D0745439

Sliding Mode Control in Electromechanical Systems

Also in the Systems and Control Series
Advances in Intelligent Control, edited by C. J. Harris
Intelligent Control in Biomedicine, edited by D. A. Linkens
Advances in Flight Control, edited by M. B. Tischler
Multiple Model Approaches to Modelling and Control, edited by R. Murray-Smith and T. A. Johansen
A Unified Algebraic Approach to Control Design R. E. Skelton, T. Iwasaki, K. M. Grigoriadis
Neural Network Control of Robot Manipulators and Nonlinear Systems, by F. L. Lewis, S. Jagannathan and A. Yesildirek
Sliding Mode Control: Theory and Applications, by C. Edwards and S. K. Spurgeon
Generalized Predictive Control with Applications to Medicine, by M. Mahfouf and D. A. Linkens
Sliding Mode Control in Electromechanical Systems, by V. I. Utkin, J. Guldner and J. Shi

Sliding Mode Control in Electromechanical Systems

VADIM UTKIN JÜRGEN GULDNER JINGXIN SHI

UK Taylor & Francis Ltd, 11 New Fetter Lane, London EC4P 4EE
USA Taylor & Francis Inc., 325 Chestnut Street, Philadelphia, PA 19106

British Library Cataloguing in Publication Data

A catalogue record for this book is available from the British Library.
ISBN 0-7484-0116-4 (cased)

Library of Congress Cataloguing in Publication Data are available

Cover design by Amanda Barragry
Typeset in Times 10/12pt by Santype International Ltd, Salisbury
Printed and bound by T.J. International Ltd, Padstow
Cover printed by Flexiprint, Lancing, West Sussex

Contents

Series Introduction ix

Preface xi

1 Introduction 1
1.1 Examples of dynamic systems with sliding modes 1
1.2 Sliding modes in relay and variable-structure systems 4
1.3 Multidimensional sliding modes 9
1.4 Outline of sliding mode control methodology 11
References 13

2 Mathematical Background 15
2.1 Problem statement 15
2.2 Regularization 18
2.3 Equivalent control method 24
2.4 Physical meaning of equivalent control 27
2.5 Existence conditions 28
References 34

3 Design Concepts 35
3.1 Introductory example 35
3.2 Decoupling 36
3.3 Regular form 39
3.4 Invariance 41
3.5 Unit control 43
References 46

4 Pendulum Systems 47
4.1 Design methodology 47
4.2 Cart–opendulum 51
4.3 Double inverted pendulum 55

4.4 Rotational inverted pendulum system 62
4.4.1 Control of inverted pendulum 63
4.4.2 Control of the base angle and the inverted pendulum 66
4.5 Simulation and experimental results for rotational inverted pendulum 68
4.5.1 Stabilization of the inverted pendulum 70
4.5.2 Stabilization of the inverted pendulum and the base 73
References 76

5 Control of Linear Systems 79
5.1 Eigenvalue placement 79
5.2 Invariant systems 82
5.3 Sliding mode dynamic compensators 83
5.4 Ackermann's formula 87
5.5 Output feedback sliding mode control 94
5.6 Control of time-varying systems 99
References 102

6 Sliding Mode Observers 103
6.1 Linear asymptotic observers 103
6.2 Observers for linear time-invariant systems 105
6.3 Observers for linear time-varying systems 106
6.3.1 Block-observable form 106
6.3.2 Observer design 109
6.3.3 Simulation results 110
References 113

7 Integral Sliding Mode 115
7.1 Motivation 115
7.2 Problem statement 116
7.3 Design principles 117
7.4 Perturbation and uncertainty estimation 118
7.5 Application examples 120
7.5.1 Linear time-invariant systems 120
7.5.2 Control of robot manipulators 122
7.5.3 Pulse-width modulation for electric drives 124
7.5.4 Robust current control for PMSMs 125
7.6 Summary 129
References 129

8 The Chattering Problem 131
8.1 Problem analysis 131
8.1.1 Example system: model 132
8.1.2 Example system: ideal sliding mode 132
8.1.3 Example system: causes of chattering 134
8.2 Boundary layer solution 139
8.3 Observer-based solution 141
8.4 Regular form solution 144
8.5 Disturbance rejection solution 147

8.6 Comparing the different solutions 151
References 153

9 Discrete-Time and Delay Systems 155
9.1 Discrete-time systems 155
9.2 Discrete-time sliding mode concept 157
9.3 Linear discrete-time systems with known parameters 161
9.4 Linear discrete-time systems with unknown parameters 163
9.5 Distributed systems and systems with delays 164
9.6 Linear systems with delays 165
9.7 Distributed systems 166
9.8 Summary 168
References 169

10 Electric Drives 171
10.1 DC motors 172
10.1.1 Introduction 172
10.1.2 Model of the DC motor 172
10.1.3 Current control 173
10.1.4 Speed control 174
10.1.5 Integrated structure for speed control 175
10.1.6 Observer design 176
10.1.7 Speed control with reduced-order model 179
10.1.8 Observer design for sensorless control 182
10.1.9 Discussion 185
10.2 Permanent-magnet synchronous motors 187
10.2.1 Introduction 187
10.2.2 Model of the PMSH motor 188
10.2.3 Current control 193
10.2.4 Speed control 201
10.2.5 Current observer 205
10.2.6 Observer design for sensorless speed control 206
10.2.7 Discussion 211
10.3 Induction motors 212
10.3.1 Introduction 212
10.3.2 Model of the induction motor 213
10.3.3 Rotor flux observer with known rotor speed 218
10.3.4 Speed control 219
10.3.5 Observer for rotor flux and rotor speed 223
10.3.6 Discussion 228
10.4 Summary 228
References 229

11 Power Converters 231
11.1 DC/DC converters 231
11.1.1 Bilinear systems 232
11.1.2 Direct sliding mode control 233
11.1.3 Observer based control 240
11.2 Boost-type AC/DC converters 251

11.2.1 Model of the boost-type AC/DC converter 252
11.2.2 Control problems 254
11.2.3 Observer for sensorless control 262
11.3 Summary 268
References 269

12 Advanced Robotics 271
12.1 Dynamic modelling 271
12.1.1 Generic inertial dynamics 272
12.1.2 Holonomic robot model 273
12.1.3 Nonholonomic robots: model of a wheel set 277
12.2 Trajectory tracking control 278
12.2.1 Componentwise control 279
12.2.2 Vector control 283
12.2.3 Continuous feedback/feedforward plus a discontinuity term 286
12.2.4 Comparing the design choices 290
12.3 Gradient tracking control 292
12.3.1 Control objectives 296
12.3.2 Holonomic robots 298
12.3.3 Nonholonomic robots 299
12.4 Application examples 302
12.4.1 Torque control for flexible robot 302
12.4.2 Collision avoidance in a known planar workspace 306
12.4.3 Collision avoidance in known workspaces of higher dimension 310
12.4.4 Automatic steering control for passenger cars 312
References 318

Index 319

Series Introduction

Control systems research has a long and distinguished tradition stretching back to nineteenth-century dynamics and stability theory. Its establishment as a major engineering discipline in the 1950s arose, essentially, from Second World War-driven work on frequency response methods by, among others, Nyquist, Bode and Wiener. The intervening 40 years have seen quite unparalleled developments in the underlying theory with applications ranging from the ubiquitous PID controller, widely encountered in the process industries, through to high-performance fidelity controllers typical of aerospace applications. And all this has been increasingly underpinned by the rapid developments in the essentially enabling technology of computing software and hardware.

This view of model-based systems and control as a mature discipline masks relatively new and rapid developments in the general area of robust control. Here an intense research effort is being directed to the development of high-performance controllers which, at least, are robust to specified classes of plant uncertainty. One measure of this effort is the fact that, after a relatively short period of work, 'near world' tests on classes of robust controllers have been undertaken in the aerospace industry. Again, this work is supported by computing hardware and software developments, such as the toolboxes available within numerous commercially marketed controller design and simulation packages.

Recently there has been increasing interest in the so-called intelligent control techniques such as fuzzy logic and neural networks. These techniques rely on learning, in a prescribed manner, the input–output behaviour of the plant to be controlled. Already it is clear there is little to be gained by applying these techniques to cases where mature model-based approaches yield high-performance control. Instead, their role, in general terms, almost certainly lies in areas where the processes encountered are ill-defined, complex, nonlinear, time-varying and stochastic. A detailed evaluation of their relative potential awaits the appearance of a rigorous supporting base such as underlying theory and implementation architectures, the essential elements of which are beginning to appear in learned journals and at conferences.

Elements of control and systems theory/engineering are increasingly finding use outside traditional numerical processing environments. One such area is intelligent command and control systems, which are central to innovative manufacturing and the management of advanced transportation systems. Another is discrete event systems which mix numerical and logical decision making.

It was in response to these exciting new developments that this series on systems and control was conceived. It publishes high-quality research texts and reference works in the diverse areas which systems and control now includes. In addition to basic theory, experimental and/or application studies are welcome, as are expository texts where theory, verification and applications come together to provide a unifying coverage of a particular topic or topics.

E. ROGERS
J. O'REILLY

Preface

The term 'sliding mode control' first appeared in the context of variable-structure systems. Soon sliding modes became the principal operational mode for this class of control systems. Practically all design methods for variable-structure systems are based on deliberate introduction of sliding modes which have played, and are still playing, an exceptional role both in theoretical developments and in practical applications. Due to its order reduction property and its low sensitivity to disturbances and plant parameter variations, sliding mode control is an efficient tool to control complex high-order dynamic plants operating under uncertainty conditions which are common for many processes of modern technology.

The development of sliding mode control theory has revealed the true potential of the original research trends. Some methods have become conventional for feedback system design whereas others have proven less promising; new research directions have been initiated due to the appearance of new classes of control problems, new mathematical methods and new control principles. It would be difficult to give a historical perspective to the events accompanying the development of the theory and the scientific analysis. Moreover, a historical survey would be rather pointless since, at each stage, monographs and survey papers have been published in which the authors summarized the vast amount of scientific results accumulated at the time. Our book is one more item in that series.

The intention of the book is to combine conventional design methodology using the available mathematical background with novel trends in sliding mode theory and the results of practical applications in a wide range of heterogeneous problems such as control in electrical, mechanical and electromechanical systems. The first chapters have a theoretical flavour and their main objective is to demonstrate the essence of the problems and the ideas underlying the sliding mode control design methodology. In situations where complex mathematical considerations are needed, we prefer to illustrate the main idea with comparatively simple examples and ask the interested reader to refer to related publications for the full-blown mathematics.

More complicated topics include such mathematical problems as deriving sliding mode equations and existence conditions, conventional and new sliding mode control design methods, and practical issues like chattering suppression methods and discrete-time sliding mode. At first sight, some of these chapters look purely theoretical, but the theory does lead to important applications. For example, integral sliding mode control is a tool for eliminating the reaching phase to achieve a higher control performance; chattering suppression methods can be used to eliminate high-frequency oscillations, once regarded as the main obstacle to sliding mode control; the results on discrete-time sliding mode are of great importance for microprocessor implementation of sliding mode control algorithms.

The application chapters embrace such challenging problems as control of electric drives, power converters, robot manipulators, mobile robots and autonomous vehicles. All these plants are governed by high-order nonlinear differential equations and operate under uncertainty conditions and in time-varying environments. The potential of sliding mode control for these processes is demonstrated in numerical and experimental results: the desired closed-loop dynamics are of reduced order and exhibit low sensitivity to different types of uncertainty due to enforcing sliding modes.

The idea to write this book stems from a long-term collaboration between the authors; it began in 1992 when V.I. Utkin was invited to work as a visiting scientist at the German Aerospace Center (DLR) in Munich. His future coauthors, J. Guldner and J. Shi, started to work with enthusiasm in the area of sliding mode control. Since then the authors have participated in joint research with colleagues at the Technical University in Munich, the University of Tokyo, the Siemens Research Department in Munich, and Ford Motor Company in the United States. Many of the chapters in this book are the result of that collaboration. Of course, the results of many other authors are included in the book, but we do not claim to cover everything that is going on in the world of sliding mode control.

It is a pleasure to acknowledge the contributions of our colleagues at the Institute of Robotics and System Dynamics of DLR, the Control Systems Lab at the Technical University in Munich, the University of Tokyo's Institute of Industrial Science, the Departments of Electrical and Mechanical Engineering of the Ohio State University, the Advanced Robotics Lab of the Siemens Research Department, and the Scientific Research Lab of Ford Motor Company. Their support of the authors' collaboration was invaluable for preparing this book. We would also like to express our sincerest thanks to our families for their endless and outstanding support during the many years of research, and especially during the 'hot phase' of preparing the final manuscript.

VADIM UTKIN
JÜRGEN GULDNER
JINGXIN SHI

CHAPTER ONE

Introduction

In the course of the entire history of automatic control theory, the intensity of investigation of systems with discontinuous control actions has been maintained at a high level. In particular, at the fist stage, relay or 'on-off' regulators ranked highly for design of feedback systems. The reason was twofold: ease of implementation and high efficiency of hardware. Monographs by Flugge-Lotz (1953) and Tsypkin (1955) were most obviously the first theoretical generalizations of the wide diversity of analysis and design methods for relay systems.

In systems with control as a discontinuous state function, so-called sliding modes may arise. The control action switches at high frequency should the sliding mode occur in the system. The study of sliding modes embraces a wide range of heterogeneous areas from pure mathematical problems to application aspects.

Systems with sliding modes have proven to be an efficient tool to control complex high-order nonlinear dynamic plants operating under uncertainty conditions, a common problem for many processes of modern technology. This explains the high level of research and publication activity in the area and the unremitting interest of practicing engineers in sliding mode control during the last two decades.

1.1 Examples of dynamic systems with sliding modes

Sliding modes as a phenomenon may appear in a dynamic system governed by ordinary differential equations with discontinuous right-hand sides. The term *sliding mode* first appeared in the context of relay systems. It may happen that the control as a function of the system state switches at high (theoretically infinite) frequency; this motion is called *sliding mode*. It may be enforced in the simplest first-order tracking relay system with the state variable $x(t)$:

$$\dot{x} = f(x) + u$$

with the bounded function $f(x)$, $|f(x)| < f_0 =$ constant and the control as a relay function (Figure 1.1) of the tracking error $e = r(t) - x$; $r(t)$ is the reference input and u is given by

$$u = \begin{cases} u_0 & \text{if } e > 0 \\ -u_0 & \text{if } e < 0 \end{cases} \quad \text{or} \quad u = u_0 \operatorname{sign}(e) \qquad u_0 = \text{const}$$

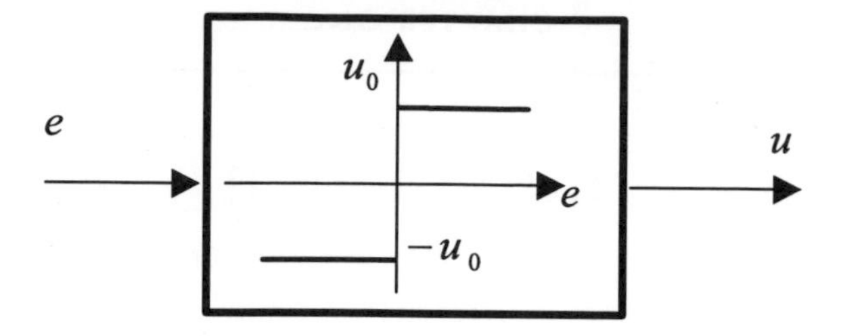

Figure 1.1 Relay control.

The values of e and $de/dt = \dot{e} = \dot{r} - f(x) - u_0 \operatorname{sign}(e)$ have different signs if $u_0 > f_0 + |\dot{r}|$. It means that the magnitude of the tracking error decays at a finite rate and the error is equal to zero identically after a finite time interval T (Figure 1.2). The argument of the control function, e, is equal to zero which is the discontinuity point. For any real-life implementation, imperfections in the switching device mean the control switches at high frequency or takes intermediate values for continuous approximation of the relay function. The motion for $t > T$ is called *sliding mode.*

Formally, sliding mode may appear not only in a control system with discontinuous control but in any dynamic system with discontinuities in the motion equations. For example, the right-hand side is a discontinuous function of the state in the simple mechanical system with Coulomb friction depicted in Figure 1.3. The motion equation is given by

$$m\ddot{x} + kx = -u_f(\dot{x})$$

where $x(t)$ is the displacement, k is the spring stiffness and the friction force is a discontinuous function of the speed:

$$u_f = u_0 \operatorname{sign}(\dot{x}) \qquad u_0 = \text{const}$$

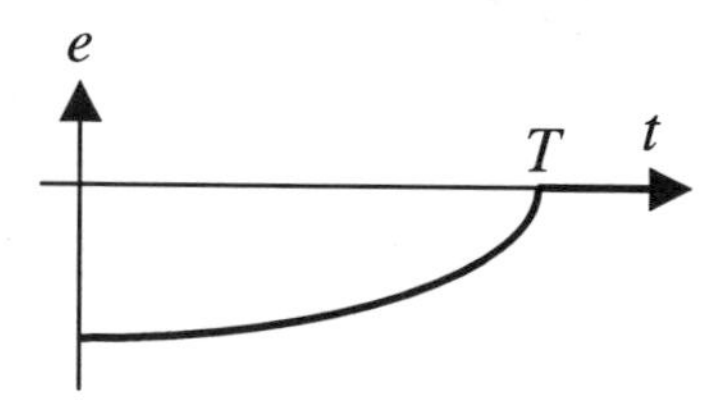

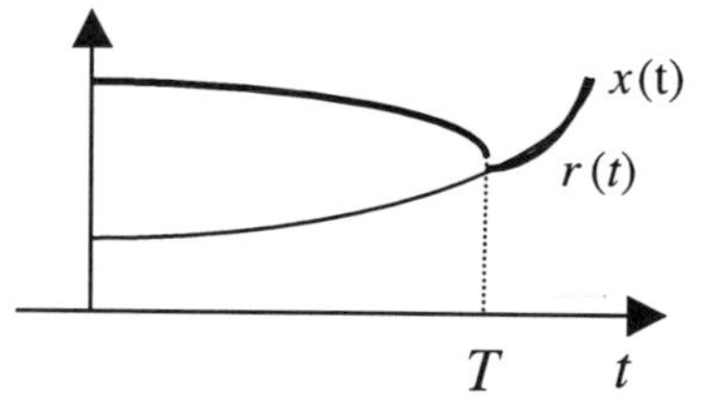

Figure 1.2 Sliding mode tracking control.

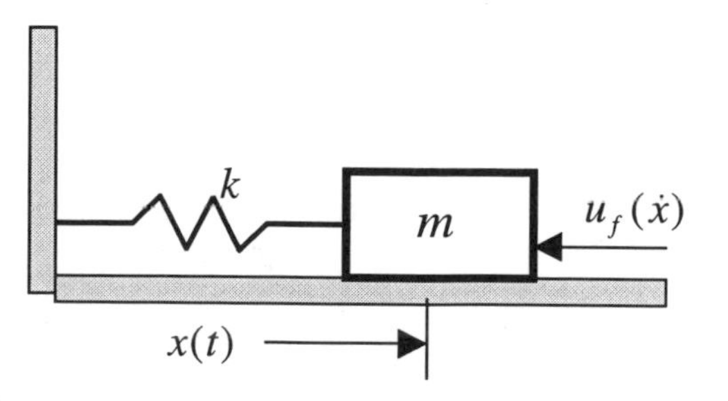

Figure 1.3 Mechanical system with Coulomb friction.

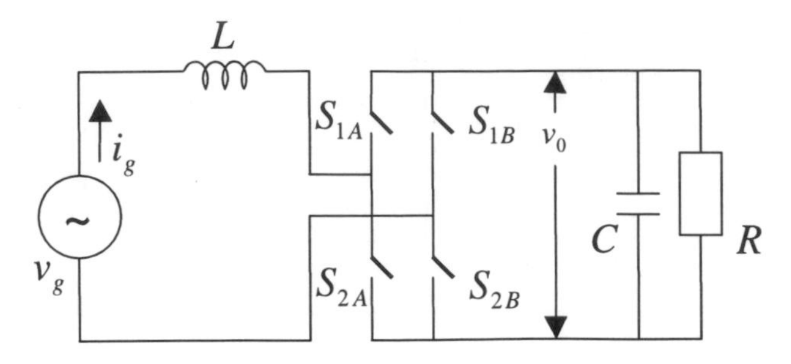

Figure 1.4 AC/DC voltage converter.

If $u_0 < k|x|$ then the friction force takes one of the extreme values and the motion is described by nonhomogeneous differential equations with the right-hand side equal to u_0 or $-u_0$. For $u_0 > k|x(t_0)|$ and $\dot{x}(t_0) = 0$, the mass sticks and $\dot{x}(t) \equiv 0$, $x(t) \equiv x(t_0)$ for $t > t_0$. This motion may be called *sliding mode* since the argument of the discontinuous function $u_f = u_0 \operatorname{sign}(\dot{x})$ is equal to zero identically, similar to the previous example.

The third example illustrates sliding motions in an electrical system, an AC/DC voltage converter (Figure 1.4) with dynamic equations

$$\frac{di_g}{dt} = \frac{V_g}{L} \sin \omega t - \frac{v_0}{L} u$$

$$\frac{dv_0}{dt} = -\frac{v_0}{RC} + \frac{i_g}{C} u$$

where the input voltage $v_g = V_g \sin \omega t$ and the switches s_{1A}, s_{2A}, s_{1B} and s_{2B} constitute the control input

$$u = \begin{cases} 1 \text{ if } S_{1A} \text{ and } S_{2B} \text{ are closed} \\ -1 \text{ if } S_{1B} \text{ and } S_{2A} \text{ are closed} \end{cases}$$

The switching logic should be found such that the output voltage v_0 is equal to the desired value $v_d(t)$. It seems reasonable to switch the function u depending on the sign of the tracking error $v_e = v_d - v_0$:

$$u = \begin{cases} 1 \text{ if } (v_d - v_0)i_g > 0 \\ -1 \text{ if } (v_d - v_0)i_g < 0 \end{cases} \quad \text{or} \quad u = \operatorname{sign}[(v_d - v_0)i_g]$$

Calculate the time derivative of the tracking error as

$$\dot{v}_e = \dot{v}_g + \frac{v_0}{RC} - \frac{|i_g|}{C} \operatorname{sign}(v_e)$$

If $|i_g| > |(v_0/R) + C\dot{v}_g|$ then v_e and $\dot{v}_e$ have different signs, hence the error v_e will vanish after finite time interval T and will be equal to zero identically afterwards. The plots $v_e(t)$, $v_0(t)$ and $v_d(t)$ are similar to those of $e(t)$, $x(t)$ and $r(t)$ in Figure 1.2. As for the first-order example, ideal tracking is provided through enforcement of this *sliding mode.*

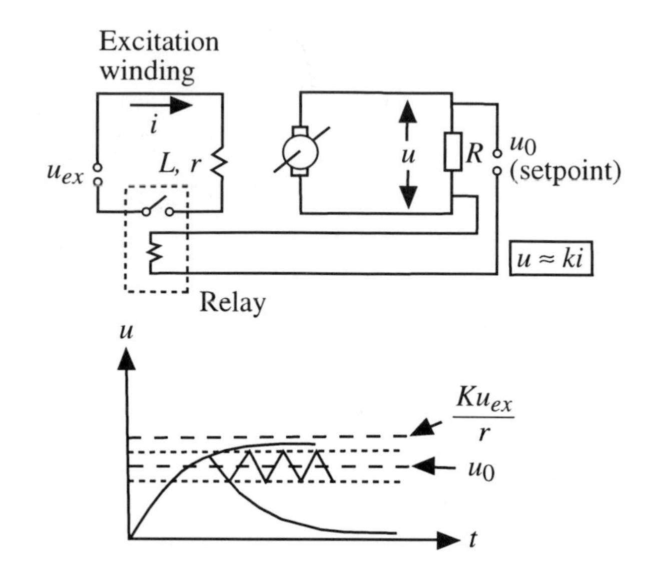

Figure 1.5 Vibration control of DC generator.

1.2 Sliding modes in relay and variable-structure systems

The ideas underlying modern analysis and design methods for sliding mode control may be found in the publications of the early 1930s. Figure 1.5 illustrates the so-called vibration control studied by V. Kulebakin (1932) in the context of voltage control for a DC generator of an aircraft. Notice that the output voltage is close to the setpoint due to discontinuous feedback and high-frequency switching in the excitation winding. It seems that 1930s 'vibration control' is just the same as our contemporary 'sliding mode control'.

A second example from the 1930s (Figure 1.6) concerns sliding mode relays for controlling the course of a ship (Nikolski, 1934). It is amazing that a paper published more than 60 years ago should be written in the language of modern control theory: phase plane, switching line and even sliding mode.

In all the examples, except for the last one, the phenomenon *sliding mode* was revealed and discussed in the time domain, although this term was not used directly. But, for analysis and design of sliding mode control, the state-space method looks much more promising.

The conventional example to demonstrate sliding modes in terms of the state-space method is a second-order time-invariant relay system

$$\begin{aligned} &\ddot{x} + a_2\dot{x} + a_1 x = u + f(t), \\ &u = -M\operatorname{sign}(s), \qquad s = \dot{x} + cx \end{aligned} \tag{1.2.1}$$

where M, a_1, a_2, c are constant parameters and $f(t)$ is a bounded disturbance.

The system behaviour may be analyzed in the state plane $(x, \dot{x})$. The state plane in Figure 1.7 is shown for $a_1 = a_2 = 0$. The control u undergoes discontinuities at the switching line $s = 0$ and the state trajectories are constituted by two families: the first family corresponds to $s > 0$ and $u = -M$ (upper semiplane); the second family corresponds to $s < 0$ and $u = M$ (lower semiplane). Within the sector $m - n$ on the

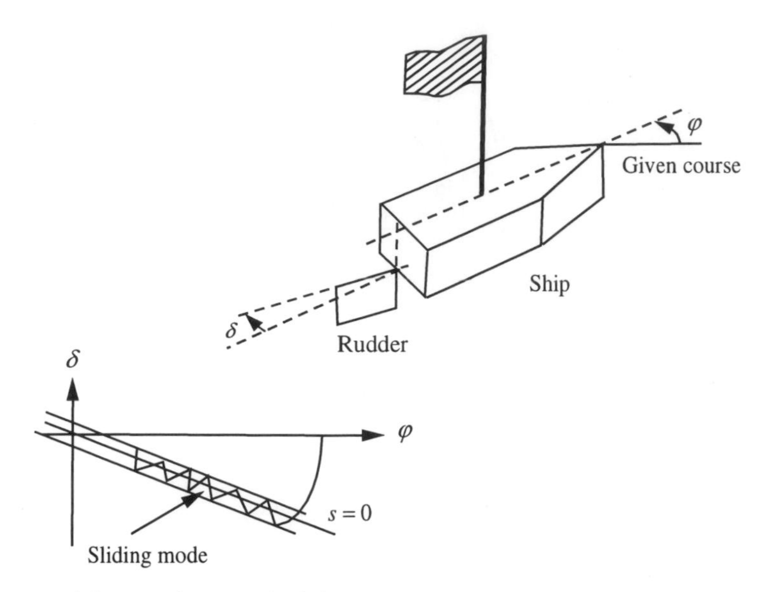

Figure 1.6 Sliding mode control of ship's course.

switching line, the state trajectories are oriented towards the line. Having reached the sector at some time t_1, the state cannot leave the switching line. This means that the state trajectory will belong to the switching line for $t > t_1$. This motion with state trajectories in the switching line is called *sliding mode.* Since, in the course of sliding mode, the state trajectory coincides with the switching line $s = 0$, its equation may be interpreted as the motion equation, i.e.

$$\dot{x} + cx = 0 \tag{1.2.2}$$

It is important that its solution $x(t) = x(t_1)e^{-c(t-t_1)}$ depends neither on the plant parameters nor the disturbance. This so-called *invariance* property looks promising for designing feedback control for dynamic plants operating under uncertainty conditions.

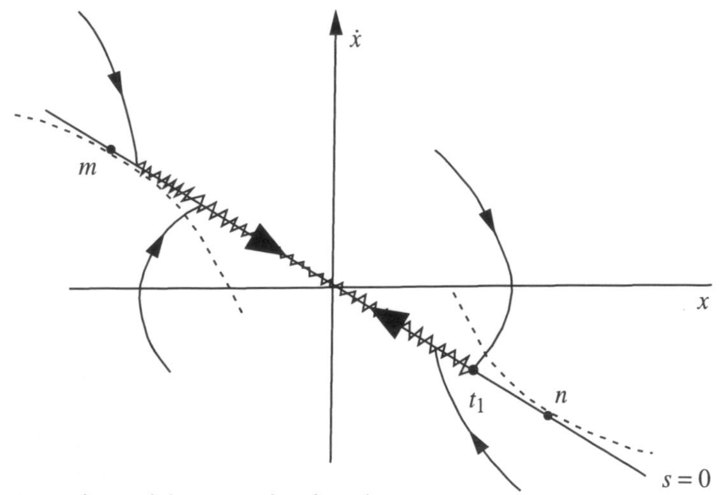

Figure 1.7 State plane of the second-order relay system.

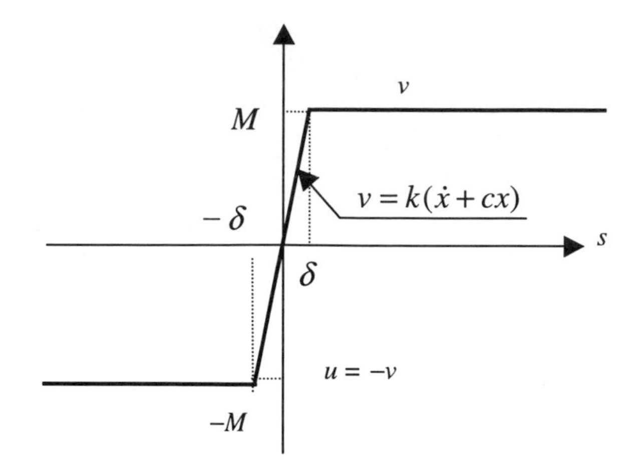

Figure 1.8 Continuous approximation of discontinuous control.

We have just described an ideal mathematical model. In real implementations, the trajectories are confined to some vicinity of the switching line. The deviation from the ideal model may be caused by imperfections of switching devices such as small delays, dead zones and hysteresis, which may lead to high-frequency oscillations as shown in Figure 1.6. The same phenomenon may appear due to small time constants of sensors and actuators having been neglected in the ideal model. This phenomenon, called *chattering*, was a serious obstacle to the use of sliding modes in control systems, and special attention will be paid to chattering suppression methods in Chapter 8. Note that the state trajectories are also confined to some vicinity of the switching line for continuous approximation of a discontinuous relay function as well (Figure 1.8). In a δ-vicinity of the line $s = 0$, control is the linear state function with a high gain k and the eigenvalues of the linear system are close to $-k$ and $-c$. This means that the motion in the vicinity consists of the fast component decaying rapidly and the slow component coinciding with solution to the ideal sliding mode equation (1.2.2).

Sliding modes became a principal operational mode in variable-structure systems or systems consisting of a set of continuous subsystems with a proper switching logic. For example, the second-order system

$$\ddot{x} - ax = u \qquad (a > 0)$$

$$u = -k|x| \operatorname{sign}(s)$$
$$s = cx + \dot{x} \qquad\qquad (k > 0, \quad c > 0)$$

consists of two unstable linear structures (Figure 1.9).

By varying the system structure along the switching lines $s = 0$ and $x = 0$ and enforcing sliding mode, the system becomes asymptotically stable (Figure 1.10). The switching line is reached for any initial conditions. If the slope of the switching line is lower than the slope of the asymptote for structure I ($c < c_0$), then the state trajectories are oriented towards the line and sliding mode may start at any point of $s = 0$. Similar to the relay system, it is governed by the first-order equation (1.2.2) with the solution $x(t) = x(t_1)e^{-c(t-t_1)}$. Again, the solution depends on neither the plant parameters nor any disturbances the plant may be subjected to.

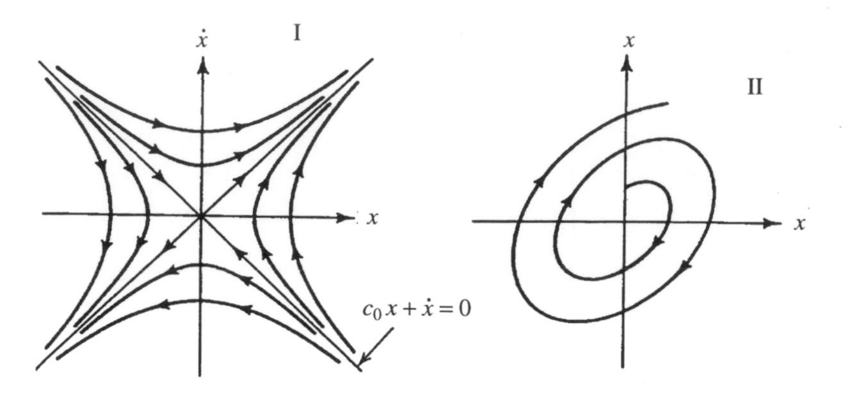

Figure 1.9 Variable-structure system consisting of two unstable subsystems.

The examples of relay and variable-structure systems demonstrated order reduction and invariance with respect to plant uncertainties of the systems with sliding modes. Use of these properties was the key idea of variable-structure theory at the first stage when only single-input single-output systems with motion equations in canonical space were studied (Emelyanov *et al.*, 1970). A control variable $x = x_1$ and its time derivatives $x^{(i-1)} = x_i$, $(i = 1, \ldots, n)$ are components of a state vector in the canonical space

$$\begin{aligned} \dot{x}_i &= x_{i+1} \qquad (i = 1, \ldots, n-1) \\ \dot{x}_n &= -\sum_{i=1}^{n} a_i(t)x_i + f(t) + b(t)u \end{aligned} \tag{1.2.3}$$

where $a_i(t)$and $b_i(t)$ are unknown parameters and $f(t)$ is an unknown disturbance.

Control undergoes discontinuities on some plane $s(x) = 0$ in the state space

$$u = \begin{cases} u^+(x, t) \text{ if } s(x) > 0 \\ u^-(x, t) \text{ if } s(x) < 0 \end{cases}$$

where $u^+(x, t)$ and $u^-(x, t)$ are continuous state functions, $u^+(x, t) \neq u^-(x, t)$, $s(x) = \sum_{i=1}^{n} c_i x_i$, $c_n = 1$ and $c_1, \ldots, c_{n-1}$ are constant coefficients.

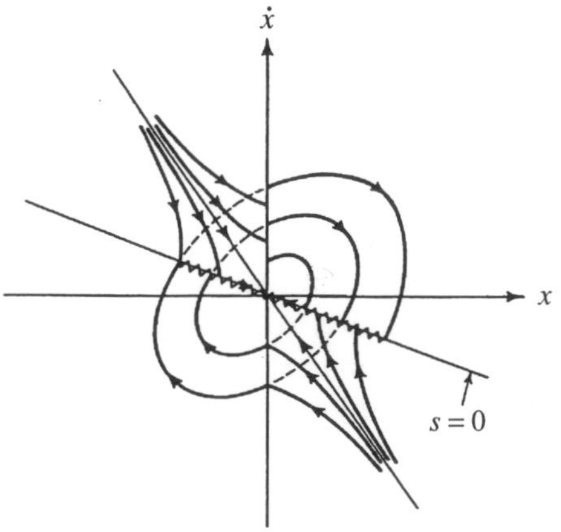

Figure 1.10 State plane of variable-structure system: $s = 0, \dot{x} + cx = 0$.

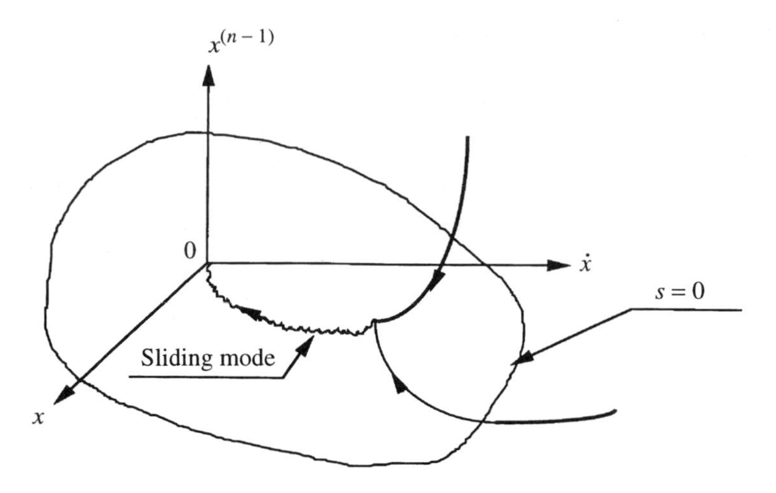

Figure 1.11 Sliding mode in canonical state space: $x^{(n-1)} + c_{n-1}x^{(n-2)} + \ldots + c_1 x = 0$.

The discontinuous control was selected such that the state trajectories are oriented towards the switching plane $s = 0$, hence sliding mode arises in this plane (Figure 1.11). Once sliding mode has begun, the motion trajectories of system (1.2) are in the switching surface,

$$x_n = -\sum_{i=1}^{n-1} c_i x_i$$

Substitution into the $(n-1)$th equation yields the sliding mode equations

$$\begin{aligned} &\dot{x}_i = x_{i+1} \qquad (i = 1, \ldots, n-2) \\ &\dot{x}_{n-1} = -\sum_{i=1}^{n-1} c_i x_i \quad \text{or} \quad x^{(n-1)} + c_{n-1}x^{(n-2)} + \ldots + c_1 = 0 \end{aligned} \tag{1.2.4}$$

The motion equation is of reduced order and depends on neither the plant parameters nor the disturbance. The desired dynamics of the sliding mode may be assigned by a proper choice of the parameters of switching plane c_i.

Although the invariance property is very useful, it did create the illusion that any control problem can be easily solved by enforcing the sliding mode in the system. The main problem is that the space of state derivatives is a mathematical idealization and ideal differentiators can hardly be implemented. As a result, another extreme appeared reflecting a certain pessimism over the possibility of implementing variable-structure systems with sliding modes. But the refusal to use sliding modes in control systems proved to be unreasonable as well.

In modern technological processes it is common that control and system output may be vector-valued quantities and only some components of the state vector are accessible for measurement. The canonical space approach did not suggest how the control may be designed in such situations. The second stage of variable-structure system studies was dedicated to the development of design methods for systems with motion equations in an arbitrary state space with vector control action and vector variables to be controlled (Utkin, 1983). The basic idea underlying the majority of control methods is the enforcement of multidimensional sliding modes.

1.3 Multidimensional sliding modes

In the previous examples of control systems with sliding modes, the control was a scalar state function and the sliding mode was governed by a differential equation with order one less than the order of the original system. So we may assume that sliding motion may appear in an intersection of several surfaces if the control is a vector-valued quantity and each component undergoes discontinuities in its own switching surface. The planar motion of a point mass m with Coulomb friction (Figure 1.12) may serve as an example.

The motion in the orthogonal frame (x, y) is governed by the fourth-order system

$$x = x_1 \qquad y = y_1$$

$$\begin{cases} \dot{x}_1 = x_2 \\ m\dot{x}_2 = -kx_1 - F_x \\ \dot{y}_1 = y_2 \\ m\dot{y}_2 = -ky_1 - F_y \end{cases}$$

where both springs have the same stiffness k, $F_x = x_2/\sqrt{x_2^2 + y_2^2}$ and $F_y = y_2/\sqrt{x_2^2 + y_2^2}$ are (x, y) components of the friction force vector $F = -Mv/\|v\|$, $M = \text{const}$, v is a speed vector with components x_2 and y_2 and $\|v\| = \sqrt{x_2^2 + y_2^2}$.

The magnitude of the friction force is equal to M for $v \neq 0$. F undergoes discontinuities when x_2 and y_2 are equal to zero simultaneously. If at initial time $v = 0$ (i.e. $x_2 = 0$ and $y_2 = 0$) and the maximum value of the friction force exceeds the spring force, $M > k\sqrt{x_1^2 + y_1^2}$, then the mass is stuck and $v \equiv 0$ for all further time.

Thus, beyond the intersections of two surfaces $x_2 = 0$ and $y_2 = 0$, the friction force is a continuous state function, and in the domain $M > k\sqrt{x_1^2 + y_1^2}$ the state trajectories $(x_1(t) = \text{const},\ y_1(t) = \text{const})$ belong to this manifold. This motion may be called *two-dimensional sliding mode* in the intersection of two discontinuity surfaces.

The next example illustrates two-dimensional sliding mode in a control system with a two-dimensional control vector:

$$\begin{aligned} \dot{x}_1 &= x_2 \\ \dot{x}_2 &= x_3 + f_1(t) + u_1 \\ \dot{x}_3 &= f_2(t) + u_2 \end{aligned} \tag{1.3.1}$$

where $f_1(t)$ and $f_2(t)$ are unknown bounded disturbances with a known range of variation.

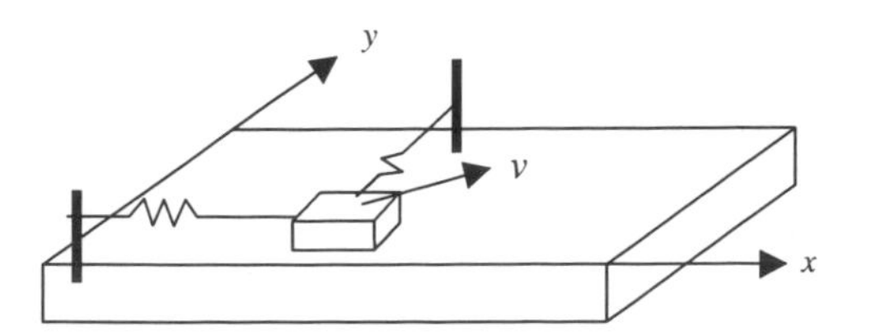

Figure 1.12 Mechanical system with Coulomb friction, on a plane.

The components of the control undergo discontinuities in two planes of the three-dimensional state:

$$u_1 = -M_1 \operatorname{sign}(s_1) \qquad s_1 = x_1 + x_2$$
$$u_2 = -M_2 \operatorname{sign}(s_2) \qquad s_2 = x_1 + x_2 + x_3$$

where M_1, M_2 are positive constant values.

If $M_2 > |x_2 + x_3 + f_1(t) + f_2(t)| + M_1$ then the values s_2 and $\dot{s}_2 = x_2 + x_3 + u_1 - M_2 \operatorname{sign}(s_2)$ have different signs. Hence the plane $s_2 = 0$ is reached after a finite time interval and then sliding mode with state trajectories in this plane will start (Figure 1.13). For this motion $x_3 = -x_1 - x_2$ and the sliding mode is governed by the second-order equation

$$\begin{cases} \dot{x}_1 = x_2 \\ \dot{x}_2 = -x_1 - x_2 + f_1(t) + u_1 \end{cases}$$

Again, for $M_1 > |x_1 + f_1(t)|$, the values s_1 and $\dot{s}_1 = -x_1 + f_1(t) - M_1 \operatorname{sign}(s_1)$ have different signs and after a finite time interval the state will reach the intersection of the planes $s_1 = 0$ and $s_2 = 0$. The further motion will be in this manifold (the straight line formed by the intersection of the two planes), and its first-order equation may be derived by substituting $-x_1$ for x_2 (since $s_1 = 0$) into the first equation to obtain

$$\dot{x}_1 = -x_1$$

The two-dimensional sliding mode is asymptotically stable, its order is two less than the order of the original system and the motion does not depend on the disturbances $f_1(t)$ and $f_2(t)$.

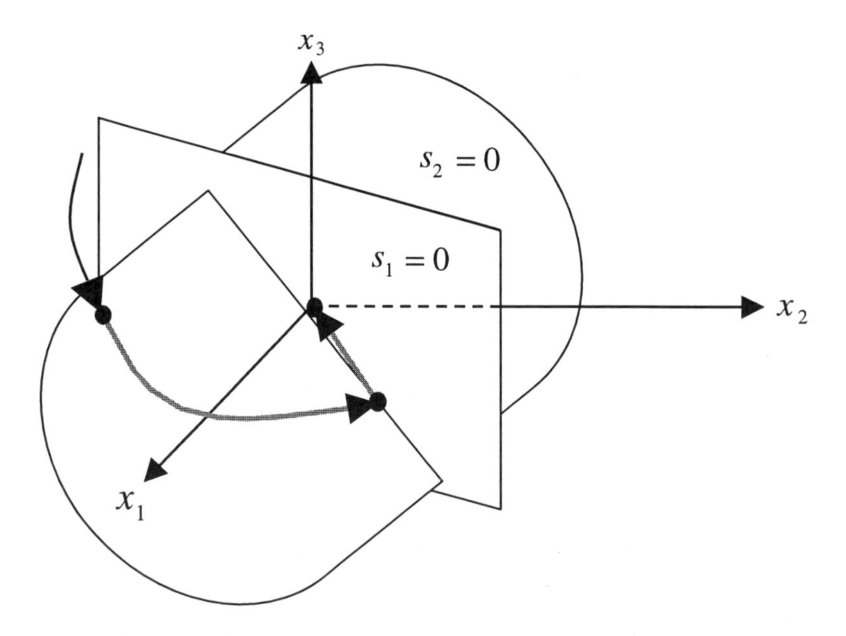

Figure 1.13 Two-dimensional sliding mode.

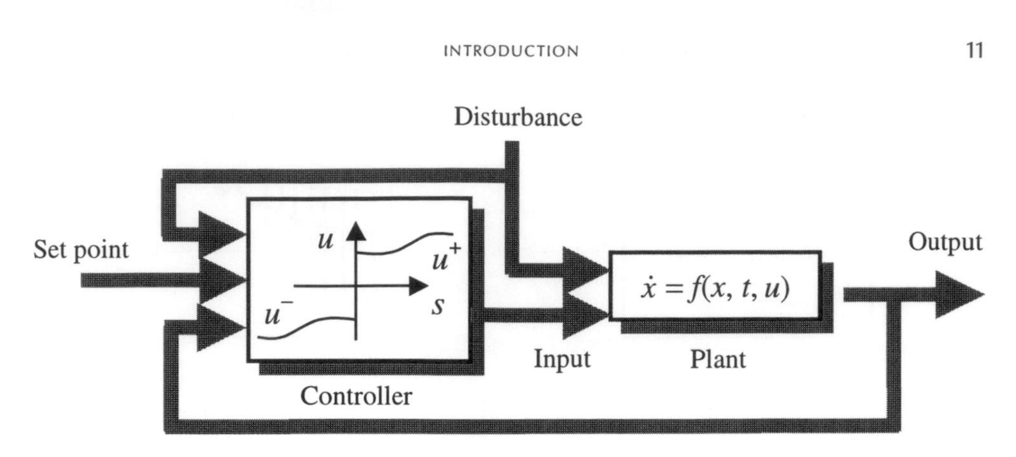

Figure 1.14 Systems with sliding mode control.

1.4 Outline of sliding mode control methodology

The examples in the previous sections allow us to outline the main reasons why enforcing sliding modes is a promising method to control high-order nonlinear dynamic plants operating under uncertainty conditions. In this book, we will deal mainly with processes described by nonlinear differential equations in an arbitrary n-dimensional state space with m-dimensional vector control actions (Figure 1.14):

$$\dot{x} = f(x, t, u) \tag{1.4.1}$$

with $x \in \Re^n$, $f \in \Re^n$, $u \in \Re^m$; t denotes the time.

The control is selected as a discontinuous function of the state. For example, each component of the control u_i may undergo discontinuities on some nonlinear surface $s_i(x) = 0$ in the state space

$$u_i = \begin{cases} u_i^+(x, t) \text{ if } s_i(x) > 0 \\ u_i^-(x, t) \text{ if } s_i(x) < 0 \end{cases} \quad (i = 1, \ldots, m) \tag{1.4.2}$$

where $u_i^+(x, t)$ and $u_i^-(x, t)$ are continuous state functions with $u_i^+(x, t) \neq u_i^-(x, t)$; the $s_i(x)$ are continuous state functions.

Similar to the example with two-dimensional sliding mode in the intersection of two discontinuity planes (Section 1.3), we may expect that sliding mode may occur in the intersection of m surfaces $s_i(x) = 0$ $(i = 1, \ldots, m)$ and the order of the motion equations is m less than the order of the original system. In connection with the control of high-dimensional plants, great interest is attached to design methods that allow the overall system motions to be decoupled into independent partial components. As we can see, enforcing sliding modes in systems with discontinuous control enables order reduction, which leads to decoupling and simplification of the design procedure.

Furthermore, the element implementing a discontinuous function $u(x)$ has the input $s(x)$ close to zero during sliding mode (Figure 1.15) whereas its output takes finite values (to be precise the average value of the output since it contains a high-frequency component). This means that the element implements high (theoretically infinite) gain, which is the conventional tool to suppress the influence of disturbances and uncertainties in the plant behaviour. Unlike continuous high-gain control systems, the invariance effect is attained using finite control actions.

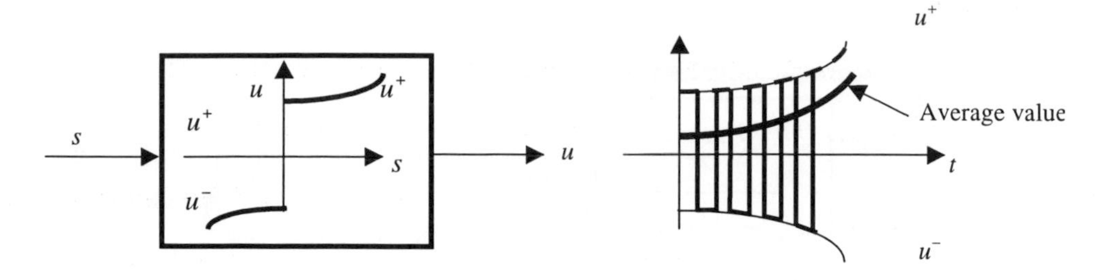

Figure 1.15 Sliding mode for high-gain implementation.

Our brief discussion of the motions in systems with sliding modes has shown two things:

- The order of the system is reduced.
- Sensitivity with respect to parameter variation and disturbances may be reduced should sliding mode occur.

As demonstrated in the previous sections, the order reduction and invariance properties are easily feasible in second-order systems with motion equations in the canonical space. The sliding mode dynamics depend on the switching surface equations and do not depend on control. Hence the design procedure should consist of two stages. First, the equation of the manifold with sliding mode is selected to design the desired dynamics of this motion in accordance with some performance criterion. Then, the discontinuous control should be found such that the state would reach the manifold and such that sliding mode exists in this manifold. As a result, the design is decoupled into two subproblems of lower dimension, and after a finite time interval preceding the sliding motion, the system will possess the desired dynamic behaviour.

We have dwelt upon the main reasons for using sliding modes in control systems and outlined the sliding mode control design methodology. The basic design concept of the control methods studied in this book will focus on enforcing sliding modes. Decoupling or invariance (or both) will be inherent in the majority of the proposed design techniques. The decoupling and invariance properties make the sliding mode methodology an efficient tool to control complex electrical and mechanical dynamic processes governed by high-order differential equation with bounded information on parameters and disturbances. Two examples are an induction motor and a multilink manipulator with unknown load torque and inertia, and with position, speed or torque to be controlled. The design methods for control of the following systems will be developed in the book:

- Purely mechanical systems with forces or torques as control actions, e.g. manipulators and mobile robots
- Purely electrical systems, e.g. power converters
- Electromechanical systems, e.g. electric motors

Widely used electrical servomechanisms are controlled by power electronic converters. When employing them, it seems reasonable to turn to control algorithms with discontinuous control actions, since only an on-off operation is admissible for such converters and discontinuities in control are dictated by the very nature of the converter elements.

References

EMELYANOV, S. *et al.*, 1970, *Theory of Variable Structure Control Systems* (in Russian), Moscow: Nauka.

FLUGGE-LOTZ, I., 1953, *Discontinuous Automatic System,* New York: Princeton University Press.

KULEBAKIN, V., 1932, On theory of vibration controller for electric machines (in Russian), *Theoretical and Experimental Electronics*, **4**.

NIKOLSKI, G., 1934, On automatic stability of a ship on a given course (in Russian), *Proceedings of the Central Communication Laboratory,* **1**, 34–75.

TSYPKIN, Y., 1955, *Theory of Relay Control Systems* (in Russian), Moscow: Gostechizdat.

UTKIN, V., 1983, Variable structure systems: present and future, *Automation and Remote Control*, **44**, 1105–20.

CHAPTER TWO

Mathematical Background

Sliding mode control is not only in the class of nonlinear control systems, but also inherently introduces discontinuities into the control loop. However, most tools for system analysis and control synthesis were developed for *continuous* linear and nonlinear systems. Consequently, these tools are not applicable to *discontinuous control* methods like sliding mode control and variable-structure systems.

This chapter provides the mathematical background of the most important tools developed for discontinuous systems, in particular for the design of sliding mode control. Since this book is mainly intended to provide sufficient tools for practical control design in real-life applications, the interested reader is referred to Utkin (1992) for a more detailed mathematical description of sliding mode techniques.

2.1 Problem statement

The sketch of design methods in Section 1.3 assumed that the properties of sliding modes in canonical spaces (1.2.3) would be preserved for arbitrary systems. These properties – order reduction and invariance – were revealed after the sliding mode equation had been derived. It was an easy problem since the equation of the switching surface was also the sliding mode equation; see equations (1.2.2) and (1.2.4). This is not the case for systems with motion equations with respect to arbitrary state variables. The analytical problems arising in such systems with sliding modes may be illustrated with the help of a linear second-order system

$$\begin{aligned} \dot{x}_1 &= a_{11}x_1 + a_{12}x_2 + b_1 u + d_1 f(t) \\ \dot{x}_2 &= a_{21}x_1 + a_{22}x_2 + b_2 u + d_2 f(t) \end{aligned} \tag{2.1.1}$$

with relay control

$$u = -M\,\mathrm{sign}(s) \qquad s = c_1 x_1 + c_2 x_2$$

All parameters a_{ij}, b_i, d_i, c_i $(i, j = 1,\ 2)$ and M are constant; $f(t)$ is a bounded disturbance.

Similar to relay systems in canonical space, the state trajectories in the state plane (x_1, x_2) may be oriented toward the switching line $s = 0$ and sliding mode arises along this line. To analyze the system behaviour in sliding mode, the question to answer is: What is the motion equation? In contrast to the second-order systems in canonical space, $x_2 = -c_2^{-1}c_1x_1$ resulting from $s = 0$ is not a motion equation. For the particular case $b_1 = 0$, substitution of $-c_2^{-1}c_1x_1$ for x_2 into the first equation of (2.1.1) allows us to derive the first-order sliding mode equation

$$\dot{x}_1 = (a_{11} - a_{12}c_2^{-1}c_1)x_1 + d_1f(t)$$

As we can see, the order reduction property takes place but invariance with respect to the disturbance does not, since the right-hand side of the motion equation depends on $f(t)$ directly. This shows that the fundamental problems related to mathematical models of sliding modes arise in systems described in the general form (1.4.1) and (1.4.2). To determine conditions for sliding mode to be insensitive to system uncertainties, special mathematical methods will need to be developed.

Next, having derived the sliding mode equations, the desired dynamics may be assigned by proper choice of the discontinuity surface equations as the first stage of the design procedure outlined in Section 1.4. The second stage implies selection of the discontinuous control inputs to enforce sliding mode in the intersection of the surfaces. To solve this problem, the conditions for sliding mode to exist should be obtained. For systems with scalar control, this condition may be interpreted easily from a geometrical viewpoint: the state trajectories should be oriented towards the discontinuity surface in its vicinity, or the variable describing deviation from the surface and its time derivative should have opposite signs. The components of the two-dimensional control in the third-order system (1.3.1) were designed using these conditions. For the general case, the problem of enforcing sliding mode in the intersection of a set of discontinuity surfaces cannot be reduced to sequential treatment of scalar subproblems. This may be illustrated by a third-order controllable system with a two-dimensional control vector:

$$\begin{aligned}\dot{x}_1 &= x_3\\ \dot{x}_2 &= -x_3 + u_1 - 2u_2\\ \dot{x}_3 &= -x_3 + 2u_1 + u_2\end{aligned} \tag{2.1.2}$$

$$\begin{aligned}u_1 &= -\text{sign}(s_1) \qquad s_1 = x_1 + x_2\\ u_2 &= -\text{sign}(s_2) \qquad s_2 = x_1 + x_3\end{aligned}$$

The analysis of the condition for sliding mode to exist in the intersection of the discontinuity surfaces may be performed in terms of motion projection on subspace (s_1, s_2):

$$\begin{aligned}\dot{s}_1 &= -\text{sign}(s_1) + 2\,\text{sign}(s_2)\\ \dot{s}_2 &= -2\,\text{sign}(s_1) - \text{sign}(s_2)\end{aligned}$$

The state trajectories are straight lines in the state plane (s_1, s_2); see Figure 2.1. It is clear from the diagram that, for any point on $s_1 = 0$ or $s_2 = 0$, the state trajectories are not oriented towards the line, therefore sliding mode does not exist at any of the switching lines taken separately. At the same time, the trajectories converge to

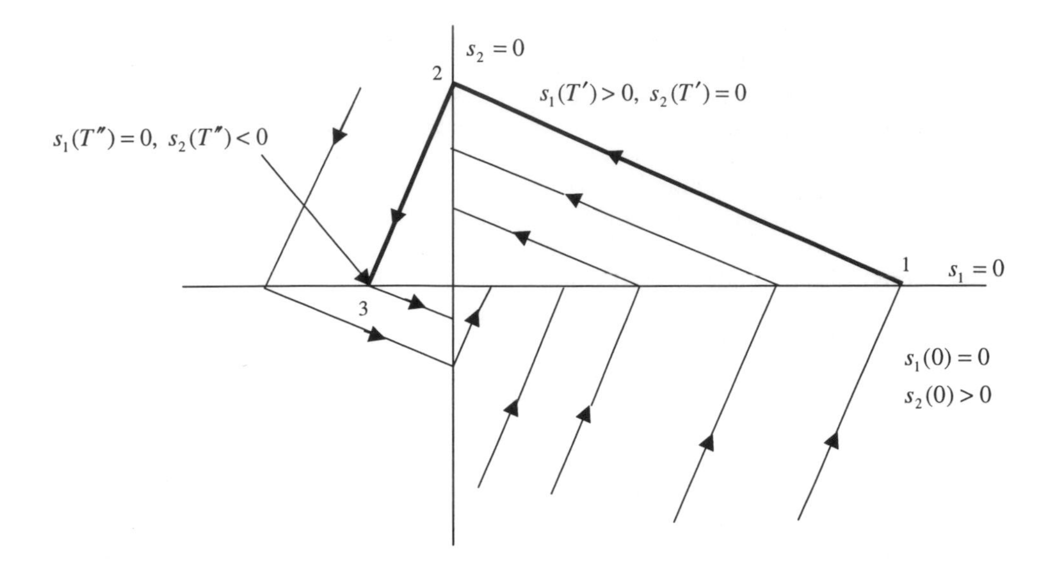

Figure 2.1 Sliding mode in the system with two-dimensional control.

the intersection of them – the origin in the subspace (s_1, s_2). Let us calculate the time needed for the state to reach the origin. For initial conditions $s_1(0) = 0, s_2(0) > 0$ (point 1)

$$\begin{aligned} \dot{s}_1 &= 1 \\ \dot{s}_2 &= -3 \end{aligned} \qquad (0 < t < T')$$

and $s_2(T') = 0,\ T' = \frac{1}{3}s_2(0),\ s_1(T') = \frac{1}{3}s_2(0)$ at point 2.

For the further motion

$$\begin{aligned} \dot{s}_1 &= -3 \\ \dot{s}_2 &= -1 \end{aligned} \qquad (T' < t < T' + T'')$$

and at point 3

$$s_1(T' + T'') = 0 \qquad T'' = \frac{1}{9}\, s_2(0) \qquad s_2(T' + T'') = -\frac{1}{9}\, s_2(0)$$

or

$$s_2(T_1) = -\frac{1}{9}\, s_2(0) \qquad T_1 = T' + T'' = \frac{4}{9}\, s_2(0)$$

This means that

$$|s_2(T_i)| = \left(\frac{1}{9}\right)^i |s_2(0)| \qquad s_1(T_i) = 0$$

$$\Delta T_i = T_i - T_{i-1} = \frac{4}{9}\, s_2(T_{i-1}) = \frac{4}{9}\left(\frac{1}{9}\right)^{i-1} s_2(0) \quad \text{for } i = 1, 2, \dots, \text{ and } T_0 = 0$$

Since

$$\lim_{i\to\infty} [s_2(T_i)] = 0, \ \lim_{i\to\infty} = s_1(T_i) = 0, \qquad \lim_{i\to\infty} \sum_{i=1}^{\infty} \Delta T_i = \frac{4}{9} s_2(0) \frac{1}{1-\frac{1}{9}} = \frac{1}{2} s_2(0)$$

the state will reach the manifold $(s_1, s_2) = 0$ after a finite time interval and thereafter sliding mode will arise in this manifold as in all the above systems with discontinuous scalar and vector controls. The example illustrates that the conditions for two-dimensional sliding mode to exist cannot be derived from analysis of scalar cases. Moreover, sliding mode may exist in the intersection of discontinuity surfaces although it does not exist on each of the surfaces taken separately.

In addition to the problems of mathematical model and invariance conditions for the general case we face one more problem with a mathematical flavour: the existence conditions for multidimensional sliding modes should be derived. The mathematical models and existence conditions for sliding modes will be studied in this chapter, and the invariance conditions will be addressed in Chapter 3.

2.2 Regularization

The first mathematical problem in the context of our plan to employ sliding modes for designing feedback control systems is a mathematical description of this motion. It arises due to discontinuities in the control inputs and hence in the right-hand sides of the motion differential equations. Discontinuous systems are not a subject of the conventional theory of differential equations dealing with continuous state functions. The conventional theory does not answer even the fundamental questions of whether the solution exists and whether the solution is unique. Formally, even for our simple examples of second-order systems in canonical form (1.2.1), our method of deriving the sliding mode equations was not legitimate. Strictly speaking, the most conventional method requires the right-hand sides of a differential equation to consist of functions $f(x)$ satisfying the Lipschitz condition $\|f(x_1) - f(x_2)\| < L\|x_1 - x2\|$ with some positive number L, known as the Lipschitz constant, for any x_1 and x_2. The condition implies that the function does not grow faster than some linear function, which is not the case for discontinuous functions if x_1 and x_2 are close to a discontinuity point.

The solution $x(t) = x(t_1)e^{-c(t-t_1)}$ should satisfy the original differential equation (1.2.1) rather than the heuristic equation (1.2.2). Direct substitution of $x(t)$ into (1.2.1) leads to $s(t) = 0$ and $(c^2 - ca_2 + a_1)x(t_1)e^{-c(t-t_1)} \overset{?}{=} -M \operatorname{sign}(0) + f(t)$. Since the function sign(·) is not defined at zero, we cannot check whether the solution $x(t)$ is correct.

In situations where conventional methods are not applicable, the common approach is to employ different methods of regularization or to replace the original problem by a closely similar one for which familiar methods are applicable. For systems with discontinuous controls, the regularization approach has a simple physical interpretation. Uncertainty of system behaviour at the discontinuity surfaces appears because the motion equations (1.4.1) and (1.4.2) are an ideal system model. The ideal model neglects nonideal factors such as small imperfections of switching devices (delay, hysteresis, small time constants), unmodelled dynamics of sensors and actuators, etc. Incorporating them into the system model makes the discontinuity point isolated in time and eliminates ambiguity in the system behaviour. Next, small parameters

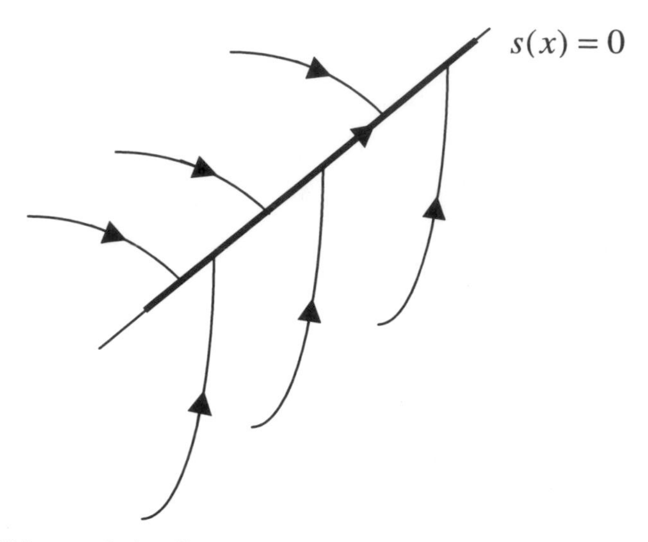

Figure 2.2 Sliding mode in a linear system.

characterizing all these factors are assumed to tend to zero. If the limit of the solutions exists with the small parameters tending to zero, then they are taken as the solutions to the equations describing the ideal sliding mode. Such a limit procedure is the *regularization* method for deriving sliding mode equations in dynamic systems with discontinuous control.

To illustrate the regularization method, we consider a linear time-invariant system with one control input, a scalar relay function of a linear combination of the state components:

$$\dot{x} = Ax + bu \qquad (x \in \Re^n) \tag{2.2.1}$$

A and b are $n \times n$ and $n \times 1$ constant matrices, $u = M\,\mathrm{sign}(s)$, M is a scalar positive constant, $s = cx$, $c = [c_1, c_2, \ldots, c_n] = \mathrm{const}$. As in the examples in Chapter 1 and Section 2.1, the state trajectories may be oriented in a direction towards the switching plane $s(x) = 0$ in the state space $x^T = (x_1, x_2, \ldots, x_n)$. Hence the sliding mode occurs in the plane (Figure 2.2) and the motion equation should be found. A similar problem was left unanswered for system (2.1.1).

Following the regularization procedure, small imperfections of a switching device should be taken into account. If a relay device is implemented with a hysteresis loop of width 2Δ (Figure 2.3), then the state trajectories oscillate in a Δ-vicinity of the switching plane (Figure 2.4). The value of Δ is assumed to be small such that the state trajectories may be approximated by straight lines with constant state velocity vectors $Ax + bM$ and $Ax - bM$ in the vicinity of some point x on the plane $s(x) = 0$.

Calculate time intervals Δt_1 and Δt_2 and increments Δx_1 and Δx_2 in the state vector for transitions from point 1 to point 2 and from point 2 to point 3, respectively:

$$\Delta t_1 = \frac{-2\Delta}{\dot{s}^+} = \frac{-2\Delta}{cAx + cbM}$$

$$\Delta x_1 = (Ax + bM)\Delta t_1 = (Ax + bM)\frac{-2\Delta}{cAx + cbM}$$

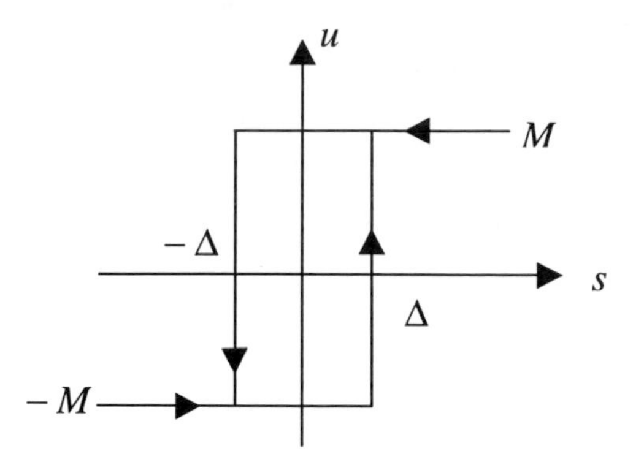

Figure 2.3 Relay with hysteresis.

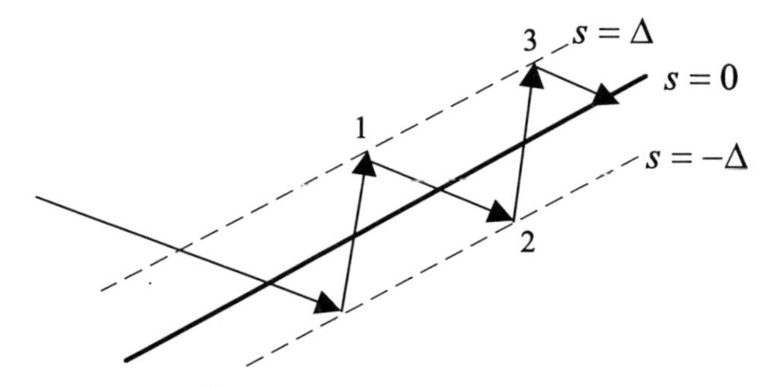

Figure 2.4 Oscillations in a vicinity of the switching surface.

Similarly for the second interval

$$\Delta t_2 = \frac{2\Delta}{\dot{s}^-} = \frac{2\Delta}{cAx - cbM}$$

$$\Delta x_2 = (Ax - bM)\Delta t_2 = (Ax - bM)\frac{2\Delta}{cAx - cbM}$$

Note that, by our assumption, sliding mode exists in the ideal system, therefore the values s and $\dot{s}$ have opposite signs, i.e. $\dot{s}^+ = cAx + cbM < 0$ and $\dot{s}^- = cAx - cbM > 0$. This implies that both time intervals Δt_1 and Δt_2 are positive. Note that the inequalities may hold if $cb < 0$. The average state velocity within the time interval $\Delta t = \Delta t_1 + \Delta t_2$ may be found as

$$\dot{x}_{av} = \frac{\Delta x_1 + \Delta x_2}{\Delta t} = Ax - (cb)^{-1}bcAx$$

The next step of the regularization procedure implies that the width of the hysteresis loop Δ should tend to zero. However, we do not need to calculate $\lim_{\Delta \to 0}(\dot{x}_{av})$: the limit procedure was performed implicitly when we assumed that state trajectories are straight lines and the state velocities are constant. This is the reason why $\dot{x}_{av}$ does not depend on Δ. And as follows from the more accurate model, the sliding mode in the plane $s(x) = 0$ is governed by

$$\dot{x} = (I_n - (cb)^{-1}bc)Ax \tag{2.2.2}$$

with initial state $s(x(0)) = 0$ and where I_n is an identity matrix. It follows from (2.2.1) and (2.2.2) that

$$\dot{s} = c(I_n - (cb)^{-1}bc)Ax \equiv 0$$

hence the state trajectories of the sliding mode are oriented along the switching plane. The condition $s(x(0)) = 0$ enables us to reduce the system order by one. To obtain the sliding mode equation of $(n-1)$th order, one of the components of the state vector, let it be x_n, may be found as a function of the other $n-1$ components and substituted into the system (2.2.2). Finally, the last equation for x_n can be disregarded.

Applying this procedure to the second-order system (2.1.1) results in a first-order sliding mode equation along the switching line $s = c_1x_1 + c_2x_2 = 0$:

$$\dot{x}_1 = (a_{11} - a_{12}c_2^{-1}c_1 - (cb)^{-1}b_1(ca^1 - ca^2c_2^{-1}c_1))x_1 + (d_1 - b_1(cb)^{-1}(cd))f$$

where $c = (c_1, c_2)$, $b^T = (b_1, b_2)$, $(a^1)^T = (a_{11}, a_{21})$, $(a^2)^T = (a_{12}, a_{22})$, $d^T = (d_1, d_2)$, and cb and c_2 are assumed to be different from zero. As we can see, for this general case of a linear second-order system, the sliding mode equation is of reduced order and depends on the plant parameters, disturbances and coefficients of switching line equations, but does not depend on control.

For the systems in canonical form (1.2.1) and (1.2.3), the above regularization method may serve as validation that the reduced-order sliding mode equations (1.2.2) and (1.2.4) depend on neither plant parameters nor disturbances.

Exactly the same equations for our examples result from regularization based on an imperfection of 'delay' type (Andre and Seibert, 1956). It is interesting to note that nonlinear systems of an arbitrary order with one discontinuity surface were studied in this paper and the motion equations proved to be the same for both types of imperfections – hysteresis and delay. This result may be easily interpreted in terms of relative time intervals for the control input to take each of two extreme values. For a system of an arbitrary order with scalar control

$$\dot{x} = f(x, u) \qquad x, f \in \Re^n, \ u(x) \in \Re \tag{2.2.3}$$

$$u(x) = \begin{cases} u^+(x) \text{ if } s(x) > 0 \\ u^-(x) \text{ if } s(x) < 0 \end{cases}$$

the components of vector f, scalar functions $u^+(x)$, $u^-(x)$ and $s(x)$ are continuous and smooth, and $u^+(x) \neq u^-(x)$. We assume that sliding mode occurs on the surface $s(x) = 0$ and try to derive the motion equations using the regularization method. Again, let the discontinuous control be implemented with some unspecified imperfections; control is known to take one of the two extreme values, $u^+(x)$ or $u^-(x)$, and the discontinuity points are isolated in time. As a result, the solution exists in the conventional sense and it does not matter whether we deal with small hysteresis, time delay or time constants that are neglected in the ideal model.

As in the system (2.2.1) with hysteresis imperfection, the state velocity vectors $f^+ = f(x, u^+)$ and $f^- = f(x, u^-)$ are assumed to be constant for some point x on the surface $s(x) = 0$ within a short time interval $(t, t + \Delta t)$. Let the time interval Δt consist of two sets of intervals Δt_1 and Δt_2 such that $\Delta t = \Delta t_1 + \Delta t_2$, $u = u^+$

for the time from the set Δt_1 and $u = u^-$ for the time from the set Δt_2. Then the increment of the state vector after time interval Δt is found as

$$\Delta x = f^+ \Delta t_1 + f^- \Delta t_2$$

and the average state velocity as

$$\dot{x}_{av} = \frac{\Delta x}{\Delta t} = \mu f^+ + (1 - \mu) f^-$$

where $\mu = \Delta t_1 / \Delta t$ is the relative time for control to take value u^+ and $(1 - \mu)$ is the relative time to take value u^-, $0 \leq \mu \leq 1$. To get the vector $\dot{x}$, the time Δt should be tended to zero. However, we do not need to perform this limit procedure, it is hidden in our assumption that the state velocity vectors are constant within time interval Δt, therefore the equation

$$\dot{x} = \mu f^+ + (1 - \mu) f^- \tag{2.2.4}$$

represents the motion during sliding mode. Since the state trajectories during sliding mode are on the surface $s(x) = 0$, the parameter μ should be selected such that the state velocity vector of the system (2.2.4) is in the tangential plane to this surface, or

$$\dot{s} = \text{grad}[s(x)] \cdot \dot{x} = \text{grad}[s(x)][\mu f^+ + (1 - \mu) f^-] = 0 \tag{2.2.5}$$

with $\text{grad}[s(x)] = [\partial s / \partial x_1 \ldots \partial s / \partial x_n]$.

The solution to (2.2.5) is given by

$$\mu = \frac{\text{grad}(s) \cdot f^-}{\text{grad}(s) \cdot (f^- - f^+)} \tag{2.2.6}$$

Substitution of (2.2.6) into (2.2.4) results in the sliding mode equation

$$\dot{x} = f_{sm}, \qquad f_{sm} = \frac{(\text{grad } s) \cdot f^-}{(\text{grad } s) \cdot (f^- - f^+)} f^+ - \frac{(\text{grad } s) \cdot f^+}{(\text{grad } s) \cdot (f^- - f^+)} f^- \tag{2.2.7}$$

representing the motion in sliding mode with initial condition $s[x(0)] = 0$. Note that sliding mode occurs on the surface $s(x) = 0$, therefore the functions s and $\dot{s}$ have different signs in the vicinity of the surface (Figure 2.5) and $\dot{s}^+ = (\text{grad } s) \cdot f^+ < 0$, $\dot{s}^- = (\text{grad } s) \cdot f^- > 0$. As follows from (2.2.6), the condition $0 \leq \mu \leq 1$ holds for parameter μ. It is easy to check the condition $\dot{s} = (\text{grad } s) \cdot f_{sm} = 0$ for the trajectories of system (2.2.7) and to show they are confined to the switching surface $s(x) = 0$. As might be expected, direct substitution of $\text{grad } s = c$, $f^+ = Ax + bu^+$ and $f^- = Ax + bu^-$ into (2.2.7) results in the sliding mode equation (2.2.2) derived for the linear system (2.2.1) with the discontinuity plane $s(x) = cx = 0$ via hysteresis regularization.

It is interesting to note that the above regularization method for deriving the sliding mode equation may be considered as a physical interpretation of the famous *Filippov* method. The method is intended for solution continuation at a discontinuity surface for differential equations with discontinuous right-hand sides (Filippov, 1988). According to this method, the ends of all state velocity vectors in the vicinity of a point on a discontinuity surface should be complemented by a minimal convex set and the state

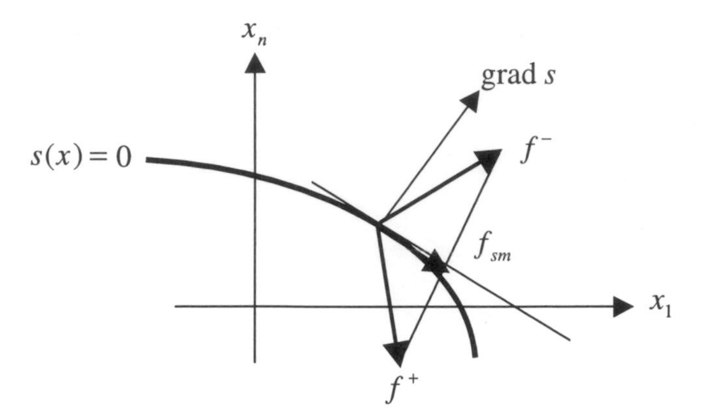

Figure 2.5 Sliding mode equation by Filippov's method.

velocity vector of the sliding motion should belong to this set. In our case there are two points, the ends of vectors f^+ and f^-, and the minimal convex set is the straight line connecting their ends. The equation of this line is exactly the right-hand side of equation (2.2.4). The intersection of the line with the tangential plane defines the state velocity vector in the sliding mode, or the right-hand side of the sliding mode equation. It is clear that the result of Filippov's method coincides with the equation derived by the regularization approach.

The regularization methods discussed above and the methods studied by Andre and Seibert (1956) were developed under rather restrictive assumptions: a special class of imperfections (delay or hysteresis) control may take only two extreme values, systems with scalar control and one discontinuity surface. The general regularization concept embracing a wider class of imperfections (such as continuous approximation of a discontinuous function) and sliding modes in the intersection of several surfaces is regularization via *boundary layer* (Utkin 1971/72, 1992). We describe the idea of the boundary layer approach for an arbitrary system with vector control:

$$\dot{x} = f(x, u) \qquad x, f \in \Re^n, \;\; u(x) \in \Re^m \tag{2.2.8}$$

$$u(x) = \begin{cases} u^+(x) \text{ for } s(x) > 0 \\ u^-(x) \text{ for } s(x) < 0 \end{cases} \quad \text{(componentwise)}$$

The components of vector $s(x)^T = [s_1(x) \ldots s_m(x)]$ are m smooth functions and the ith component of the control undergoes discontinuities on the ith surface $s_i(x) = 0$. Similar to sliding mode in the intersection of two planes (2.1.2), sliding mode may occur in the manifold $s(x) = 0$. To obtain the sliding mode equations, the ideal control in (2.2.8) is replaced by a new control $\tilde{u}$ such that the solution to (2.2.8) with this control exists in the conventional sense. Due to the substitution, the trajectories are not confined to the manifold $s(x) = 0$ but run in its boundary layer of width $\Delta > 0$ (Figure 2.6):

$$\|s(x)\| \le \Delta \qquad \|s\| = (s^T s)^{1/2}$$

The imperfections taken into account in the control $\tilde{u}$ are not specified and it is only known that the solution to (2.2.8) with the new control exists in the conventional sense. As a rule, real-life imperfections belong to this class (including hysteresis, time delay and small time constants neglected in the ideal model).

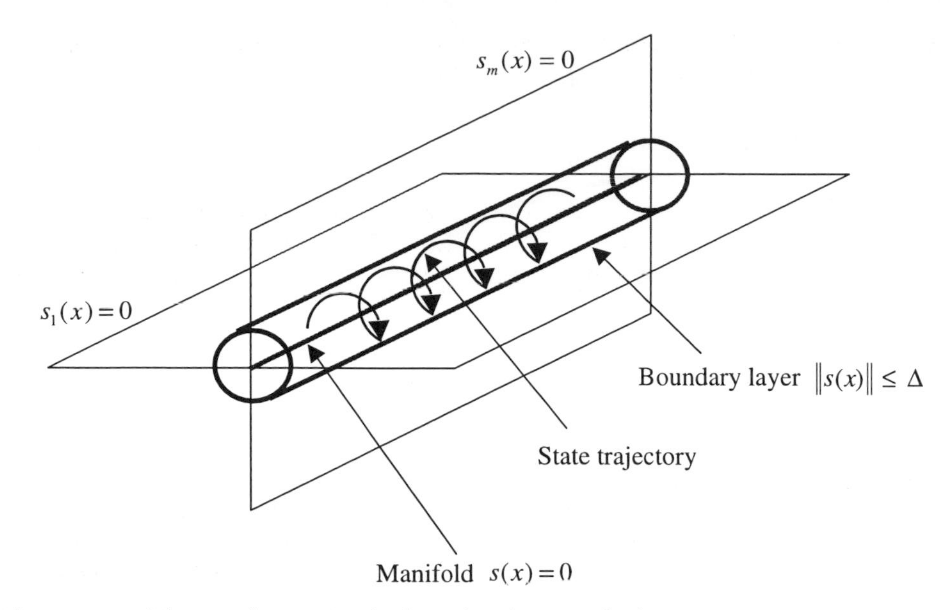

Figure 2.6 Sliding mode equation by boundary layer method.

The core idea of the boundary layer regularization method is as follows. If the limit of the solution to (2.2.8), with $u = \tilde{u}$ and the width of the boundary layer tending to zero, exists and is unique and does not depend on the type of imperfections and the way in which Δ tends to zero, i.e. it is independent of the limit procedure in

$$\lim_{\Delta \to 0} x(t, \Delta) = x^*(t) \tag{2.2.9}$$

then the function $x^*(t)$ is taken as the solution to (2.2.8) with ideal sliding mode. Otherwise, we should recognize that the motion equations beyond the discontinuity manifold do not let us derive unambiguously the equations for the motion in the manifold. The particular cases of regularization for the systems with scalar discontinuous control handled in this section have shown that equation (2.2.7) is the 'right' model of sliding mode. More general cases will be discussed in the next section using the boundary layer approach.

2.3 Equivalent control method

From a methodological viewpoint it is convenient to develop a procedure for deriving the sliding mode equations for system (2.2.8) by beginning with a heuristic method and then, using the boundary layer regularization approach, going on to analyze whether the equations can be taken as the motion model.

We assume that the initial state vector of system (2.2.8) is in the intersection of all discontinuity surfaces, i.e. in the manifold $s(x) = 0$, and sliding mode occurs with the state trajectories confined to this manifold for $t > 0$.

Since motion in the sliding mode implies $s(x) = 0$ for $t > 0$, we may assume that $ds/dt = \dot{s} = 0$ as well. Hence, in addition to $s(x) = 0$, the time derivative $\dot{s}(x) = 0$ may be used to characterize the state trajectories during sliding mode. The fast switching control u is an obstacle for using conventional methods, so disregard the control discontinuities and calculate the vector u such that the time derivative of vector on the state trajectories of (2.2.8) is equal to zero:

$$\dot{s}(x) = G \cdot f(x, u) = 0 \tag{2.3.1}$$

where $G = (\partial s/\partial x)$ is an $m \times n$ matrix with gradients of functions $s_i(x)$ as rows. Let a solution to the algebraic equation (2.3.1) exist. The solution $u_{eq}(x)$ will be called the *equivalent control*. This continuous function is substituted for the discontinuous control u into the original system (2.2.8):

$$\dot{x} = f(x, u_{eq}) \tag{2.3.2}$$

It is evident that for initial conditions $s(x(0)) = 0$, in compliance with (2.3.1), further motion governed by (2.3.2) will be along the state trajectories in the manifold $s(x) = 0$, just as for sliding mode in system (2.2.8). Equation (2.3.2) is taken as the equation of sliding mode in the intersection of m discontinuity surfaces $s_i(x) = 0$, $(i = 1, \ldots, m)$. The procedure for deriving the equation will be called the *equivalent control method.*

From a geometrical viewpoint, the equivalent control method means replacement of discontinuous control in the intersection of switching surfaces by a continuous control such that the state velocity vector lies in the tangential manifold. For example, in the system with scalar control (2.2.3) this vector may be found as the intersection of the tangential plane and the locus $f(x, u)$ with control u running from u^- to u^+ (Figure 2.7). The intersection point defines the equivalent control u_{eq} and the right-hand side $f(x, u_{eq})$ in the sliding mode equation (2.1.2).

Note that the right-hand side $f(x, u_{eq})$ of the motion equation resulting from the equivalent control method does not coincide with that of Filippov's method (f_{sm} in (2.7) and in Figure 2.5). They are equal if system (2.2.3) with scalar control is linear with respect to control $f(x, u) = f_0(x) + b(x)u$ (f_0 and b are n-dimensional vectors). Then the locus $f(x, u)$ of the equivalent control method (Figure 2.7) coincides with a minimal convex set (the straight line connecting the ends of vectors f^+ and f^-) of *Filippov's* method. The discrepancy reflects the fact that different ways of regularization lead to different sliding mode equations in systems with nonlinear functions of control input in motion equations (Utkin, 1971/72).

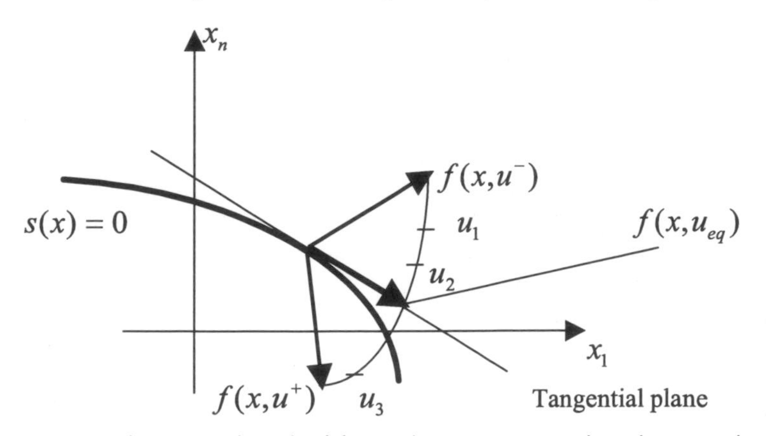

Figure 2.7 Equivalent control method for nonlinear systems with scalar control.

We apply the equivalent method procedure to so-called affine systems, i.e. nonlinear systems with right-hand sides in the motion equations (2.2.8) as linear functions of the control input u:

$$\dot{x} = f(x) + B(x)u \qquad x,\ f(x) \in \Re^n,\ \ B(x) \in \Re^{n\times m},\ \ u(x) \in \Re^m \tag{2.3.3}$$

$$u(x) = \begin{cases} u^+(x) \text{ for } s(x) > 0 \\ u^-(x) \text{ for } s(x) < 0 \end{cases} \quad \text{(componentwise)} \qquad s(x)^T = [s_1(x) \ldots s_m(x)]$$

Similar to system (2.2.8), each surface $s_i(x) = 0$ is the set of discontinuity points for corresponding control component u_i.

Equation (2.3.1) of the equivalent control method for system (2.3.2) is of form

$$\dot{s} = Gf + GBu_{eq} = 0 \quad \text{where} \quad G = (\partial s/\partial x) \tag{2.3.4}$$

Assuming that matrix GB is nonsingular for any x, find the equivalent control $u_{eq}(x)$ as the solution to (2.3.4):

$$u_{eq}(x) = -(G(x)B(x))^{-1}G(x)f(x)$$

and substitute $u_{eq}(x)$ into (2.3.3) to yield the sliding mode equation as

$$\dot{x} = f(x) - B(x)\,(G(x)B(x))^{-1}G(x)f(x) \tag{2.3.5}$$

Equation (2.3.5) is taken as the equation of sliding mode in the manifold $s(x) = 0$. The equation has been postulated. According to our concept, the question of whether it is a 'right' model of the motion in sliding mode, may be answered by involving the regularization method based on introducing a boundary layer of the manifold $s(x) = 0$. For the affine systems (2.3.3), the sliding mode equation is found uniquely in the framework of the method and it coincides with (2.3.5) resulting from the equivalent control method. This statement is substantiated in Utkin (1971/72) under general assumptions related to smoothness and growth rate for the functions f, B, u^+, u^- and s. According to these results, condition (2.2.9) holds, which means that any solution in the boundary layer $x(t, \Delta)$ tends to a solution $x^*(t)$ of equation (2.3.5) regardless what kind of imperfection caused the motion in the boundary layer and regardless how the boundary layer is reduced to zero.

Formally, the equivalent control method may be applied to systems which are nonlinear with respect to control as well. The result differs from the equations of Filippov's method even for systems with scalar control.

Attempts to show whether this method or that method is 'right' by employing the regularization approach have been unsuccessful, since the sliding mode equations that result from the limit procedure depend on the nature of the introduced imperfections and the way in which they are tended to zero. For example, sliding modes in relay systems with small delay or hysteresis in a switching device are governed by the equation of Filippov's method; but for a piecewise smooth continuous approximation of a discontinuous function they are governed by the equation of the equivalent control method. Utkin (1992) gives details of sliding mode analysis in nonlinear systems and explains the reasons for the ambiguity.

Qualitatively, the result for affine systems may be explained in terms of a system block diagram (Figure 2.8). In compliance with the equivalent control method, the time derivative $\dot{s}$ is formally set equal to zero (2.3.4). For motion in a boundary layer,

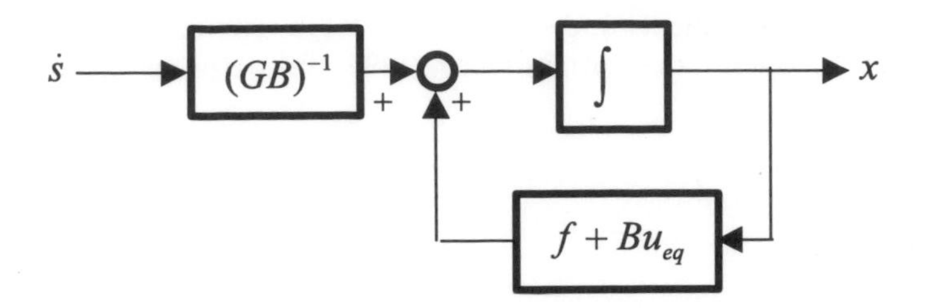

Figure 2.8 Equivalent control method for affine systems.

s is a small value of order Δ but $\dot{s}$ takes finite values and does not tend to zero with Δ. This means that real control does not satisfy equation (2.3.4) and may be found as

$$u = u_{eq} + (GB)^{-1}\dot{s}$$

The motion equation in the boundary layer is governed by

$$\dot{x} = f + Bu_{eq} + (GB)^{-1}\dot{s}$$

It differs from the ideal sliding equation (2.16) by the additional term $(BG)^{-1}\dot{s}$.

In terms of block diagrams any dynamic system may be represented as a series of integrators and it is natural to assume that an output of each of them may be estimated by an upper bound of the integral of an input.

In our case the input is $\dot{s}$ and its integral s tends to zero with $\Delta \to 0$. Therefore the response to this input tends to zero. This explains why the solution with the boundary layer reduced to zero tends to the same result as the solution to the equation derived using equivalent control. Of course, this is not the case for systems with nonlinear functions of control in motion equations; indeed, generally speaking $\int_0^t h(\dot{s})\,d\tau$ does not tend to zero even if s does and $h(0) = 0$. For example, if $s = -(1/\omega)\cos\omega t$, $\dot{s} = \sin\omega t$, then s tends to zero with $\omega \to \infty$, but $\int_0^t (\dot{s})^2 d\tau$ does not. Uniqueness of sliding mode equations in affine systems explains why the major attention is paid to this class in theory of sliding mode control. In practical applications the most common systems are those which are nonlinear with respect to a state vector and linear with respect to a control input.

2.4 Physical meaning of equivalent control

The motion in sliding mode was regarded as a certain idealization. It was assumed that the control changes at high, theoretically infinite, frequency such that the state velocity vector is oriented precisely along the intersection of discontinuity surfaces. However, in reality, various imperfections make the state oscillate in some vicinity of the intersection and components of control are switched at finite frequency, alternately taking the values $u_i^+(x)$ and $u_i^-(x)$. These oscillations have high frequency and slow components. The high frequency is filtered out by a plant under control while its motion in sliding mode is determined by the slow component. On the other hand, sliding mode equations were obtained by substitution of equivalent control for the real control. It is reasonable to assume that the equivalent control is close to the slow component of the real control which may be derived by filtering out the high-frequency component using a lowpass filter. Its time constant should be sufficiently small to preserve the slow component undistorted but large enough to eliminate the high-frequency component. As shown in Utkin (1992), the output of the lowpass filter

$$\tau\dot{z} + z = u$$

tends to the equivalent control

$$\lim_{\tau\to 0,\, \Delta/\tau\to 0} z = u_{eq}$$

This way of tending z to u_{eq} is not something complicated, but naturally follows from physical properties of the system. Indeed, the vicinity of a discontinuity manifold of width Δ, where the state oscillates, should be reduced to make the real motion close to ideal sliding mode. For reduction of Δ, the switching frequency f of the control should be increased, otherwise the amplitude of oscillations would exceed Δ since $\Delta \approx 1/f$. To eliminate the high-frequency component of the control in sliding mode, the frequency should be much higher than $1/\tau$, or $1/f \ll \tau$, hence $\Delta \ll \tau$. Finally, the time constant of the lowpass filter should be made to tend to zero because the filter should not distort the slow component of the control. Thus the conditions $\tau \to 0$ and $\Delta/\tau \to 0$ (which implies $\Delta \to 0$) should be fulfilled to extract the slow component equal to the equivalent control and to filter out the high-frequency component.

It is interesting that the equivalent control depends on plant parameters and disturbances which may be unknown. For example, let us assume that sliding mode exists on the line $s = 0$ in system (2.1.1):

$$\ddot{x} + a_2\dot{x} + a_1 x = u + f(t)$$
$$u = -M\,\text{sign}(s) \qquad s = \dot{x} + cx$$

where M, a_1, a_2, c are constant parameters and $f(t)$ is a bounded disturbance. Equivalent control is the solution to the equation $\dot{s} = -a_2\dot{x} - a_1 x + u + f(t) = 0$ with respect to u under condition $s = 0$, or $\dot{x} = -cx$:

$$u_{eq} = (-a_2 c + a_1)x - f(t)$$

The equivalent control depends on parameters a_1, a_2 and disturbance $f(t)$. Extracting equivalent control by a lowpass filter, this information may be obtained and used for improvement of feedback control system performance. Furthermore, this opportunity will be used in Chapter 6 for designing state observers with sliding modes and in Chapter 8 for chattering suppression.

2.5 Existence conditions

The methods developed in the previous section enable us to write down the sliding mode equation should sliding mode occur in a system. If the sliding mode exhibits the desired dynamic properties the control should be designed such that this motion is enforced. Hence the conditions for sliding mode to exist should be derived – the second mathematical problem in the analysis of sliding mode as a phenomenon. For the systems with scalar control studied in Chapters 1 and 2, the conditions were obtained from geometrical considerations: the deviation from the switching surface

s and its time derivative should have opposite signs in the vicinity of a discontinuity surface $s = 0$, or (Barbashin, 1967)

$$\lim_{s \to +0} \dot{s} < 0, \quad \text{and} \quad \lim_{s \to -0} \dot{s} > 0 \tag{2.4.1}$$

For system (1.2.1) the domain of sliding mode (sector $m - n$ on the switching line in Figure 1.7) was found based on geometric considerations. It may be found analytically from (2.4.1) as

$$\dot{s} = (-c^2 + a_2 c - a_1)x - M\,\text{sign}(s) + f(t)$$

and the domain of sliding for bounded disturbance $|f(t)| < f_0$ is given by

$$|x| < \frac{M - f_0}{|-c^2 + a_2 c - a_1|}$$

As demonstrated by system (2.1.2), for the existence of sliding mode in an intersection of a set of discontinuity surfaces $s_i(x) = 0$, $(i = 1, \ldots, m)$ it is not necessary to fulfill inequalities (2.4.1) for each of them. This system showed that the trajectories converge to the manifold $s^T = [s_1 \ldots s_m] = 0$ and reach it after a finite time interval, similar to the systems with scalar control. The term 'converge' means that we deal with the problem of stability of the origin in m-dimensional subspace $(s_1, \ldots, s_m)$, therefore the existence conditions may be formulated in terms of stability theory.

In addition, *finite time convergence* should take place. This nontraditional condition is important to distinguish systems with sliding modes from continuous systems with state trajectories converging to some manifold asymptotically. For example, the state trajectories of the system $\ddot{x} - x = 0$ converge to the manifold $s = \dot{x} - x = 0$ asymptotically since $\dot{s} = -s$; however, it would hardly be reasonable to call the motion in $s = 0$ 'sliding mode'.

In the sequel, we examine the conditions for sliding mode to exist for affine systems (2.3.3). To derive the conditions, we need to analyze the stability of the motion projection on subspace s governed by the differential equation

$$\dot{s} = Gf + GBu \tag{2.4.2}$$

The control (2.3.3)

$$u(x) = \begin{cases} u^+(x) \text{ for } s(x) > 0 \\ u^-(x) \text{ for } s(x) < 0 \end{cases} \quad \text{(componentwise)} \qquad s(x)^T = [s_1(x) \ldots s_m(x)]$$

may be represented as

$$u(x) = u_0(x) + U(x)\,\text{sign}(s) \tag{2.4.3}$$

where $u_0(x) = \frac{1}{2}(u^+(x) + u^-(x))$, $U(x)$ is a diagonal matrix with elements $U_i(x) = \frac{1}{2}(u_i^+(x) - u_i^-(x))$ for $i = 1, \ldots, m$ and the discontinuous control sign(s) takes the form of a componentwise sign function

$$\text{sign}(s)^T = [\text{sign}(s_1) \ldots \text{sign}(s_m)] \tag{2.4.4}$$

Then the motion projection on subspace s is governed by

$$\dot{s} = d(x) - D(x)\,\text{sign}(s) \quad \text{with} \quad d = Gf + GBu_0, \; D = -GBU \tag{2.4.5}$$

To find the stability conditions of the origin $s = 0$ for nonlinear system (2.4.5), i.e. the conditions for sliding mode to exist, we will follow the standard methodology for stability analysis of nonlinear systems – try to find a *Lyapunov* function. At the same time, we should remember that the right-hand side in the motion equation is discontinuous and not defined in the points where arguments of the sign functions are equal to zero. To illustrate that the problem needs subtle treatment, let us turn to system (2.1.2) in Section 2.1 with the equation of motion projection on subspace (s_1, s_2) given by

$$\dot{s}_1 = -\text{sign}\ s_1 + 2\,\text{sign}\ s_2$$
$$\dot{s}_2 = -2\,\text{sign}\ s_1 - \text{sign}\ s_2$$

The time derivative of the positive definite Lyapunov function candidate $V = |s_1| + |s_2|$ along the system trajectories

$$\dot{V} = \frac{\partial V}{\partial s_1}\dot{s}_1 + \frac{\partial V}{\partial s_{21}}\dot{s}_2 = \text{sign}(s_1)(-\text{sign}\ s_1 + 2\,\text{sign}\ s_2)$$
$$+\text{sign}(s_2)(-2\,\text{sign}\ s_1 - \text{sign}\ s_2) = -2$$

is negative definite and we may conclude that the origin of the state plane (s_1, s_2) is asymptotically stable. Figure 2.1 shows that this is the case. The time needed for the state to reach the origin is calculated as $T = V(0)/|\dot{V}| = V(0)/2 = |s_2(0)|/2$ for initial condition $s_1(0) = 0$, $s_2(0) \neq 0$. The result coincides with the reaching time found by the point-to-point transform method in Section 2.1.

However, the conclusion on asymptotic stability made for system (2.1.2) is not always correct. The time derivative of the positive definite Lyapunov function candidate

$$V = 4|s_1| + |s_2|$$

along the trajectories of another system

$$\dot{s}_1 = -2\,\text{sign}\ s_1 - \text{sign}\ s_2$$
$$\dot{s}_2 = -2\,\text{sign}\ s_1 + \text{sign}\ s_2$$

is negative

$$\dot{V} = \frac{\partial V}{\partial s_1}\dot{s}_1 + \frac{\partial V}{\partial s_2}\dot{s}_2 = -7 - 6\,\text{sign}(s_1)\text{sign}(s_2)$$

everywhere except at the discontinuity surfaces. But it does not testify to stability. The state plane in Figure 2.9 shows that the state trajectories reach the plane $s_1 = 0$ on which sliding mode occurs (the existence conditions (2.4.1) hold in this plane). Following the equivalent control method, the motion equation may be obtained by finding $[\text{sign}(s_1)]_{eq}$ from equation $\dot{s}_1 = 0$ and substituting it into the second equation. This results in sliding equation $\dot{s}_2 = 2\text{sign}(s_2)$ with unstable solution and s_2 tending to infinity.

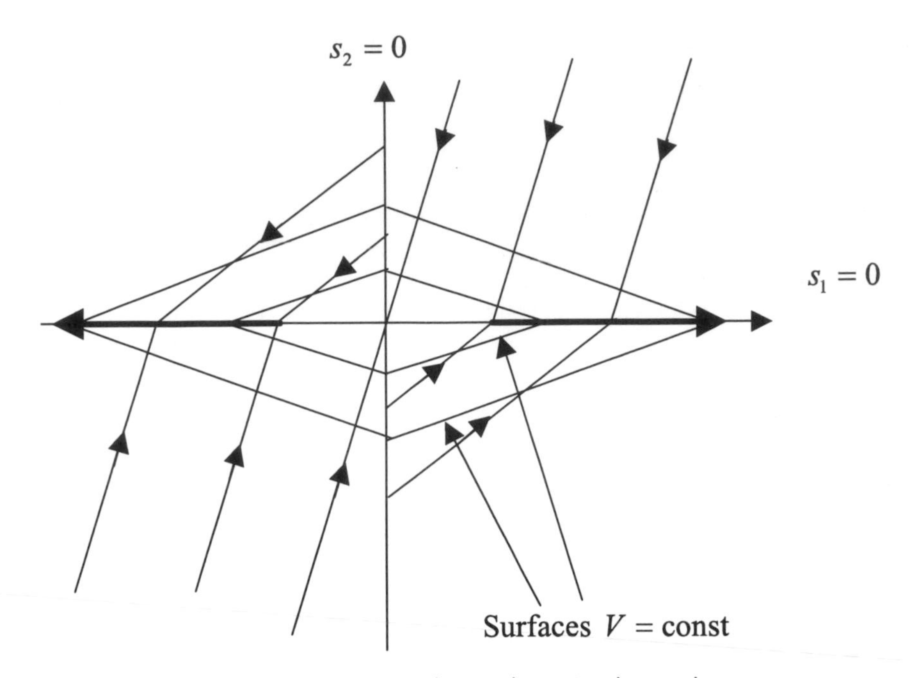

Figure 2.9 The s-plane for the system with two-dimensional control.

Instability of the motion along the plane $s_1 = 0$ means that no sliding mode occurs in the intersection of the discontinuity surfaces $s_1 = 0$ and $s_2 = 0$. The trajectories intersect the surfaces V = const from outside everywhere except for the corner points with $s_1 = 0$, and the trajectories of sliding mode in $s_1 = 0$ diverge through these points. In Figure 2.1, which shows the state trajectories for system (2.1.2), sliding mode occurs on none of the discontinuity surfaces, therefore the discontinuity points of the right-hand sides in the motion equations are isolated. This is the reason for decreasing the Lyapunov function; it also explains the difference in the signs of V and $\dot{V}$, suggesting asymptotic stability of the origin in subspace (s_1, s_2), and the existence of sliding mode in the intersection of planes $s_1 = 0$ and $s_2 = 0$. However, as shown by Figure 2.9, knowledge of the signs of a piecewise smooth function and its derivative, generally speaking, is not sufficient to ascertain the existence of sliding mode.

To be able to use a Lyapunov function in the form of the sum of absolute values whenever sliding mode occurs on some of the discontinuity surfaces, the corresponding components of discontinuous control should be replaced by their equivalent ones and only then should the time derivative of the Lyapunov function be found.

Definition 2.1

The set $S(x)$ in the manifold $s(x) = 0$ is the domain of sliding mode if, for the motion governed by equation (2.4.5), the origin in the subspace s is asymptotically stable with finite convergence time for each x from $S(x)$. □

Definition 2.2

Manifold $s(x) = 0$ is called a *sliding manifold* if sliding mode exists at each point, or $S(x) = \{x : s(x) = 0\}$. □

Theorem 2.1

If the matrix D in the equation

$$\dot{s} = -D\,\mathrm{sign}(s) \tag{2.4.6}$$

is positive definite

$$D + D^T > 0 \tag{2.4.7}$$

then the origin $s = 0$ is an asymptotically stable equilibrium point with finite convergence time. □

Proof

Let the sum of absolute values of s_i

$$V = \mathrm{sign}(s)^T s > 0 \tag{2.4.8}$$

be a Lyapunov function candidate.

Partition vector s into two subvectors $s^T = (s^k)^T(s^{m-k})^T$, assuming that sliding mode occurs in the intersection of k discontinuity surfaces, $s^k = 0$, and the components of vector s^{m-k} are different from zero.

According to the equivalent control method, vector $\mathrm{sign}(s^k)$ in the motion equation must be replaced by the function $(\mathrm{sign}(s^k))_{eq}$ such that $\dot{s}^k = 0$. Since $s^k = 0$ in sliding mode, the time derivative of V in (2.4.8) consists of $(m-k)$ terms:

$$\begin{aligned}\dot{V} &= \frac{d}{dt}((\mathrm{sign}(s^{m-k}))^T s^{m-k}) \\ &= (\mathrm{sign}(s^{m-k}))^T \dot{s}^{m-k}\end{aligned}$$

The value of $(\mathrm{sign}(s^k))_{eq}^T \dot{s}^k$ is equal to zero in sliding mode, therefore

$$\dot{V} = (\mathrm{sign}(s^k))_{eq}^T \dot{s}^k + (\mathrm{sign}(s^{m-k}))^T \dot{s}^{m-k}$$

Replacing vector $\dot{s}$ with its value from (2.4.6) with vector $\mathrm{sign}(s^k)$ substituted by $(\mathrm{sign}(s^k))_{eq}$, we have

$$\dot{V} = -z^T D z = -z^T \frac{D + D^T}{2} z$$

where $z^T = (\mathrm{sign}(s^k))_{eq}^T (\mathrm{sign}(s^{m-k}))^T$. Since matrix D is positive definite (2.4.7), $\dot{V} \le -\lambda_{\min}\|z\|^2$, $\lambda_{\min} > 0$ is the minimal eigenvalue of matrix $\frac{1}{2}(D + D^T)$, $\|z\| = \sqrt{z^T z}$. (Generally speaking, $Q \ne \frac{1}{2}(Q + Q^T)$ but $z^T Q z = z^T[\frac{1}{2}(Q + Q^T)]z$.)

At the initial time, at least one of the components of vector s is different from zero, therefore one of the components of vector z is equal to $+1$ or -1, $\|z\| \ge 1$ and the Lyapunov function (2.4.8) decays at a finite rate

$$\dot{V} \le -\lambda_{\min} \tag{2.4.9}$$

Condition (2.4.9) means that V and vector s vanish after a finite time interval and the origin $s = 0$ is an asymptotically stable equilibrium point with finite convergence time. ■

Two second-order systems were studied in this section. For both of them the time derivatives of positive definite Lyapunov functions of 'sum of absolute values' type were negative definite everywhere except at the discontinuity surfaces. For the first second-order system in this section

$$D = \begin{pmatrix} 1 & -2 \\ 2 & 1 \end{pmatrix}$$

and matrix $D + D^T > 0$ is positive definite, which testifies to stability (Figure 2.1). The second system with

$$D = \begin{pmatrix} 2 & 1 \\ 2 & -1 \end{pmatrix}$$

is unstable (Figure 2.9) and matrix

$$D + D^t = \begin{pmatrix} 4 & 3 \\ 3 & -2 \end{pmatrix}$$

is not positive definite.

The sufficient stability (or sliding mode existence) condition formulated in Theorem 2.1 for the system (2.4.6) can be easily generalized for the cases (2.4.2) or (2.4.5). For $s(x) = 0$ to be a sliding manifold, it is sufficient that for any $x \in S$, $S\{x : s(x) = 0\}$

$$\begin{aligned} &D(x) + D^T(x) > 0 \\ &\lambda_0 > d_0\sqrt{m} \quad \text{with} \quad \lambda_{\min}(x) > \lambda_0 > 0 \\ &\|d(x)\| < d_0 \end{aligned} \tag{2.4.10}$$

where $\lambda_{\min}(x)$ is the minimal eigenvalue of $\frac{1}{2}(D + D^T)$, λ_0 is a constant positive value, and d_0 is an upper estimate of vector $d(x)$ in (2.4.5).

The time derivative of Lyapunov function (2.4.8) along the trajectories of (2.4.2) has a form similar to (2.4.9):

$$\begin{aligned} \dot{V} &= z^T d(x) - z^T \frac{D + D^T}{2} z \\ &\leq \|z\| \, \|d(x)\| - \lambda_{\min} \end{aligned}$$

The components of vector z are either $\text{sign}(s_i)$ or $(\text{sign}(s_i))_{eq}$. As shown in Section 2.3, the equivalent control is an average value of discontinuous control, and this value is a continuous function varying between the two extreme values of the discontinuous control. This means that $|(\text{sign}(s_i))_{eq}| \leq 1$, hence the norm of m-dimensional vector z with components from the range ± 1 does not exceed $\sqrt{m}$ and

$$\dot{V} \leq d_0\sqrt{m} - \lambda_0 < 0 \tag{2.4.11}$$

Since the Lyapunov function decays at a finite rate, it vanishes and sliding mode occurs after a finite time interval.

Remark 2.1

If inequality (2.4.11) holds for any x then it is simultaneously *the reaching condition*, i.e. the condition for the state to reach the sliding manifold from any initial point. □

Remark 2.2

Upper and lower estimates of $d(x)$ and $\lambda_{\min}(x)$, respectively, may depend on x. Then the condition for a sliding manifold to exist and the reaching condition is of the form

$$\dot{V} \leq \|d(x)\|\sqrt{m} - \lambda_{\min}(x) < -v_0 \tag{2.4.12}$$

where v_0 is a positive constant. □

Remark 2.3

Functions $d(x)$ and $D(x)$ depend on control (2.4.3) and (2.4.5). The value of $\lambda_{\min}(x)$ may be increased by increasing the difference between $u_i^+(x)$ and $u_i^-(x)$ without varying $u_0(x)$ and $d(x)$. Then condition (2.4.11) or (2.4.12) can be fulfilled. □

General Remark for Chapter 2

When studying the equations of sliding modes and conditions for this motion to exist, only time-invariant systems were handled, but all results are valid for time-varying systems as well. The difference for a time-varying switching manifold $s(x, t) = 0$ is that to find the equivalent control, equation (2.3.4) should be complemented by an additional term $\partial s(x, t)/\partial t$:

$$\dot{s} = Gf + GBu_{eq} + \partial s(x, t)/\partial t = 0$$

and its solution $u_{eq} = -(GB)^{-1}(Gf + \partial s/\partial t)$ should be substituted into (2.3.3) regardless whether functions f and B in motion equation (2.3.3) depend on time or are time-invariant. □

References

ANDRE, J. and SEIBERT, P., 1956, Über Stückweise Lineare Differential-Gleichungen, die bei Regelungsproblemen Auftreten I, II (in German), *Archiv der Math.*, **7**, 2–3.

BARBASHIN, E., 1967, *Introduction to the Theory of Stability* (in Russian), Moscow: Nauka.

FILIPPOV, A., 1988, *Differential Equations with Discontinuous Right-Hand Sides*, Dordrecht: Kluwer.

UTKIN, V., 1971/72, Equations of slipping regimes in discontinuous systems I, II, *Automation and Remote Control*, **32**, 1897–1907 and **33**, 211–19.

UTKIN, V., 1992, *Sliding Modes in Control and Optimization*, Berlin: Springer-Verlag.

CHAPTER THREE

Design Concepts

The key idea of the design methodology for sliding mode control was outlined in Section 1.4. According to this idea, any design procedure should consist of two stages. As established in Chapter 2, sliding modes are governed by a reduced-order system depending on the equations of some discontinuity surfaces. The first stage of design is the selection of the discontinuity surfaces such that sliding motion would exhibit desired properties. The methods of conventional control theory, such as stabilization, eigenvalue placement and dynamic optimization, may be applied at this stage. The second stage is to find discontinuous control to enforce sliding mode in the intersection of the surfaces selected at the first stage. The second problem is of reduced order as well, since its dimension is equal to the number of discontinuity surfaces, which is usually equal to the dimension of control.

Partitioning of the overall motion into two motions of lower dimensions – the first motion precedes sliding mode within a finite time interval and the second motion is sliding mode with the desired properties – may simplify the design procedure considerably. In addition, sliding modes may be insensitive with respect to unknown plant parameters and disturbances, although the invariance property does take place for any system, as demonstrated for system (2.1.1) in Sections 2.1 and 2.2. In this chapter, different sliding mode control design methods based on the decoupling principle will be developed and special attention will be paid to the class of systems with invariant sliding motions.

3.1 Introductory Example

As an example of sliding mode control design, a multilink manipulator may be considered under the assumption that each link is subjected to a control force or torque. The system motion is represented by a set of interconnected second-order equations

$$M(q)\ddot{q} + f(\dot{q}, q, t) = u$$

where q and u are vectors of the same dimension of generalized states and force or torque control components respectively, $M(q)$ is a positive definite inertia matrix, and $f(q, \dot{q}, t)$ is a function depending on the system geometry, the velocity vector, unknown parameters and disturbances. The motion equation may be represented as

$$\dot{p} = v \qquad M(p)\dot{v} = -f(p, v, t) + u$$

with $q = p, \dot{q} = v$.

If sliding mode is enforced in the manifold $s = cp + v = 0$ then

$$\dot{p} = -cp$$

(Formally, equivalent control u_{eq} should be found from equation $\dot{s} = 0$, substituted into the second equation and then v should be replaced by $-cp$; it will result in the above equation with respect to p.) Assigning the eigenvalues of the sliding mode equation by a proper choice of matrix c, the desired rate of convergence of $p = q$ and $v = \dot{q}$ (note that $s = cp + v = 0$) to zero may be determined. To enforce the sliding mode with the desired dynamics, convergence of the motion projection on subspace s

$$\dot{s} = c\dot{p} + \dot{v} = cv - M^{-1}f + M^{-1}u$$

should be provided. The inertia matrix and its inverse M^{-1} are positive definite. It follows from Theorem 2.1 of Section 2.4 that the discontinuous control

$$u = -U(q, \dot{q})\text{sign}(s) \qquad U(q, \dot{q}) > 0$$

with sufficiently high value of scalar function $U(q, \dot{q})$ enforces sliding motion in manifold $s = 0$. Only an upper bound of function $f(q, \dot{q}, t)$ and a lower estimate for the minimal eigenvalue of M^{-1} of (see Remarks 2.2 and 2.3) are needed for the design of the control stabilizing the mechanical system operating under the uncertainty condition with the desired rate of convergence. Remember the solution has been obtained in the framework of an ideal model with known state vector (p^T, v^T) and assuming that the control forces may be implemented as discontinuous state functions.

3.2 Decoupling

In the sequel, we will deal with affine systems

$$\dot{x} = f(x, t) + B(x, t)u \qquad x, f \in \Re^n, \quad B(x) \in \Re^{n\times m}, \quad u(x) \in \Re^m \tag{3.2.1}$$

$$u(x) = \begin{cases} u^+(x, t) \text{ if } s(x) > 0 \\ u^-(x, t) \text{ if } s(x) < 0 \end{cases} \quad \text{(componentwise)} \qquad s(x)^T = [s_1(x) \dots s_m(x)]$$

with the right-hand side of (3.2.1) being a linear function of control.

To obtain the equation of sliding mode in manifold $s(x) = 0$ under the assumption that matrix GB (matrix $G = \{\partial s/\partial x\}$ with rows as gradients of the components of vector s) is nonsingular, the equivalent control

$$u_{eq}(x, t) = -(G(x)B(x, t))^{-1}G(x)f(x, t)$$

should be substituted into (3.2.1) for the control $u(x)$ to yield

$$\dot{x} = f_{sm}(x, t)$$
$$f_{sm}(x, t) = f(x, t) - B(x, t)(G(x)B(x, t))^{-1}G(x)f(x, t) \tag{3.2.2}$$

Since $s(x) = 0$ in sliding mode, this system of m algebraic equations may be solved with respect to m components of the state vector constituting subvector x_2:

$$x_2 = s_0(x_1), \;\; x_2 \in \Re^m, \;\; x_1 \in \Re^{n-m} \qquad x^T = [x_1^T \quad x_2^T] \;\; \text{and} \;\; s(x) = 0$$

Replacing x_2 by $s_0(x_1)$ in the first $n - m$ equations of (3.2.2) yields a reduced-order sliding mode equation

$$\dot{x}_1 = f_{1sm}(x_1, s_0(x_1), t) \tag{3.2.3}$$

where $f_{sm}^T(x, t) = f_{sm}^T(x_1, x_2, t) = [f_{1sm}^T(x_1, x_2, t) \quad f_{2sm}^T(x_1, x_2, t)]$

The motion equation (3.2.3) depends on function $s_0(x_1)$, i.e. on the equation of the discontinuity manifold. Function $s_0(x_1)$ may be handled as m-dimensional control for the reduced-order system. Note that the design problem is not a conventional one, since the right-hand sides in (3.2.2) and (3.2.3) depend not only on the discontinuity manifold equation but on the gradient matrix G as well. If a class of functions $s(x)$ is preselected, e.g. linear functions or functions in the form of finite series, then both $s(x)$, G and therefore the right-hand sides in (3.2.3) depend on the set of parameters to be selected when designing the desired dynamics of sliding motion.

The second-order system (2.1.1) with a scalar control

$$\dot{x}_1 = a_{11}x_1 + a_{12}x_2 + b_1u + d_1f(t)$$
$$\dot{x}_2 = a_{21}x_1 + a_{22}x_2 + b_2u + d_2f(t)$$

$$u = -M\,\text{sign}(s) \qquad s = c_1x_1 + c_2x_2$$

may serve as an example. As shown in Section 2.2, sliding mode along the switching line $s = c_1x_1 + c_2x_2$ is governed by the first-order equation

$$\dot{x}_1 = (a_{11} - a_{12}c_2^{-1}c_1 - (cb)^{-1}b_1(ca^1 - ca^2c_2^{-1}c_1))x_1 + (d_1 - b_1(cb)^{-1}(cd))f(t)$$

where $c = [c_1 \quad c_2]$, $b^T = [b_1 \quad b_2]$, $(a^1) = [a_{11} \quad a_{21}]^T, [a^2] = [a_{12} \quad a_{22}]^T, d^T = [d_1 \quad d_2]$ and cb and c_2 are assumed to be different from zero. The equation may be rewritten in the form

$$\dot{x}_1 = (a_{11} - a_{12}c_1^* - (c^*b)^{-1}b_1(c^*a^1 - c^*a^2c_1^*))x_1 + (d_1 - b_1(c^*b)^{-1}(c^*d))f(t)$$

with $c^* = [c_1^* \quad 1]$, and $c_1^* = c_2^{-1}c_1$. Hence only one parameter c_1^* should be selected to provide the desired motion of the first-order dynamics in our second-order example.

The second stage of the design procedure is selection of discontinuous control enforcing sliding mode in manifold $s(x) = 0$ which has been chosen at the first stage. The conditions for sliding mode to exist are equivalent to the stability condition of the motion projection on subspace s:

$$\dot{s} = Gf + GBu \tag{3.2.4}$$

with a finite convergence time (Section 2.4).

Generally speaking, matrix $-(GB + (GB)^T)$ is not positive definite, therefore the stability cannot be provided by increasing the elements of matrix U, as recommended in Remark 2.3 of Section 2.4 for control (3.2.1).

Let the positive definite function

$$V = 0.5s^T s > 0$$

be a Lyapunov function candidate. Its time derivative along the system trajectories is of the form

$$\dot{V} = s^T Gf + s^T GBu \tag{3.2.5}$$

Assuming that matrix GB is nonsingular, select the control as a discontinuous function

$$u = -U(x)\text{sign}(s^*) \quad \text{with} \quad s^* = (GB)^T s \tag{3.2.6}$$

where $U(x)$ is a scalar positive function of the state. Then (3.2.5) is of form

$$\dot{V} = s^T Gf - U|s^*|$$

where $|s^*| = (s^*)^T \text{sign}(s^*)$, or

$$\dot{V} = (s^*)^T (GB)^{-1} Gf - U|s^*| \tag{3.2.7}$$

Since $|s^*| \geq \|s^*\|$ due to

$$\sum_{i=1}^{m} |s_i^*| \geq \left(\sum_{i=1}^{m} (s_i^*)^2 \right)^{\frac{1}{2}}$$

it follows from (3.2.7) that

$$\dot{V} \leq |s^*||(GB)^{-1}Gf| - U|s^*| \tag{3.2.8}$$

If an upper estimate $F \geq |(GB)^{-1}Gf|$ is known, then $\dot{V} < 0$ for $U > F$, the motion is asymptotically stable and sliding mode is enforced in the system. Later it will be proven that the time interval preceding sliding mode is finite and may be decreased by increasing the magnitude $U(x)$ of the discontinuous control. Sliding mode occurs in the manifold $s^* = 0$. The transformation (3.2.6) is nonsingular, therefore the manifolds $s = 0$ and $s^* = 0$ coincide and sliding mode takes place in the manifold $s = 0$, which was selected to design sliding motion with the desired properties.

The design procedure has been decoupled into two independent subproblems of lower dimensions m and $n - m$. Decoupling is feasible because the sliding mode equations do not depend on the control but they do depend on the sliding manifold equation. When designing a switching manifold, one constraint should be taken into account: matrix GB should be nonsingular. Exact knowledge of plant parameters and disturbances (vector f and matrix B) is not needed, only knowledge of an upper bound F is sufficient to enforce sliding mode in manifold $s^* = 0$. Matrix $B(x, t)$ is needed to calculate vector s^* in (3.2.6). However, the range of parameter variation in matrix $B(x, t)$ may be found such that sliding mode can be enforced without exact knowledge of these parameters.

First, we will show that any $m \times m$ transformation matrix Q in

$$s^* = Q(x)s$$

fits if $(Q^{-1})^T GB = L(x)$ is a matrix with a dominant diagonal

$$|l_{ii}| > \sum_{\substack{j=1 \\ j \neq i}}^{m} |l_{ij}| \quad \text{or} \quad \alpha_i = |l_{ii}| - \sum_{\substack{j=1 \\ j \neq i}}^{m} |l_{ij}| \qquad (\alpha_i > 0 \text{ for any } i = 1, \ldots, m)$$

Indeed for control $u = -U(x)(\text{sign}(L))(\text{sign}(s^*))$ with $\text{sign}(L)$ being a diagonal matrix with elements $\text{sign}(l_{ii})$

$$\dot{V} = (s^*)^T (Q^{-1})^T Gf - U(x)(s^*)^T L(\text{sign}(L))(\text{sign}(s^*))$$

$$= \sum_{i=1}^{m} s_i^* q_i - U \sum_{i=1}^{m} \left(|s_i^*| (|l_{ii}| + \sum_{\substack{j=1 \\ j \neq i}}^{m} l_{ij} \, \text{sign}(l_{ij} s_i^* s_j^*) \right)$$

where q_i are elements of vector $(Q^{-1})^T Gf$. The time derivative of the Lyapunov function is negative, i.e. sliding mode is enforced in $s^* = 0$ if

$$U(x) > \max_i |q_i(x, t)| / \alpha_i$$

To illustrate the design method for the system (3.2.1), assume that matrix B consists of a known nominal part and unknown variation $B = B_0 + \Delta B$. Then for $Q = (GB_0)^T$ matrix L is of form $L = I_m + \Delta L$, $\Delta L = (GB_0)^{-1} G \Delta B$ (I_m is the $m \times m$ identity matrix). This form enables us to find an admissible range of variations in matrix B: the sum of absolute values in any row of matrix ΔL should not exceed 1. Hence sliding mode can be enforced with control (3.2.6) in systems with unknown parameters in the input matrix $B(x, t)$.

3.3 Regular form

The two-stage design procedure – selection of a switching manifold and then finding of control that enforces sliding mode in this manifold – becomes simpler for systems in so-called *regular form*. The regular form for an affine system (3.2.1) consists of two blocks

$$\begin{aligned} \dot{x}_1 &= f_1(x_1, x_2, t) \\ \dot{x}_2 &= f_2(x_1, x_2, t) + B_2(x_1, x_2, t)u \end{aligned} \tag{3.3.1}$$

where $x_1 \in \Re^{n-m}$, $x_2 \in \Re^m$ and B_2 is an $m \times m$ nonsingular matrix, i.e. $\det B_2 \neq 0$.

The first block does not depend on control, and the dimension of the second block coincides with the dimension of the control. The design is performed in two stages as well. First, m-dimensional state vector x_2 is handled as the control of the first block and designed as a function of the state x_1 of the first block in correspondence with some performance criterion

$$x_2 = -s_0(x_1) \tag{3.3.2}$$

Again we deal with a reduced-order design problem. At the second stage, discontinuous control is to be selected to enforce sliding mode in the manifold

$$s(x_1, x_2) = x_2 + s_0(x_1) = 0 \tag{3.3.3}$$

After sliding mode occurs in the sliding manifold (3.3.3), condition (3.3.2) holds and the further motion in the system will be governed by the differential equation

$$\dot{x}_1 = f_1(x_1, -s_0(x_1), t) \tag{3.3.4}$$

with the desired dynamic properties.

The design of the discontinuous control may be performed using the methods of Section 3.1 with $x^T = [x_1^T \;\; x_2^T]$, $f^T = [f_1^T \;\; f_2^T]$, $B^T = [0_{m\times(n-m)} \;\; B_2^T]$, $G = [G_1 \;\; I_m]$; $G_1 = \{\partial s_0/\partial x_1\}$ is an $m \times (n-m)$ matrix.

Note the following characteristics for the design in regular form:

1. In contrast to (3.2.2) and (3.2.3), the sliding mode equation does not depend on gradient matrix G, which makes the design problem at the first stage a conventional one–design of m-dimensional control x_2 in an $(n-m)$-dimensional system with state vector x_1.
2. Calculation of the equivalent control to find the sliding mode equation is not needed.
3. The condition $\det(GB) = \det(B_2) \neq 0$ holds. (Recall that this condition is needed to enforce sliding mode in the preselected manifold $s(x) = 0$.)
4. Sliding mode is invariant with respect to functions f_2 and B_2 in the second block.

These characteristics suggest that we should find a coordinate transformation reducing the original affine system (3.2.1) to the regular form (3.3.1) before designing sliding mode control. We will confine ourselves to systems with scalar controls. The methods related to systems with vector control may be found in Luk'yanov and Utkin (1981).

We assume that in a system

$$\begin{aligned} &\dot{x} = f(x, t) + b(x, t)u \\ &x \in \Re^n, \quad u \in \Re, \quad f^T = [f_1, \ldots, f_n] \end{aligned} \tag{3.3.5}$$

$b(x, t)$ is an n-dimensional vector with components $b_i(x, t)$, $i = 1, \ldots, n$. Assume that at least one of them, let it be $b_n(x, t)$, is different from zero for any x and t:

$$b_n(x, t) \neq 0$$

Let a solution to an auxiliary system of $(n-1)$th order

$$dx_i/dx_n = b_i/b_n \qquad i = 1, \ldots, n-1 \tag{3.3.6}$$

be a set of functions

$$x_i = \varphi_i(x_n, t) \qquad (i = 1, \ldots, n-1) \tag{3.3.7}$$

Let us introduce the nonsingular coordinate transformation

$$y_i = x_i - \varphi_i(x_n, t) \qquad (i = 1, \ldots, n-1) \tag{3.3.8}$$

According to equations (3.3.5) to (3.3.8), the motion equations with respect to new state vector $(y_1, \ldots, y_{n-1}, x_n)$ are of the form

$$\begin{aligned}\dot{y}_i &= \dot{x}_i - \frac{d\varphi_i(x_n, t)}{dx_n}\dot{x}_n\\ &= \dot{x}_i - \frac{dx_i}{dx_n}\dot{x}_n = f_i + b_i u - \frac{b_i}{b_n}(f_n + b_n u)\\ &= f_i - \frac{b_i}{b_n} f_n \qquad (i = 1, \ldots, n-1)\\ \dot{x}_n &= f_n + b_n u.\end{aligned}$$

Replacing x_i by $y_i + \varphi_i(x_n)$ leads to motion equations

$$\begin{aligned}\dot{y} &= f^*(y, x_n, t)\\ \dot{x} &= f_n^*(y, x_n) + b_n^*(y, x_n, t)u\end{aligned} \tag{3.3.9}$$

where y and f^* are $(n-1)$-dimensional vectors, and f_n^* and b_n^* are scalar functions.

The system with respect to y and x_n is in the regular form (3.3.1) with one $(n-1)$th order and one first-order block. For a particular case with b_i depending only on one coordinate x_n, the state transformation may be found in the explicit form

$$y_i = x_i - \int_0^{x_n} \frac{b_i(\gamma, t)}{b_n(\gamma, t)} d\gamma \tag{3.3.10}$$

The system with respect to the new variables is in the regular form (3.3.9) as well.

3.4 Invariance

Consider a control system with time-varying parameters operating in the presence of disturbances. Given reference inputs may be treated as disturbances if the deviations of control variables from the inputs are included in a state vector. The possibility of designing systems with invariant sliding motions in canonical spaces was discussed in the introduction to Chapter 1.

Let variable x_i and its time derivatives $x_1^{i-1} = x_i$, $i = 2, \ldots, n$ be components of a state vector in the canonical space, then the motion equations of a single-input single-output system in canonical space are of the form

$$\begin{aligned}\dot{x}_i &= x_{i+1} \qquad (i = 1, \ldots, n-1)\\ \dot{x}_n &= -\sum_{i=1}^{n} a_i(t)x_i + f(t) + b(t)u\end{aligned} \tag{3.4.1}$$

where $a_i(t)$ and $b(t)$ are bounded parameters with known range $|a_i(t)| \le a_{i0}$, $|b(t)| \ge b_0$; $f(t)$ is a bounded disturbance $|f(t)| \le f_0$ with a_{i0}, b_0, f_0 being known scalars.

Let the control be a discontinuous state function

$$u = -(\alpha|x| + M)\text{sign}(s) \qquad |x| = \sum_{i=1}^{n} |x_i|, \quad s = \sum_{i=1}^{n} c_i x_i$$

where α, M, c_i are constant values and $c_n = 1$. Calculate the time derivative of function s as

$$\dot{s} = \sum_{i=1}^{n}(c_{i-1} - a_i)x_i - b(\alpha|x| + M)\text{sign}(s) \quad \text{with} \quad c_0 = 0$$

The condition for the state to reach plane $s = 0$ in the state space and for sliding mode to exist, see (2.4.1), are fulfilled if

$$b_0\alpha > \max(c_{i-1} - a_{i0}) \qquad i = 1, \ldots, n, \quad \text{and} \quad b_0 M > f_0$$

After a finite time interval, sliding mode occurs in the plane $s = 0$. To obtain the sliding mode equation, $x_n = -\sum_{i=1}^{n-1} c_i x_i$ should be substituted into the $(n-1)$th equation of system (3.4.1) and the last one should be disregarded:

$$\dot{x}_i = x_{i+1} \qquad (i = 1, \ldots, n-2)$$
$$\dot{x}_{n-1} = -\sum_{i=1}^{n-1} c_i x_i$$

The sliding mode equation is invariant to the plant parameter variations and the disturbance, and its dynamics are determined by the roots of the characteristic equation

$$p^{n-1} + c_{n-1}p^{n-2} + \ldots + c_2 p + c_1 = 0$$

which may be shaped by a proper choice of coefficients c_i in the equations of the discontinuity surface. However, technical difficulties involved in obtaining time derivatives of the plant output x_1 are the major obstacles for implementation of such specific sliding modes. At the same time, by means of both scalar and vector control, invariant sliding modes can be enforced in the spaces whose coordinates may not only be derivatives, but arbitrary physical variables as well.

Let us formulate the invariance conditions for arbitrary affine systems of the form (3.2.1)

$$\dot{x} = f(x, t) + B(x, t)u + h(x, t) \tag{3.4.2}$$

where vector $h(x, t)$ characterizes disturbances and parameter variations which should not affect the feedback system dynamics. In compliance with the equivalent control method (Section 2.3) the solution to $\dot{s} = G(f + Bu + h) = 0$ with respect to control,

$$u_{eq} = -(GB)^{-1}G(f + h)$$

should be substituted into the system equation (3.4.2) to yield

$$\dot{x} = f - B(GB)^{-1}Gf + (I_n - B(GB)^{-1}G)h \tag{3.4.3}$$

Let range $(B(x, t))$ be a subspace formed by the base vectors of matrix $B(x, t)$ for each point (x, t). Sliding mode is invariant with respect to vector $h(x, t)$ if

$$h(x, t) \in \text{range}(B(x, t)) \tag{3.4.4}$$

Condition (3.4.4) means there exists vector $\gamma(x, t)$ such that

$$h(x, t) = B(x, t)\gamma(x, t) \tag{3.4.5}$$

Direct substitution of vector $h(x, t)$ in the form (3.4.5) into (3.4.3) demonstrates that the sliding motion in any manifold $s(x) = 0$ does not depend on perturbation vector $h(x, t)$. Following from the design methods in Sections 3.2 and 3.3, an upper estimate of this vector is needed to enforce the sliding motion. Condition (3.4.5) generalizes the invariance condition obtained in Drazenovic (1969) for linear systems.

3.5 Unit control

The objective of this section is to demonstrate a design method for discontinuous control that enforces sliding mode in some manifold without individual selection of each component of control as a discontinuous state function. The approach implies design of control based on a Lyapunov function selected for a nominal (feedback or open-loop) system. The control is to be found such that the time derivative of the Lyapunov function is negative along the trajectories of the system with perturbations caused by uncertainties in the plant model and environmental conditions.

The roots of the above approach may be found in papers by Leitmann and Gutman published in the 1970s (Gutman and Leitmann, 1976; Gutman, 1979). The design idea may be explained for an affine system

$$\dot{x} = f(x, t) + B(x, t)u + h(x, t) \tag{3.5.1}$$

with state and control vectors $x \in \Re^n$, $u \in \Re^m$, state-dependent vectors $f(x, t)$, $h(x, t)$ and control input matrix $B(x, t) \in \Re^{n \times m}$. The vector $h(x, t)$ represents the system uncertainties and its influence on the control process should be rejected.

The equation

$$\dot{x} = f(x, t) \tag{3.5.2}$$

represents an open-loop nominal system which is assumed to be asymptotically stable with a known Lyapunov function candidate

$$V(x) > 0$$

$$W_o = dV/dt|_{h=0,u=0} = \operatorname{grad}(V)^T f < 0 \qquad \operatorname{grad}(V)^T = \left[\frac{\partial V}{\partial x_1} \cdots \frac{\partial V}{\partial x_n}\right] \tag{3.5.3}$$

The perturbation vector $h(x, t)$ is assumed to satisfy the matching conditions (3.4.4), hence there exists vector $\gamma(x, t) \in R^m$ such that

$$h(x, t) = B(x, t)\gamma(x, t) \tag{3.5.4}$$

$\gamma(x, t)$ may be an unknown vector with known upper scalar estimate $\gamma_0(x, t)$,

$$\|\gamma(x, t)\| < \gamma_o(x, t) \tag{3.5.5}$$

Calculate the time derivative of Lyapunov function $V(x)$ along the trajectories of the perturbed system (3.5.2) to (3.5.5) as

$$W = dV/dt = W_o + \text{grad}(V)^T B(u + \gamma) \tag{3.5.6}$$

For control u depending on the upper estimate of the unknown disturbance, chosen as

$$u = -\rho(x, t)\frac{B^T \text{grad}(V)}{\|B^T \text{grad}(V)\|} \tag{3.5.7}$$

with a scalar function $\rho(x, t) > \gamma_0(x, t)$ and

$$\|B^T \text{grad}(V)\|^2 = (\text{grad}(V)^T B)(B^T \text{grad}(V))$$

the time derivative of the Lyapunov function

$$\begin{aligned} W &= W_o - \rho(x, t)\|B^T \text{grad}(V)\| + \text{grad}(V)^T B\gamma(x, t) \\ &< W_o - \|B^T \text{grad}(V)\|[\rho(x, t) - \gamma_o(x, t)] \\ &< 0 \end{aligned}$$

is negative. This implies that the perturbed system with control (3.5.7) is asymptotically stable as well.

Two important features should be underlined for the system with control (3.5.7):

1. Control (3.5.7) undergoes discontinuities in $(n - m)$-dimensional manifold $s(x) = B^T \text{grad}(V) = 0$ and is a continuous state function beyond this manifold. This is the principal difference between control (3.5.7) and all the control inputs in the previous sections with individual switching functions for each control component.
2. The disturbance $h(x, t)$ is rejected *due to enforcing sliding mode* in the manifold $s(x) = 0$. Indeed, if the disturbance (3.5.4) is rejected, then control u should be equal to $-\gamma(x, t)$, which is not generally the case for the control (3.5.7) beyond the discontinuity manifold $s(x) = B^T \text{grad}(V) \neq 0$. It means that sliding mode occurs in the manifold $s = 0$ and the equivalent value of control u_{eq} is equal to $-\gamma(x, t)$.

Note that the norm of control (3.5.7) with the gain $\rho(x, t) = 1$

$$\left\| \frac{B^T \text{grad}(V)}{\|B^T \text{grad}(V)\|} \right\|$$

is equal to 1 for any value of the state vector. It explains the term 'unit control' for (3.5.7).

Later on, unit control was used directly with a Lyapunov function at the second stage of the conventional two-stage design procedure for sliding mode control: selection of a sliding manifold $s(x) = 0$ and enforcing sliding mode in this manifold

(Ryan and Corless, 1984; Dorling and Zinober, 1986). The manifold $s(x) = 0$ was selected in compliance with some performance criterion and the control was designed in the form (3.5.7):

$$u = -\rho(x, t)\frac{D^T s(x)}{\|D^T s(x)\|} \tag{3.5.8}$$

with $D = GB$, $G = \{\partial s/\partial x\}$ and D was assumed to be nonsingular.

The equation of the motion projection of the system (3.5.1) on the subspace s is of form

$$\dot{s} = G(f + h) + Du \tag{3.5.9}$$

The conditions for the trajectories to converge to the manifold $s(x) = 0$ and for sliding mode to exist in this manifold may be derived based on a Lyapunov function candidate

$$V = \tfrac{1}{2}s^T s > 0 \tag{3.5.10}$$

Find the time derivative of Lyapunov function (3.5.10) along the trajectories of system (3.5.9) with control (3.5.8):

$$\begin{aligned}\dot{V} &= s^T G(f + h) - \rho(x, t)\|D^T s(x)\| \\ &= [s^T D][D^{-1}G(f + h)] - \rho(x, t)\|D^T s(x)\| \\ &\leq \|D^T s(x)\|[\|D^{-1}G(f + h)\| - \rho(x, t)]\end{aligned}$$

For $\rho(x, t) > \|D^{-1}G(f + h)\|$ the value of $\dot{V}$ is negative, therefore the state will reach the manifold $s(x) = 0$. Next we will show that if $\rho(x, t) - \|D^{-1}G(f + h)\| \geq \rho_0 > 0$ (ρ_0 is a constant), then $s(x)$ vanishes and sliding mode occurs after a finite time interval. Preliminarily we estimate $\|D^T s(x)\|$:

$$\|s\| = \|(D^T)^{-1}D^T s\| \leq \|(D^T)^{-1}\|\|D^T s\| \quad \text{and} \quad \|D^T s\| \geq \|(D^T)^{-1}\|^{-1}\|s\|.$$

Thus $\dot{V} \leq -\|(D^T)^{-1}\|^{-1}\rho_0\|s\|$ and since $V = \frac{1}{2}\|s\|^2$, $\|s\| = \sqrt{2V}$ leads to

$$\dot{V} < -\eta V^{1/2} \qquad \eta = \sqrt{2}\|(D^T)^{-1}\|^{-1}\rho_0$$

The solution to the differential inequality $V(t)$ is nonnegative and is bounded by

$$V(t) < \left(-\frac{\eta}{2}t + \sqrt{V_0}\right)^2 \qquad V_0 = V(0)$$

Since the solution vanishes after some $t_s < (2/\eta)\sqrt{V_0}$, the vector s vanishes as well and sliding mode starts after a finite time interval.

Remember the principal difference in motions preceding the sliding mode in $s(x) = 0$ for the conventional componentwise control and the unit control design methods. Conventional control undergoes discontinuities should any of the components of vector s change sign, whereas unit control is a continuous state function until the manifold $s(x) = 0$ is reached.

References

DORLING, C. M. and ZINOBER, A. S. I., 1986, Two approaches to sliding mode design in multivariable variable structure control systems, *International Journal of Control*, **44**, 65–82.

DRAZENOVIC, B., 1969, The invariance conditions for variable structure systems, *Automatica*, **5**, 287–95.

GUTMAN, S., 1979, Uncertain dynamic systems – a Lyapunov min-max approach, *IEEE Transactions on Automatic Control*, **AC-24**, 437–49.

GUTMAN, S. and LEITMANN, G., 1976, Stabilizing feedback control for dynamic systems with bounded uncertainties, *Proceedings of the IEEE Conference on Decision and Control*, pp. 94–99.

LUK'YANOV, A. and UTKIN, V., 1981, Methods of reducing equations for dynamic systems to a regular form, *Automation and Remote Control*, **42**, 413–20.

RYAN, E. P. and CORLESS, M., 1984, Ultimate boundness and asymptotic stability of a class of uncertain systems via continuous and discontinuous feedback control, *IMA Journal of Mathematics, Control and Information*, **1**, 222–42.

CHAPTER FOUR

Pendulum Systems

The design of the sliding mode control for nonlinear multivariable systems has been extensively studied in many books and papers. The design procedure of such high-order nonlinear control systems may be complicated and varies from case to case. The objective of this chapter is to develop design methods for nonlinear mechanical systems governed by a set of second-order equations. The proposed approach assumes that control systems can be transformed into a regular form (Section 3.3), which enables one to decouple the controller design. Control laws are illustrated for different inverted pendulum systems.

4.1 Design methodology

When controlling mechanical systems, we deal with a set of interconnected second-order nonlinear differential equations

$$J(q)\ddot{q} = f(q, \dot{q}) + Bu \tag{4.1.1}$$

where $q \in \Re^n$, $u \in \Re^m$, u is a vector of control forces and torques, elements of matrix B are equal to either 0 or 1, and $\text{rank}(B) = m$. In particular, for rotational mechanical systems $J(q)$ is an inertia matrix. The system may be underactuated, i.e. it has fewer inputs than degrees of freedom, and/or is unstable.

The system (4.1.1) may be represented in the form of $2n$ equations of first order with respect to vectors q and $\dot{q}$. Then the regular form approach can be applied. Here we will generalize the approach for systems consisting of blocks governed by the second-order equations. Then it can be applied directly to nonlinear mechanical systems (4.1.1).

The inertia matrix $J(q)$ in mechanical systems is nonsingular and B is a full-rank matrix, hence $J^{-1}(q)B$ is a full-rank matrix. The components of vector q may be reordered such that in the motion equations

$$\ddot{q}_1 = \tilde{f}_1(q, \dot{q}) + \tilde{B}_1(q)u$$
$$\ddot{q}_2 = \tilde{f}_2(q, \dot{q}) + \tilde{B}_2(q)u$$
$$q = \begin{bmatrix} q_1 \\ q_2 \end{bmatrix}, \; q_1 \in \Re^{n-m}, \; q_2 \in \Re^m \qquad \begin{bmatrix} \tilde{f}_1 \\ \tilde{f}_2 \end{bmatrix} = J^{-1}f \qquad \begin{bmatrix} \tilde{B}_1 \\ \tilde{B}_2 \end{bmatrix} = J^{-1}B, \; \det(\tilde{B}_2) \neq 0 \tag{4.1.2}$$

According to the regular form technique, as discussed in Section 3.3, a coordinate transformation $z = \phi(q) \in \Re^{n-m}$, $y = q_2$ with continuously differentiable function $\phi(q)$ should be found such that the condition

$$\frac{\partial \phi(q)}{\partial q} J^{-1} B = 0$$

is fulfilled. Then

$$\dot{z} = \frac{\partial \phi(q)}{\partial q} \dot{q} \qquad \ddot{z} = \frac{\partial}{\partial q}\left(\frac{\partial \phi(q)}{\partial q} \dot{q}\right) \dot{q} + \frac{\partial \phi(q)}{\partial q} J^{-1}(f + Bu)$$

and the mechanical system equation is reduced to the regular form consisting of a set of second-order equations

$$\begin{aligned} \ddot{z} &= f_1(z, y, \dot{z}, \dot{y}) \\ \ddot{y} &= f_2(z, y, \dot{z}, \dot{y}) + B_2(z, y)u \qquad \det(B_2) \neq 0 \end{aligned} \tag{4.1.3}$$

For the regular form with blocks consisting of first-order equations (Section 3.3) the state of the low block was handled as control in the top block. The desired dependence between the two subvectors was provided by enforcing sliding mode.

In our case the top block equation in (4.1.3) depends on both vectors y and $\dot{y}$. This fact introduces some peculiarities which should be taken into account when designing sliding mode control. Furthermore, stabilization tasks for different types of mechanical systems will be studied. It is assumed that the origin in a system state space is an equilibrium point of an open-loop system:

$$\begin{aligned} f_1(0, 0, 0, 0) &= 0 \\ f_2(0, 0, 0, 0) &= 0 \end{aligned}$$

Case 1

First, the stability of the system zero dynamics with vector y as an output is checked. They are governed by the first equation in (4.1.3) with $y = 0$, $\dot{y} = 0$:

$$\ddot{z} = f_1(z, 0, \dot{z}, 0) \tag{4.1.4}$$

If they are stable then sliding mode is enforced in the manifold $s = \dot{y} + cy = 0$ with scalar parameter $c > 0$. It is a simple task since $\text{rank}(B_2) = m$ and any method of enforcing sliding modes in Sections 3.2 and 3.4 is applicable. After sliding mode starts in the manifold $s = 0$, the state y tends to zero as a solution to $\dot{y} + cy = 0$, and due to stability of the zero solution of (4.1.4), z therefore decays as well.

Case 2

Now stability of the system zero dynamics with vector z as an output is checked. If $z(t) \equiv 0$ then the zero dynamics equations are obtained from the top block of (4.1.3):

$$f_1(0, y, 0, \dot{y}) = 0 \tag{4.1.5}$$

Here the zero dynamics are a set of first-order equations whereas in Case 1 they are a set of second-order equations. If the zero dynamics are stable then sliding mode is enforced in the manifold

$$s = f_1 + c_1 z + c_2 \dot{z} = 0$$

After sliding mode starts,

$$f_1 = -c_1 z - c_2 \dot{z} \tag{4.1.6}$$

and the equation for z in (4.1.3) is of form

$$\ddot{z} = -c_1 z - c_2 \dot{z} \tag{4.1.7}$$

For positive scalar parameters c_1 and c_2, the solution to (4.1.7) tends to zero and then $y(t)$ as a solution to (4.1.6) tends to zero as well. This stabilization method for systems with stable zero dynamics is applicable if

$$\text{rank}\left(\frac{\partial f_1}{\partial \dot{y}} B_2\right) \geq \dim(s) = \dim(z) \tag{4.1.8}$$

Then $\dot{s} = F(z, y, \dot{z}, \dot{y}) + (\partial f_1 / \partial \dot{y}) B_2 u$ (F is a function independent of the control u) and, as shown in Section 3.2, sliding mode can be enforced. Generally speaking, this condition holds if $\dim(z) \leq \dim(y)$.

Case 3

We assume that function f_1 in (4.1.3) does not depend on $\dot{y}$, i.e. $f_1 = f_1(z, y, \dot{z})$. If condition (4.1.6) holds, then z and $\dot{z}$ tend to zero. After z decays, y is found from the algebraic equation $f_1(0, y, 0) = 0$. Since the origin of the state space is the equilibrium point, $f_1(0, 0, 0) = 0$, coordinate y tends to zero as well. To provide condition (4.1.6), the switching manifold is selected as

$$s = \dot{s}_1 + \alpha s_1 = 0 \quad (\alpha > 0)$$

with $s_1 = f_1 + c_1 z + c_2 \dot{z}$. The time derivatives of s_1 and s are of form

$$\dot{s}_1 = \frac{\partial f_1}{\partial y} \dot{y} + F_1(z, y, \dot{z})$$

where function $F_1(z, y, \dot{z})$ depends neither on $\dot{y}$ nor on control and

$$\dot{s} = F(z, y, \dot{z}, \dot{y}) + \frac{\partial f_1}{\partial y} B_2 u$$

where function $F(z, y, \dot{z}, \dot{y})$ does not depend on control.

As for Case 2, sliding mode can be enforced in the manifold $s = 0$ if condition (4.1.8) holds. In sliding mode, $s_1(t)$ decays as a solution to the equation $\dot{s}_1 + \alpha s_1 = 0$. This means that condition (4.1.6) holds and $z(t)$, $\dot{z}(t)$ and $y(t)$ tend to zero.

Case 4

Let us assume that condition (4.1.8) holds and consider the special case of function

$$f_1 = f_{11}(y)\dot{y} + f_{12}(z, y, \dot{z})$$

which is linear with respect to $\dot{y}$ and the zero dynamics governed by $f_{11}(y)\dot{y} + f_{12}(0, y, 0) = 0$ are unstable (otherwise the design method of Case 2 is applicable). Then the first equation of (4.1.3) with respect to new variables

$$z_1 = \dot{z} - g(y), \quad z_2 = z \qquad (z_1, z_2 \in \Re^{n-m})$$

is transformed to

$$\begin{aligned}
&\dot{z}_1 = f'_{12}(z_1, z_2, y) + f_{11}(y)\dot{y} - \frac{\partial g}{\partial y}\dot{y} \\
&\dot{z}_2 = z_1 + g(y) \\
&\frac{\partial g}{\partial y} = \left\{\frac{\partial g_i}{\partial y_j}\right\} \qquad (i = 1, \ldots, n-m;\ j = 1, \ldots, m) \\
&f'_{12}(z_1, z_2, y) = f_{12}(z_2, y, z_1 + g(y))
\end{aligned} \tag{4.1.9}$$

If the function $g(y)$ is a solution to the partial differential equation

$$\frac{\partial g}{\partial y} = f_{11}(y) \tag{4.1.10}$$

then the system (4.1.9) is reduced to

$$\dot{p} = F(p, y) \qquad p = \begin{bmatrix} z_1 \\ z_2 \end{bmatrix} \qquad F = \begin{bmatrix} f'_{12}(z_1, z_2, y) \\ z_1 + g(y) \end{bmatrix} \tag{4.1.11}$$

In the reduced-order system (4.1.11), the state of the second block in (4.1.3), y, is handled as $(n-m)$-dimensional control. For instance, it is possible to choose

$$y = -s_0(p) \tag{4.1.12}$$

such that the system

$$\dot{p} = F(p, -s_0(p))$$

is asymptotically stable. The relationship (4.1.12) is valid if sliding mode is enforced in the manifold

$$s = y + s_0(p) = 0$$

Similar to Case 2 this can be done since (4.1.8) holds by our assumption.

Remark 4.1

The design procedures for Cases 2 to 4 were developed under assumption (4.1.8). If this condition does not hold, a multistep procedure may be applied similar to that described in Luk'yanov (1993) and Luk'yanov and Dodds (1996). □

An example of a two-step design will be described in Section 5.6 for linear time-varying systems. Three nonlinear pendulum systems based on the above design procedures will be studied in this section. The specific design procedures for selecting a sliding manifold and discontinuous control for a cart–pendulum, a double inverted pendulum and a rotational inverted pendulum will be demonstrated. The theoretical studies will be complemented by simulation and experimental results.

4.2 Cart–pendulum

Figure 4.1 shows the physical model of a cart–pendulum system. M and m are the masses of the cart and the inverted pendulum, respectively; l is the distance from the centre of gravity of the link to its attachment point. The coordinate x represents the position of the cart on the horizontal axis to a fixed point, and θ is the rotational angle of the pendulum. Using the method of *Lagrangian* equations, one can easily show that the dynamic equations of the cart–pendulum (Mori *et al.*, 1976) are

$$\begin{cases} (M+m)\ddot{x} + ml\ddot{\theta}\cos\theta - ml\dot{\theta}^2\sin\theta = u \\ \frac{4}{3}ml\ddot{\theta} + m\ddot{x}\cos\theta - mg\sin\theta = 0 \end{cases} \tag{4.2.1}$$

The goal of the control is to stabilize the system in a position where the pendulum is in the unstable vertical position $\theta = 0$ and the cart is at a given point on the straight line $x = 0$ under the action of the control force u. It is assumed that the parameters and the state vector are available. For convenience of design, we first rewrite the given system (4.2.1) with respect to the second derivatives of the coordinates x and θ as

$$\begin{cases} \ddot{x} = \dfrac{1}{k}(-mg\cos\theta\sin\theta + \dfrac{4}{3}u^*) \\ \ddot{\theta} = \dfrac{1}{kl}((M+m)g\sin\theta - u^*\cos\theta) \end{cases} \tag{4.2.2}$$

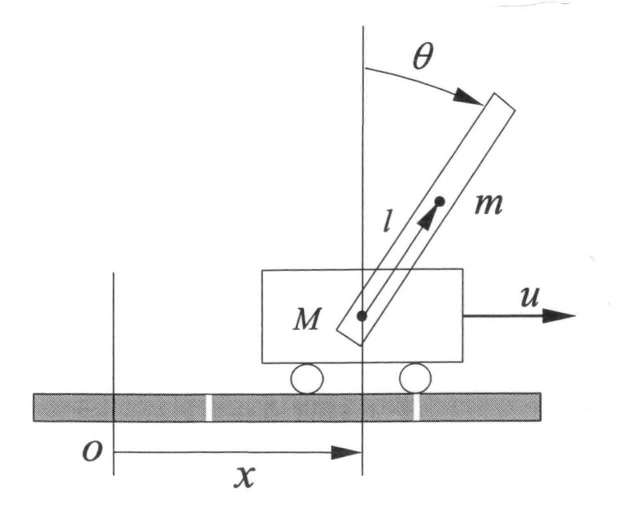

Figure 4.1 The cart–pendulum system.

where

$$k = \frac{4}{3}(M+m) - m\cos^2\theta > 0$$

and

$$u^* = u + ml\dot{\theta}^2 \sin\theta \tag{4.2.3}$$

The system (4.2.2) with scalar control u^* is in the form (4.1.2) with

$$\begin{bmatrix} \tilde{B}_1 \\ \tilde{B}_2 \end{bmatrix} = \begin{bmatrix} 4/3k \\ -(\cos\theta)/kl \end{bmatrix}$$

To reduce the system to the regular form (4.1.3), coordinate transformation

$$y = \phi(x, \theta)$$

should be found such that the second-order differential equation with respect to y does not depend on the control u^*. The solution to the problem is given in Section 3.3. According to (3.3.5) to (3.3.8),

$$\begin{aligned} y &= \phi(x, \theta) \\ &= x - \varphi(\theta) \end{aligned} \tag{4.2.4}$$

where $\varphi(\theta)$ is a solution to the equation

$$\frac{d\varphi}{d\theta} = \frac{\tilde{B}_1}{\tilde{B}_2}$$

or

$$\frac{d\varphi}{d\theta} = -\frac{4l}{3\cos\theta}$$

The coordinate transformation (4.2.4) with the solution to this equation

$$\varphi(\theta) = -\frac{4l}{3}\ln\frac{1+\tan(\theta/2)}{1-\tan(\theta/2)}$$

results in

$$\dot{y} = \dot{x} + \frac{4l}{3}\frac{\dot{\theta}}{\cos\theta}$$

and

$$\ddot{y} = \ddot{x} + \frac{4l}{3}\frac{\ddot{\theta}}{\cos\theta} + \frac{4l}{3}\frac{\sin\theta\dot{\theta}^2}{\cos^2\theta} \tag{4.2.5}$$

The regular form of the system obtained from equations (4.2.2) and (4.2.5) is in the form

$$\begin{aligned} \ddot{y} &= G(\theta, \dot{\theta}) \tan\theta \\ \ddot{\theta} &= v(\theta, u^*) \end{aligned} \tag{4.2.6}$$

where

$$G(\theta, \dot{\theta}) = \frac{g}{k}\left(\left(\frac{4}{3} - \cos^2\theta\right)m + \frac{4}{3}M\right) + \frac{4l}{3}\frac{\dot{\theta}^2}{\cos\theta}$$

and

$$v(\theta, u^*) = \frac{1}{kl}((M+m)g\sin\theta - u^*\cos\theta) \tag{4.2.7}$$

Control of the nonlinear system (4.2.6) may be found as follows.

Step 1

Consider the first equation of system (4.2.6). The function denoted as $G(\theta, \dot{\theta})$ is positive for any values of the arguments if $-\pi/2 < \theta < \pi/2$. The function $\tan\theta$ is handled as control based on the choice proposed in (4.1.6). For this intermediate control, select $\tan\theta$ as a linear combination of y and $\dot{y}$

$$\tan\theta = -\alpha_1 s_2 \qquad (\alpha_1 > 0, \quad s_2 = y + \dot{y}) \tag{4.2.8}$$

then the upper equation of system (4.2.6) is represented as

$$\begin{aligned} \dot{y} &= -y + s_2 \\ \dot{s}_2 &= -y - (\alpha_1 G(\theta, \dot{\theta}) - 1)s_2 \end{aligned}$$

The time derivative of the Lyapunov function candidate

$$V = \frac{1}{2}(y^2 + s_2^2)$$

with $V = 0$ at the origin $(y, s_2) = (0, 0)$ is

$$\dot{V} = -y^2 - (\alpha_1 G - 1)s_2^2$$

Since $G(\theta, \dot{\theta}) > 0$ for $-\pi/2 < \theta < \pi/2$, we observe that $\dot{V} < 0$ for $\alpha_1 G > 1$, i.e. design parameter α_1 should be chosen such that $\alpha_1 > 1/G > 0$ for any t. Hence the equilibrium point is asymptotically stable with $y \to 0$ and $s_2 \to 0$ as $t \to \infty$. Consequently, $(x, \theta) \to (0, 0)$ as $t \to \infty$, as follows from (4.2.4) and (4.2.8).

To implement intermediate control (4.2.8), control u^* will be designed such that the function $s_1 = \tan\theta + \alpha_1(y + \dot{y}) \to 0$ as $t \to \infty$, then $\tan\theta \to -\alpha_1(y + \dot{y})$.

Step 2

The function s_1 tends to zero asymptotically, if it is a solution to the differential equation

$$\dot{s}_1 = -\frac{\alpha}{\cos^2\theta} s_1$$

or

$$s(\theta, \dot{\theta}, y, \dot{y}) = (\cos^2\theta)\dot{s}_1 + \alpha s_1 = 0$$

with

$$s_1 = \tan\theta + \alpha_1(y + \dot{y}) \quad \text{and} \quad \dot{s}_1 = \frac{1}{\cos^2\theta}\dot{\theta} + \alpha_1(\dot{y} + G\tan\theta)$$

Step 3

In order to assign the control law such that

$$s = \dot{\theta} + \alpha_1 \cos^2\theta(\dot{y} + G\tan\theta) + \alpha s_1 = 0$$

calculate the time derivative of the function s along the solutions of (4.2.6)

$$\dot{s} = \Psi(\theta, \dot{\theta})v + F(\theta, \dot{\theta}, \dot{y}) \tag{4.2.9}$$

where

$$\Psi(\theta, \dot{\theta}) = 1 + \frac{8}{3}\alpha_1 l(\sin\theta)\dot{\theta}$$

and

$$\begin{aligned} F(\theta, \dot{\theta}, y) = {} & \alpha_1(\cos^2\theta)G\tan\theta + \alpha\dot{s}_1 + \alpha_1(G - 2(\cos\theta\sin\theta)\dot{y} - 2G\sin^2\theta)\dot{\theta} \\ & + \alpha_1\cos\theta\sin\theta\left(\frac{2gm}{k}\cos\theta\sin\theta + \frac{4}{3}l\dot{\theta}^2\frac{\sin\theta}{\cos^2\theta}\right)\dot{\theta} \end{aligned}$$

The state reaches the surface $s = 0$ for any initial conditions and sliding mode exists at any point of the surface if the deviation from the surface s and its time derivative have opposite signs. This condition is satisfied if

$$v = -v_o\,\text{sign}(s\Psi(\theta, \dot{\theta})) \tag{4.2.10}$$

where

$$v_o \geq \frac{1}{|\Psi|_{\min}}|F|_{\max}$$

Finally, the real control is obtained from (4.2.10), (4.2.7) and (4.2.3):

$$u = \frac{1}{\cos\theta}(((M + m)g\sin\theta - ml\cos\theta\sin\theta\dot{\theta}^2) + klv_o\,\text{sign}(s\Psi))$$

Note that sliding mode may disappear if $\Psi = 0$, since $\dot{s}$ in (4.2.9) does not depend on control v for $\Psi = 0$. On one hand, the function Ψ is positive for the domain $8\alpha_1 l(\sin\theta)\dot{\theta} > -3$ including the origin. On the other hand for the domain

$$\frac{9}{64\alpha_1^2 l^2}\frac{\cos\theta}{\sin^2\theta} > (\sin\theta)v_o \tag{4.2.11}$$

a system trajectory may intersect the surface $\Psi = 0$ once only. To derive this condition, calculate the time derivative of the function Ψ on system trajectories for the points on the surface $\Psi = 0$:

$$\dot{\Psi} = \frac{8}{3}\alpha_1 l((\cos\theta)\dot{\theta}^2 + (\sin\theta)\ddot{\theta}) = \frac{8}{3}\alpha_1 l\left(\frac{9}{64\alpha_1^2 l^2}\frac{\cos\theta}{\sin^2\theta} - (\sin\theta)v_o\,\mathrm{sign}(s\Psi)\right)$$

It is clear that $\dot{\Psi} > 0$ if the condition (4.2.11) holds and shows that the domain of sliding mode, including the equilibrium point, exists.

4.3 Double inverted pendulum

Consider a more complicated system consisting of two rotational inverted pendulums as shown in Figure 4.2. Each link of the pendulum is composed of mass m_i and inertia I_i. The total length of each link is L_i and the distance from the centre of gravity of each link to its pivot point is l_i $(i = 1, 2)$; θ refers to the rotational angle from the vertical axis. It is assumed that only one control force, the motor torque $F_1 = u$, is available. The motion of the dynamic system is governed by the following equation (Ledgerwood and Misawa, 1992):

$$H\begin{bmatrix}\ddot{\theta}_1\\ \ddot{\theta}_2\end{bmatrix} + P\begin{bmatrix}\dot{\theta}_1\\ \dot{\theta}_2\end{bmatrix} + W = \begin{bmatrix}F_1\\ 0\end{bmatrix} \tag{4.3.1}$$

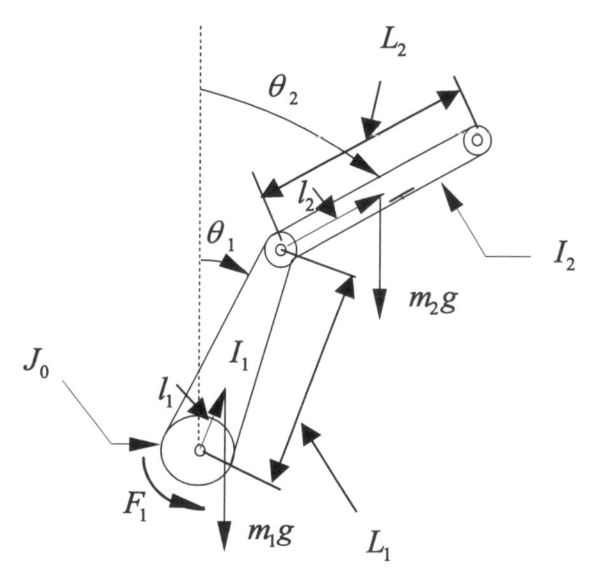

Figure 4.2 The rotational inverted pendulum system.

with matrices

$$H = \begin{bmatrix} J_0 + I_1 + m_1 l_1^2 + m_2 L_1^2 & m_2 L_1 l_2 \cos(\theta_1 - \theta_2) \\ m_2 L_1 l_2 \cos(\theta_1 - \theta_2) & m_2 l_2^2 + I_2 \end{bmatrix}$$

$$W = \begin{bmatrix} -g(m_1 l_1 + m_2 L_1) \sin \theta_1 \\ -m_2 g l_2 \sin \theta_2 \end{bmatrix}$$

and

$$P = \begin{bmatrix} 0 & V\dot{\theta}_2 \\ -V\dot{\theta}_1 & 0 \end{bmatrix} \quad \text{with} \quad V = m_2 L_1 l_2 \sin(\theta_1 - \theta_2)$$

The control objective is to drive the mechanical system from a perturbed state to the desired equilibrium point $\theta_1 = \theta_2 = 0$ and maintain it there. First, we rewrite the given plant equations as

$$\begin{aligned} \ddot{\theta}_1 &= f_1(\theta_1, \theta_2, \dot{\theta}_1, \dot{\theta}_2) + \frac{1}{\Delta} H_{22} u \\ \ddot{\theta}_2 &= f_2(\theta_1, \theta_2, \dot{\theta}_1, \dot{\theta}_2) - \frac{1}{\Delta} H_{12} u \end{aligned} \tag{4.3.2}$$

where

$$\begin{aligned} f_1 &= \frac{1}{\Delta}(-H_{22}(P_{11}\dot{\theta}_1 + P_{12}\dot{\theta}_2 + W_1) + H_{12}(P_{21}\dot{\theta}_1 + P_{22}\dot{\theta}_2 + W_2 - F_2)) \\ f_2 &= \frac{1}{\Delta}(H_{12}(P_{11}\dot{\theta}_1 + P_{12}\dot{\theta}_2 + W_1) - H_{11}(P_{21}\dot{\theta}_1 + P_{22}\dot{\theta}_2 + W_2 - F_2)) \\ \Delta &= H_{11}H_{22} - H_{12}^2 > 0 \quad \text{since inertia matrix } H \text{ is positive definite} \end{aligned}$$

and H_{ij}, P_{ij}, W_i $(i, j = 1, 2)$ are elements of matrices H, P and vector W.

First rewrite the motion equation (4.3.2) in the form

$$\begin{aligned} \ddot{\theta}_2 &= f_2 - \frac{1}{\Delta} H_{12} u \\ \ddot{y} &= f_1 - f_2 + \frac{1}{\Delta}(H_{22} + H_{12}) u \\ y &= \theta_1 - \theta_2 \end{aligned} \tag{4.3.3}$$

The system (4.3.3) is in the form (4.1.2) with

$$\begin{aligned} \tilde{B}_1 &= -\frac{1}{\Delta} H_{12} = A_1 \cos(y) \\ \tilde{B}_2 &= \frac{1}{\Delta}(H_{22} + A_1 \cos(y)) \\ A_1 &= m_2 L_1 l_2 \end{aligned}$$

Similar to the cart–pendulum in Section 4.2, a nonlinear transformation should be found to reduce (4.3.3) to the regular form

$$z = \theta_2 - \varphi(y)$$

where $\varphi(y)$ is a solution to the equation

$$\frac{d\varphi}{dy} = \frac{\tilde{B}_1}{\tilde{B}_2}$$

or

$$\frac{d\varphi}{dy} = -\frac{\cos y}{a + \cos y} \quad \text{with} \quad a = \frac{m_2 l_2^2 + I_2}{m_2 L_1 l_2}$$

We confine our study to the case $a > 1$. Then

$$\varphi(y) = -y - \frac{2a}{\sqrt{a^2 - 1}} \tan^{-1}\left(\sqrt{\frac{a-1}{a+1}} \tan\left(\frac{y}{2}\right)\right)$$

$$z = \theta_2 + y + \frac{2a}{\sqrt{a^2 - 1}} \tan^{-1}\left(\sqrt{\frac{a-1}{a+1}} \tan\left(\frac{y}{2}\right)\right)$$

$$\dot{z} = \dot{\theta}_2 + \frac{\cos y}{a + \cos y} \dot{y}$$

$$\ddot{z} = \ddot{\theta}_2 + \frac{\cos y}{a + \cos y} \ddot{y} - \frac{a \sin y}{(a + \cos y)^2} \dot{y}^2$$

and the system in transformed coordinates is represented by

$$\begin{aligned} \ddot{z} &= G_1(z, y, \dot{z}, \dot{y}) \\ \ddot{y} &= G_2(z, y, \dot{z}, \dot{y}) + \tilde{B}_2(y)u \end{aligned} \tag{4.3.4}$$

where

$$G_1 = \frac{\sin y}{a + \cos y}\left(\left(\dot{z} + \frac{a}{a + \cos y}\dot{y}\right)^2 - \frac{a}{a + \cos y}\dot{y}^2\right) + \frac{A_2}{A_1(a + \cos y)}\sin(z - \varphi(y))$$

$$G_2 = \frac{-1}{\Delta}\left((H_{22} + A_1 \cos y)\left(A_1 \sin y\left(\dot{z} - \frac{\cos y}{a + \cos y}\dot{y}\right)^2 - A_3 \sin(z - \varphi(y)\right)\right.$$

$$\left. + (H_{11} + A_1 \cos y)\left(A_1 \sin y\left(\dot{z} + \frac{a}{a + \cos y}\dot{y}\right)^2 + A_2 \sin(z - \varphi(y)\right)\right)$$

The constant coefficients $A_2 = m_2 g l_2$, $A_3 = g(m_1 l_1 + m_2 L_1)$ and $\tilde{B}_2$ are positive for $-\pi/2 < y < \pi/2$. The sliding mode control will again be designed using the proposed decoupling approach.

Step 1

We first consider the same choice of the sliding manifold as for the previous cart–pendulum example. If the last term of the function G_1 can be forced to

$$\sin(z - \varphi(y)) = -k_1 z - k_2 \dot{z} \tag{4.3.5}$$

then the upper second-order dynamics of the system (4.3.4) are given by

$$\ddot{z} = \gamma - G(k_1 z + k_2 \dot{z}) \tag{4.3.6}$$

where

$$G = \frac{A_2}{A_1(a + \cos y)} > 0$$

and

$$\gamma = \frac{\sin y}{a + \cos y}\left(\left(\dot{z} + \frac{a}{a + \cos y}\dot{y}\right)^2 - \frac{a}{a + \cos y}\dot{y}^2\right).$$

The stability analysis of the system (4.3.6) is more complicated than for our first example in Section 4.2. We will show qualitatively that there exists an attraction domain including the origin in the state space $(z, \dot{z}, y, \dot{y})$.

First let us show that the equilibrium point $z = 0$, $\dot{z} = 0$ is asymptotically stable in large for the system (4.3.6) with $\gamma = 0$. The motion equations with respect to z and $s_2 = k_1 z + k_2 \dot{z}$ are

$$\begin{aligned} \dot{z} &= -\frac{k_1}{k_2} z + \frac{1}{k_2} s_2 \\ \dot{s}_2 &= -\frac{k_1^2}{k_2} z + \frac{k_1}{k_2} s_2 - k_2 G s_2 \end{aligned} \tag{4.3.7}$$

For the Lyapunov function candidate

$$V = \frac{1}{2}(k_1^2 z^2 + s_2^2)$$

calculate the time derivative along the system trajectories given by (4.3.7):

$$\dot{V} = -\frac{k_1^3}{k_2} z^2 - \left(k_2 G - \frac{k_1}{k_2}\right) s_2^2$$

The function $G > A_2/A_1(a + 1)$ – see equation (4.3.6) – therefore positive k_1 and k_2 can be found such that

$$\frac{k_1}{k_2^2} < \frac{A_2}{A_1(a + 1)} \qquad \text{hence} \quad \dot{V} < 0$$

which implies that the equilibrium point of the system (4.3.6) with $\gamma = 0$ is asymptotically stable in large.

The state function γ is a small value of third order with respect to small deviations from the origin. Hence there exists a bounded domain of initial conditions for which the origin in the state space $(z, \dot{z}, y, \dot{y})$ is asymptotically stable as well.

Step 2

The condition (4.3.5) is equivalent to,

$$s_1 = \sin(z - \varphi(y)) + k_1 z + k_2 \dot{z} = 0$$

The function s_1 satisfies the first-order linear equation

$$\dot{s}_1 = -\lambda s_1 \tag{4.3.8}$$

with stable solutions for a positive constant $\lambda > 0$, if

$$\dot{y} = \frac{a + \cos y}{\cos y \cos(z - \varphi(y))}(\cos(z - \varphi(y))\dot{z} + k_1 \dot{z} + k_2 \ddot{z} + \lambda s_1) \tag{4.3.9}$$

since

$$\dot{s}_1 = \frac{-\cos y}{a + \cos y}\cos(z - \varphi(y))\dot{y} + \cos(z - \varphi(y))\dot{z} + k_1 \dot{z} + k_2 \ddot{z}$$

(We assume that the magnitudes of y and $z - \varphi(y)$ are less then $\pi/2$.)

Step 3

In order to satisfy equation (4.3.9), the control u will be selected to enforce sliding mode on the surface

$$s = \dot{y}\frac{a + \cos y}{\cos y \cos(z - \varphi(y))}(\cos(z - \varphi(y))\dot{z} + k_1 \dot{z} + k_2 \ddot{z}) - \lambda s_1 = 0$$

Since only derivatives of both $\dot{y}$ and $\ddot{z}(z, y, \dot{z}, \dot{y})$ depend on the control force u, we can obtain

$$\begin{aligned}\dot{s} &= \psi_1(y, z, \dot{y}, \dot{z})\ddot{y} - \psi_2(y, z, \dot{y}, \dot{z}) = \psi_1(y, z, \dot{y}, \dot{z})(G_2 + \tilde{B}_2 u) - \psi_2(y, z, \dot{y}, \dot{z}) \\ &= \psi_1(y, z, \dot{y}, \dot{z})\tilde{B}_2 u - \tilde{\psi}_2(y, z, \dot{y}, \dot{z})\end{aligned}$$

$$\text{where} \quad \psi_1 = 1 - \frac{2ak_2 \tan y}{(a + \cos y)\cos(z - \varphi(y))}\left(\dot{z} - \frac{\cos y}{a + \cos y}\dot{y}\right)$$

with ψ_2 and $\tilde{\psi}_2$ continuous functions independent of control.

Again, for some bounded domain, the functions $s, \psi_1, \tilde{\psi}_2$ exist and $\psi_1 > 0$. The function $\tilde{B}_2$ is always positive, therefore sliding mode exists for this domain in the system with the discontinuous control

$$\begin{aligned} &u = -u_o \operatorname{sign}(s) \\ &u_o > \frac{1}{(\psi_1 \tilde{B}_2)_{\min}}|\tilde{\psi}_2'|_{\max} \end{aligned} \tag{4.3.10}$$

since the functions s and $\dot{s}$ have opposite signs. After sliding mode occurs, the function s_1 tends to zero as a solution to (4.3.8), which means that (4.3.5) holds. Hence the state components z and y, and therefore θ_1 and θ_2, tend to zero.

4.3.1 Simulation results

The proposed control algorithms for the second example, a double inverted pendulum, were verified by the following computer simulation results. Parameters of the pendulum system are taken from a real model used in Ledgerwood and Misawa (1992) and listed in Table 4.1. The design method for sliding mode control is valid for the parameter $a > 1$. For the set of parameters in Table 4.1, $a = 1.15$.

Table 4.1 Parameters of the pendulum system

Parameter	Parameter
$m_1 = 0.132\,\mathrm{kg}$	$I_1 = 0.003\,62\,\mathrm{kg\,m^2}$
$m_1 = 0.088\,\mathrm{kg}$	$I_2 = 0.001\,14\,\mathrm{kg\,m^2}$
$L_1 = 0.2032\,\mathrm{m}$	$J_o = 0.000\,06\,\mathrm{kg\,m^2}$
$L_2 = 0.2540\,\mathrm{m}$	
$l_1 = 0.1574\,\mathrm{m}$	
$l_2 = 0.1109\,\mathrm{m}$	$g = 9.8\,\mathrm{m\,s^{-2}}$

Figure 4.3 shows the results using the sliding mode control law (4.3.9) with the gains of $k_1 = k_2 = 0.174$, $\lambda = 35$ and $u_o = 1$. Both links as shown in Figure 4.3(a) are balanced in the upright position $\theta_1 = \theta_2 = 0$. The angular velocities are shown

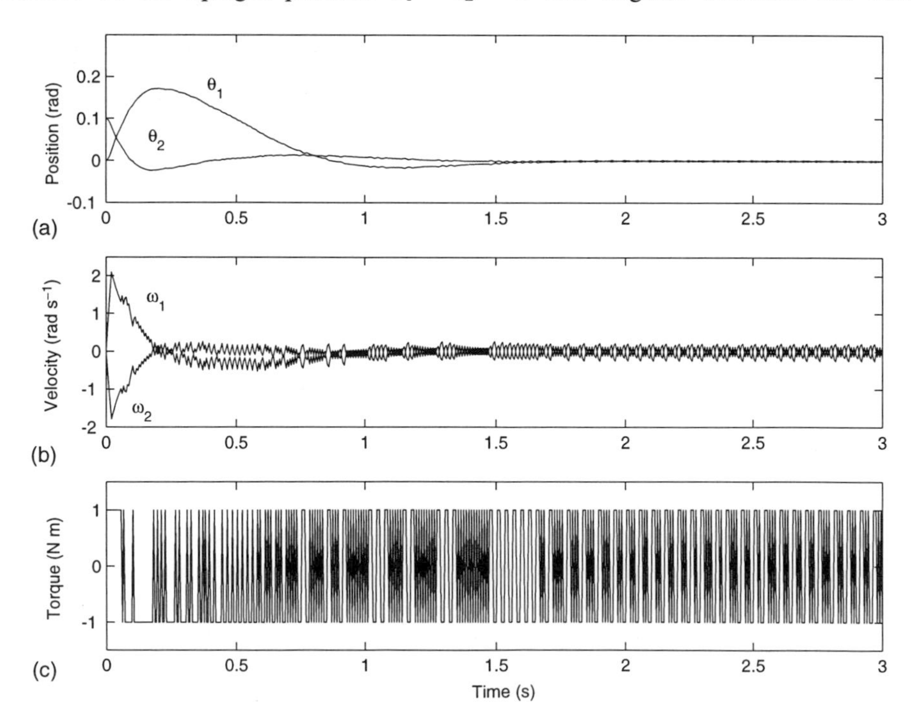

Figure 4.3 Results using SMC with $\theta_2(0) = 0.1\,\mathrm{rad}$.

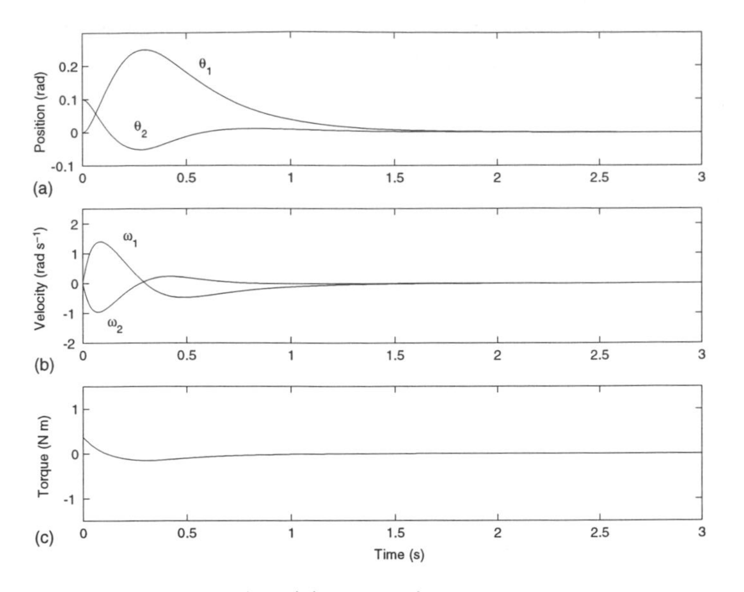

Figure 4.4 Results using LQR with $\theta_2(0) = 0.1$ rad.

in Figure 4.3(b). The control torque is shown in Figure 4.3(c). Simulation results obtained using linear quadratic regulator (LQR) control $u = 0.0001\theta_1 + 3.7388\theta_2 + 0.3218\dot{\theta}_1 + 0.5597\dot{\theta}_2$ are shown in Figure 4.4. Both the LQR and sliding mode (SMC) controllers can maintain the pendulum at the unstable equilibrium point $(\theta_1, \theta_2) = (0, 0)$ for a small value of the initial condition $\theta_2(0) = 0.1$ rad. Then both of the controllers were tested for a wider initial perturbed state of θ_2. As can be seen in Figure 4.5, the sliding mode control can handle the nonlinear control system for an initial value up to $\theta_2(0) = 0.49$ rad $\approx 28°$. The linear approach (LQR) can stabilize the system within the range $\theta_2(0) \leq 0.32$ rad $\approx 18.3°$ (Figure 4.6).

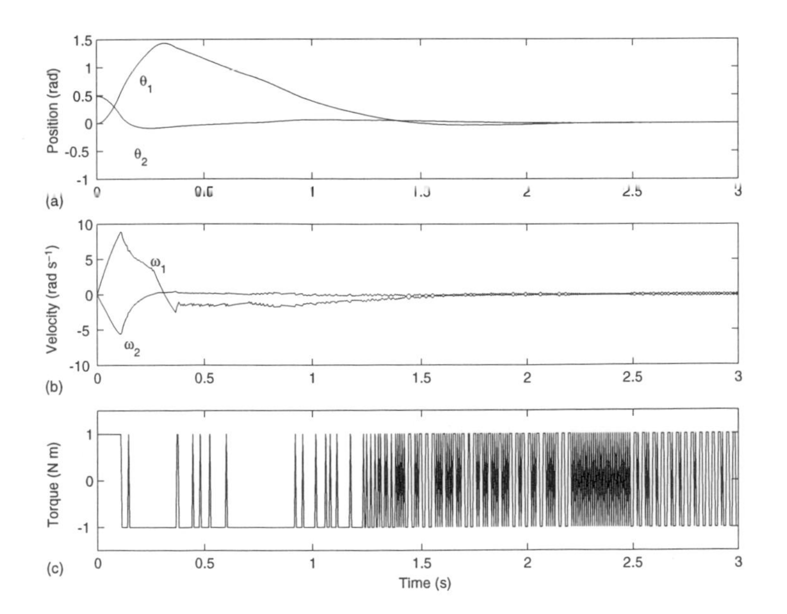

Figure 4.5 Results using SMC with $\theta_2(0) = 0.49$ rad.

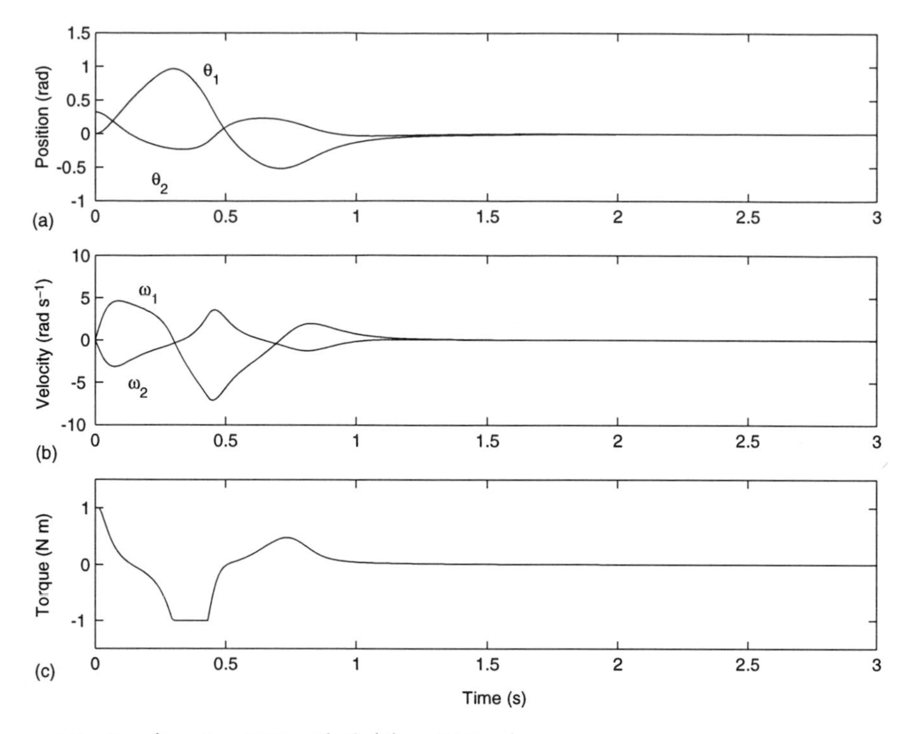

Figure 4.6 Results using LQR with $\theta_2(0) = 0.32\,\text{rad}$.

4.4 Rotational inverted pendulum system

A rotational inverted pendulum system as described by Widjaja (1994) is considered in this section. Figure 4.7 shows the plant consisting of a rotating base and a pendulum. Parameters m_1 and J_1 are the mass and inertia of the pendulum, l_1 is the distance from the centre of gravity of the link to its pivot point, g is the gravitational acceleration, and C_1 is the frictional constant between the pendulum and the rotating base. The coordinate θ_o represents the rotational angle of the base with respect to some horizontal axis (usually defined as the starting position) and θ_1 is the rotational angle of the pendulum with respect to the vertical axis. $\theta_1 = 0$ refers to the unstable equilibrium point.

The dynamic equations of the system are represented by

$$\begin{aligned} \ddot{\theta}_0 &= -a_p\dot{\theta}_o + K_p u \\ \ddot{\theta}_1 &= -\frac{C_1}{J_1}\dot{\theta}_1 + \frac{m_1 g l_1}{J_1}\sin\theta_1 + \frac{K_1}{J_1}\ddot{\theta}_o \end{aligned} \tag{4.4.1}$$

The upper equation is a simplified model of the permanent-magnet DC motor used to drive the rotating base with constants a_p and K_p. The lower equation of system (4.4.1) represents the dynamics of the pendulum; K_1 is a proportionality constant. The sign of K_1 depends on the position of the pendulum: $K_1 < 0$ for the inverted position and $K_1 > 0$ for the noninverted position. The applied armature voltage u is the only control input of the system.

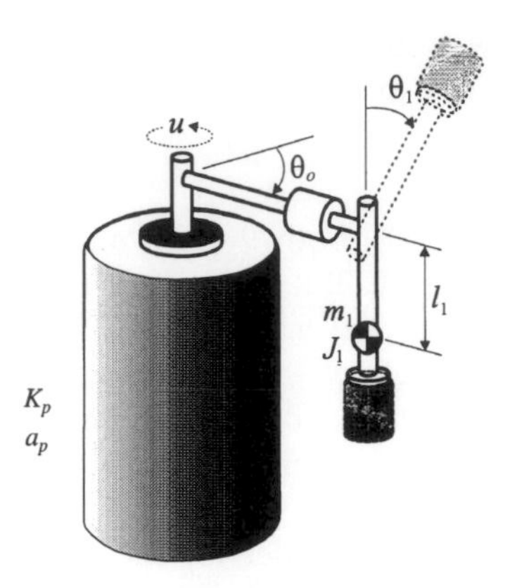

Figure 4.7 Inverted pendulum with rotating base.

As addressed in Widjaja (1994), the inverted pendulum system includes several control problems: swing-up, balancing, and both swing-up and balancing. In this section we will concentrate on a sliding mode controller for balancing the pendulum. The swing-up algorithm in the experiments will be directly taken from Widjaja (1994). First we will try to stabilize the system such that the pendulum is in the unstable vertical position $\theta_1 = 0$ and allow the base to be at an arbitrary fixed position. Then the design method will be generalized to drive both the pendulum and the rotating base to the equilibrium point $\theta_1 = \theta_0 = 0$ and maintain it there.

4.4.1 Control of the inverted pendulum

Notice, first, that in the system (4.4.1) rewritten in the form

$$\ddot{\theta}_0 = -a_p\ddot{\theta}_o + K_p u$$
$$\ddot{\theta}_1 = -\frac{C_1}{J_1}\dot{\theta}_1 + \frac{m_1 g l_1}{J_1}\sin\theta_1 - \frac{K_1}{J_1}a_p + \frac{K_1}{J_1}K_p u$$

the control u is multiplied by constant coefficients. Since $B(x)$ in this case is a constant matrix, a linear transformation is needed to reduce the system to the regular form. Let

$$y = \theta_o - \frac{J_1}{K_1}\theta_1 \tag{4.4.2}$$

Differentiating equation (4.4.2) results in

$$\dot{y} = \dot{\theta}_o - \frac{J_1}{K_1}\dot{\theta}_1 \tag{4.4.3}$$

and the motion equations in the regular form

$$\begin{aligned} \ddot{y} &= \frac{C_1}{K_1}\dot{\theta}_1 - \frac{m_1 g l_1}{K_1}\sin\theta_1 \\ \ddot{\theta}_1 &= -\frac{a_p K_1}{J_1}\dot{y} - \left(\frac{C_1}{J_1} + a_p\right)\dot{\theta}_1 + \frac{m_1 g l_1}{J_1}\sin\theta_1 + \frac{K_1 K_p}{J_1}u \end{aligned} \tag{4.4.4}$$

Let us first consider the lower subsystem of the regular form (4.4.4) and try to stabilize the system with respect to $\theta_1 = 0$. If the discontinuous control

$$u = -M \operatorname{sign}(s)$$

is applied with $s = \dot{\theta}_1 + \alpha\theta_1$ $(\alpha > 0)$, both $\dot{\theta}_1 \to 0$ and $\theta_1 \to 0$ as $t \to \infty$ if sliding mode is enforced in the plane $s = 0$. But the zero dynamics of the pendulum from the upper equation of (4.4.4) are governed by $\ddot{y} = 0$, hence $y \to \infty$ as $t \to \infty$, and the system is unstable. Therefore, the conventional design approach (Case 1 in Section 4.1) does not work for the pendulum system if the control should stabilize the inverted pendulum in the unstable vertical position with an arbitrary fixed position of the rotating base, $\theta_o = \text{const}$.

Now we design a sliding mode controller for the pendulum system based on the procedure of Case 2 in Section 4.1. Consider the upper equation of system (4.4.4). According to (4.1.6), the sliding manifold should be selected as

$$\frac{C_1}{J_1}\dot{\theta}_1 - \frac{m_1 g l_1}{J_1}\sin\theta_1 = -\alpha_1(y + \dot{y})$$

Hence the upper subsystem is stable:

$$\ddot{y} = -\alpha_1(y + \dot{y})$$

for a positive constant α_1. Both $y \to 0$ and $\dot{y} \to 0$ as $t \to \infty$; however, as follows from the upper equation of (4.4.4), the zero dynamics of the reduced-order system,

$$\dot{\theta}_1 = \frac{m_1 g l_1}{C_1}\sin\theta_1 \qquad \frac{m_1 g l_1}{C_1} > 0$$

are unstable. Case 2 in Section 4.1 is not applicable; it does not work either.

We now combine the ideas of Cases 3 and 4 to stabilize the pendulum.

Step 1

Following the approach of Case 4, introduce a new variable

$$x = \dot{y} - \frac{C_1}{K_1}\theta_1 \tag{4.4.5}$$

such that the right-hand side of the upper block in the motion equations would not depend on the time derivative of the state variable of the lower block. Since $\dot{x} = \ddot{y} - (C_1/K_1)\dot{\theta}_1$, substitution of $\ddot{y}$ from system (4.4.4) yields

$$\begin{aligned}
\dot{x} &= -\frac{m_1 g l_1}{K_1}\sin\theta_1 \\
\ddot{\theta}_1 &= -\frac{a_p K_1}{J_1}x - \frac{a_p C_1}{J_1}\theta_1 - \left(\frac{C_1}{J_1} + a_p\right)\dot{\theta}_1 + \frac{m_1 g l_1}{J_1}\sin\theta_1 + \frac{K_1 K_p}{J_1}u
\end{aligned} \tag{4.4.6}$$

The right-hand side of the upper equation in system (4.4.6) does not depend on $\dot{\theta}$. Then following the approach of Case 3, select the control such that the condition

$$\frac{m_1 g l_1}{K_1}\sin\theta_1 = \alpha_1 x \tag{4.4.7}$$

holds. The reduced-order system becomes

$$\dot{x} = -\alpha_1 x$$

with $x \to 0$ as $t \to \infty$ for positive constant α_1. In addition, since x decays exponentially, we can conclude from equations (4.4.7), (4.4.5) and (4.4.3) that functions θ_1, $\dot{y}$, $\dot{\theta}_1$ and $\dot{\theta}_o$ all decay exponentially as well. As a result, the desired system dynamics with $(\dot{\theta}_o, \theta_1) \to (0, 0)$ as $t \to \infty$ are obtained and the rotating base remains at a fixed position ($\theta_o = \text{const}$).

Step 2

The condition (4.4.7) holds if the function

$$s_1 = \frac{m_1 g l_1}{K_1} \sin\theta_1 - \alpha_1 x = 0 \tag{4.4.8}$$

The derivative of s_1 does not depend on the control u,

$$\dot{s}_1 = \frac{m_1 g l_1}{K_1} \cos\theta_1 \dot{\theta}_1 + \alpha_1 \frac{m_1 g l_1}{K_1} \sin\theta_1$$

but decays to zero if

$$\dot{s}_1 = -\alpha s_1$$

or

$$\frac{m_1 g l_1}{K_1} \cos\theta_1 \dot{\theta}_1 + \alpha_1 \frac{m_1 g l_1}{K_1} \sin\theta_1 = -\alpha s_1 \qquad (\alpha > 0)$$

Step 3

This condition is satisfied if sliding mode is enforced on the surface

$$s = \dot{s}_1 + \alpha s_1 = \frac{m_1 g l_1}{K_1} \cos\theta_1 \dot{\theta}_1 + \alpha_1 \frac{m_1 g l_1}{K_1} \sin\theta_1 + \alpha s_1 = 0 \tag{4.4.9}$$

Sliding mode exists if the functions s and $\dot{s}$ have opposite signs. Since only the derivative of $\dot{\theta}_1$ depends on the control force u, the function $\dot{s}$ may be represented in the form

$$\dot{s} = \frac{m_1 g l_1}{K_1} \cos\theta_1 \ddot{\theta}_1 + \psi_1(x, \theta_1, \dot{\theta}_1) = \frac{K_p m_1 g l_1}{J_1} \cos\theta_1 u + \tilde{\psi}_1(x, \theta_1, \dot{\theta}_1)$$

where ψ_1 and $\tilde{\psi}_1$ are functions of the system states. Notice that the function $\cos\theta_1$ is positive for the pendulum angle $-\pi/2 < \theta_1 < \pi/2$. The condition for existence of the sliding mode is satisfied if

$$u = -u_o \operatorname{sign}(s) \tag{4.4.10}$$

where

$$u_o \geq \frac{J_1}{K_p m_1 g l_1 \cos\theta_1} |\psi_1'|_{\max}$$

Once the state trajectories of sliding mode are confined to the switching manifold $s = 0$ after a finite time interval $s_1 \to 0$ and $x \to 0$ as $t \to \infty$. The desired dynamical behaviour with $\theta_o \to$ const and $\theta_1 \to 0$ as $t \to \infty$ is guaranteed.

4.4.2 Control of the base angle and the inverted pendulum

We have just shown that the system can be stabilized with respect to $\theta_1 = 0$ and $\dot{\theta}_o = 0$ by introducing a new variable x. Design of the control system for stabilizing both the pendulum and the rotating base at the equilibrium point $(\theta_o, \theta_1) = (0, 0)$ is performed as follows:

Step 1

The first equation of (4.4.6) and the first equation of (4.4.5) constitute a system similar to (4.1.11) in the design method of Case 4:

$$\begin{aligned} \dot{x} &= -\frac{m_1 g l_1}{K_1} \sin\theta_1 \\ \dot{y} &= x + \frac{C_1}{K_1}\theta_1 \end{aligned} \tag{4.4.11}$$

The state component θ_1 in the system (4.4.11) is handled as control. If the last term of the upper equation satisfies

$$\sin\theta_1 = -\lambda_1(x + y) \tag{4.4.12}$$

with constant λ_1, then the system is equivalent to

$$\begin{aligned} \dot{x} &= -a_1\lambda_1(x + y) \\ \dot{y} &= x + a_2\lambda_1 h(\theta_1)(x + y) \end{aligned} \tag{4.4.13}$$

where constants a_1 and a_2 for the interval $-\pi/2 < \theta_1 < \pi/2$ are positive since they are defined as

$$a_1 = -\frac{m_1 g l_1}{K_1} > 0 \qquad a_2 = -\frac{C_1}{K_1} > 0 \qquad (K_1 < 0) \tag{4.4.14}$$

Parameter h is a function of the pendulum angle θ_1:

$$h(\theta_1) = \theta_1 / \sin\theta_1$$

The stability of the system (4.4.13) is analyzed using the Lyapunov function candidate

$$V = \frac{1}{2}(x + y)^2 + \frac{1}{2}x^2$$

with the time derivative along the solutions of system (4.4.13)

$$\dot{V} = (x+y)(\dot{x}+\dot{y}) + x\dot{x} = -(a_1 - a_2 h(\theta_1))\lambda_1(x+y)^2 + x(x+y) - a_1\lambda_1 x(x+y)$$

The function $\dot{V}(t)$ is negative semidefinite,

$$\dot{V} = -(a_1 - a_2 h(\theta_1))\lambda_1(x+y)^2 \leq 0$$

$$\text{if} \quad \lambda_1 = \frac{1}{a_1} > 0$$

and the coefficient

$$(a_1 - a_2 h(\theta_1)) > 0 \tag{4.4.15}$$

The function $h(\theta_1)$ satisfies the inequalities

$$1 \leq h(\theta_1) < \pi/2 \tag{4.4.16}$$

for pendulum angle $-\pi/2 < \theta_1 < \pi/2$. Combining the inequalities (4.4.15) and (4.4.16) and substituting a_1 and a_2 from (4.4.14), we obtain a sufficient condition for the pendulum system to be stable as

$$\frac{m_1 g l_1}{C_1} > \pi/2$$

From a practical viewpoint, since the inverted pendulum is designed to rotate freely around its pivot, the frictional constant C_1 is much less than the torque $(m_1 g l_1)$ of the pendulum itself. Therefore, condition (4.4.15) holds for a real pendulum system. Moreover, if $\dot{V} = 0$ or $x + y = 0$, it follows from (4.4.13) that x is a constant value but $y \to \infty$ as $t \to \infty$ if this constant value is different from zero. Therefore, the system (4.4.13) can maintain the condition $\dot{V} = 0$ only at the equilibrium point $(x, y) = (0, 0)$. It is shown that the equilibrium point is asymptotically stable in the large with $x \to 0$ and $y \to 0$ as $t \to \infty$. Consequently, as follows from (4.4.12) and (4.4.2) $(\theta_o, \theta_1) \to (0, 0)$ as $t \to \infty$, which is the control objective.

Step 2

Following the same procedure as described in the previous case, equation (4.4.12) holds if the function

$$s_1 = \sin\theta_1 + \lambda_1(x+y) = 0$$

The function s_1 satisfies the linear first-order differential equation

$$\dot{s}_1 = -\lambda s_1 \qquad (\lambda > 0)$$

if

$$\cos\theta_1\dot{\theta}_1 + \lambda_1(\dot{x}+\dot{y}) = -\lambda s_1 \tag{4.4.17}$$

since in this case

$$\dot{s}_1 = \cos\theta_1\dot{\theta}_1 + \lambda_1(\dot{x} + \dot{y})$$

Step 3

In order to satisfy equation (4.4.17), sliding mode should be enforced on the switching surface

$$s = \dot{s}_1 + \lambda s_1 = \cos\theta_1\dot{\theta}_1 + \lambda_1(\dot{x} + \dot{y}) + \lambda s_1 = 0$$

The time derivative of the function s is of the form

$$\dot{s} = \cos\theta_1\ddot{\theta}_1 + \psi_2(x, y, \theta_1, \dot{\theta}_1) = \frac{K_1 K_p}{J_1}\cos\theta_1 u + \tilde{\psi}_2(x, y, \theta_1, \dot{\theta}_1)$$

where ψ_2 and $\tilde{\psi}_2$ are functions of the system states. The function $\cos\theta_1$ is positive and parameter K_1 is negative for the pendulum angle $-\pi/2 < \theta_1 < \pi/2$. The condition for existence of sliding mode (the functions s and $\dot{s}$ need to have opposite signs) is satisfied if

$$u = u_o \operatorname{sign}(s) \quad \text{with} \quad u_o \geq \frac{-J_1}{K_1 K_p \cos\theta_1}|\tilde{\psi}_2|_{\max}$$

After sliding mode occurs on the surface $s = 0$, $s_1 \to 0$ and $(x, y) \to (0, 0)$ as $t \to \infty$. Finally, we obtain the desired dynamic behaviour, $(\theta_o, \theta_1) \to (0, 0)$ as $t \to \infty$.

4.5 Simulation and experimental results

Both simulation and experimental results for stabilizing the rotational inverted pendulum system will be presented in this section. The simulation results for the two pendulum systems in Figure 4.2 have been presented in Section 4.3, so special emphasis will be placed on robustness by investigating the ability of the sliding mode controllers for significant plant parameter variations.

The experimental setup was developed, and it is currently available for both undergraduate and graduate control system laboratories at The Ohio State University. Figure 4.8 illustrates the complete hardware setup configuration of the inverted pendulum system. The real-time control system mainly consists of three parts: the controller, the interface circuits and the pendulum system. Two optical encoders are used to measure the angular positions of both the pendulum and the base. All parameters of the inverted pendulum system are listed in Table 4.2, and they were determined experimentally by identification techniques; Widjaja (1994) gives more details.

The inverted pendulum system allows the user to change the system parameters, or add disturbances by attaching containers of various sizes and contents to the end of the pendulum. A container of metal bolts and water will later be added to the pendulum in the set of experiments. The mass of the container and its contents significantly change the system parameters, and the motion of the water within the container acts as a disturbance to the system.

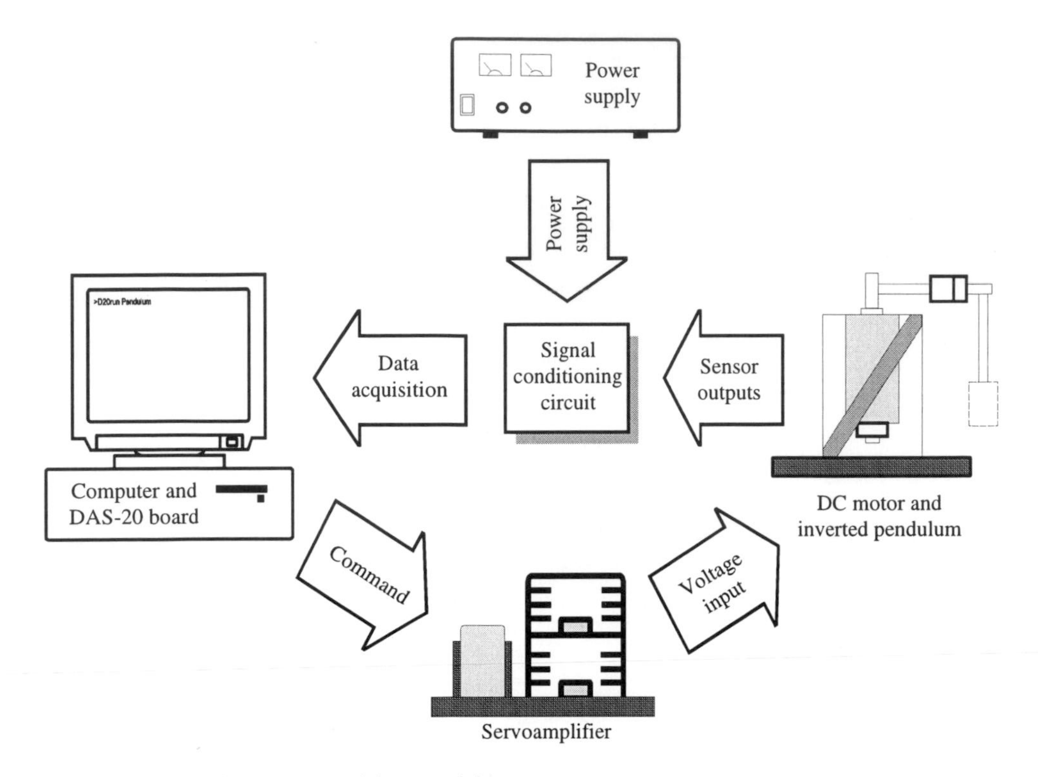

Figure 4.8 Hardware setup of the pendulum system.

Table 4.2 Parameters of the rotational inverted pendulum system

Parameter	Parameter
$l_1 = 0.113\,\mathrm{m}$	$m_1 = 8.6184 \times 10^{-2}\,\mathrm{kg}$
$g = 9.8066\,\mathrm{m\,s^{-2}}$	$J_1 = 1.301 \times 10^{-3}\,\mathrm{N\,m\,s^2}$
$a_p = 33.04$	$C_1 = 2.979 \times 10^{-3}\,\mathrm{N\,m\,s/rad^{-1}}$
$K_p = 74.89$	$K_1 = \begin{cases} -1.9 \times 10^{-3}, & \text{if } -\pi/2 < \theta_1 < \pi/2 \\ 1.9 \times 10^{-3} & \text{otherwise} \end{cases}$

Figure 4.9 shows the simulation results for stabilizing both the pendulum and the rotating base using the linear quadratic regulator (LQR) technique with

$$u = 0.7\theta_o + 1.0\dot{\theta}_o + 10.8\theta_1 + 0.7\dot{\theta}_1$$

The pendulum is first swung up with the swing-up algorithm, and then the LQR begins to take over the control when the rotational angle of the pendulum is within the range $|\theta_1| \leq 0.3\,\mathrm{rad}$. The experimental results for nominal conditions using the LQR technique may be found in Widjaja (1994) and Ordonez *et al.* (1997).

We will focus on the performance of the inverted pendulum system using our previously developed sliding mode controllers. Two case studies of the control objectives will be presented. Case 1 stabilizes the pendulum at $\theta_1 = 0$ with $\dot{\theta}_o = 0$. Case 2 stabilizes both the pendulum and the rotating base with respect to the equilibrium point $\theta_1 = \theta_o = 0$.

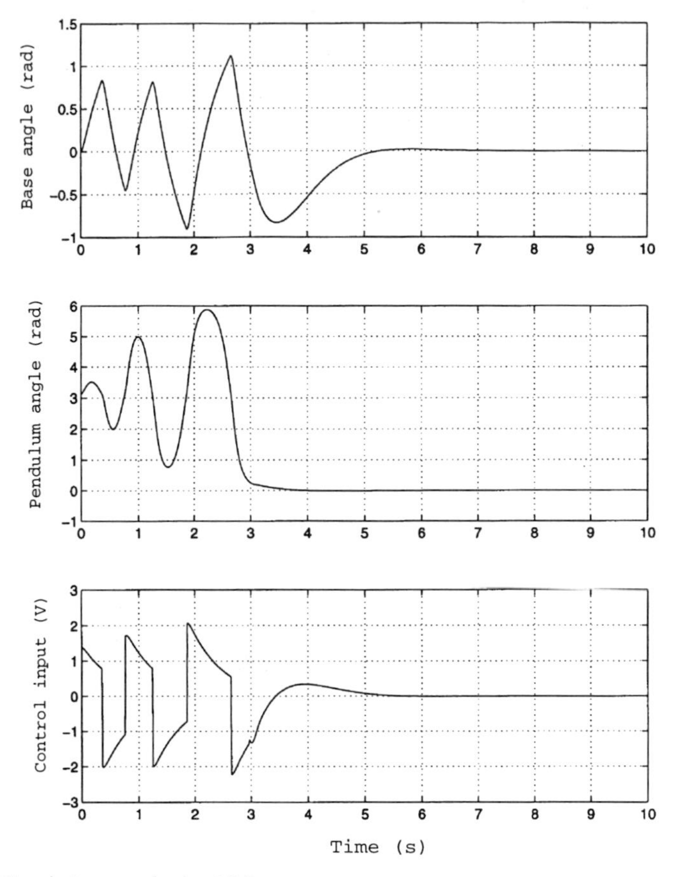

Figure 4.9 Simulation results by LQR.

4.5.1 Stabilization of the inverted pendulum

The simulation results using the control laws developed in this section are shown in Figure 4.10. The required information for calculating the control input is equations (4.4.2), (4.4.5), (4.4.8), (4.4.9) and (4.4.10). As can be seen, the pendulum angle is driven to zero, and the rotating base at the same time remains at a fixed position (its angular velocity equals to zero) with the selected input gains $\lambda_1 = 0.08$, $\lambda = 100$ and $u_o = 3$.

The discontinuous controller was implemented for real-time control of the pendulum. We observed that due to the sampling issue of the discrete-time control system, in practice, the ideal sliding mode control cannot be implemented. Besides, as presented in many publications (Utkin, 1992; Kwatny and Siu, 1987; Bartolini 1989), the chattering, which appears as a high-frequency oscillation at the vicinity of the desired manifold, may be excited by unmodelled high-frequency dynamics of the system. In order to suppress the chattering, the saturating continuous approximation (Slotine and Sastry, 1983; Burton and Zinober, 1986) will be used to replace the ideal switching at the vicinity of the switching manifold (see also Chapters 8 and 9). This results in a trade-off between accuracy and robustness.

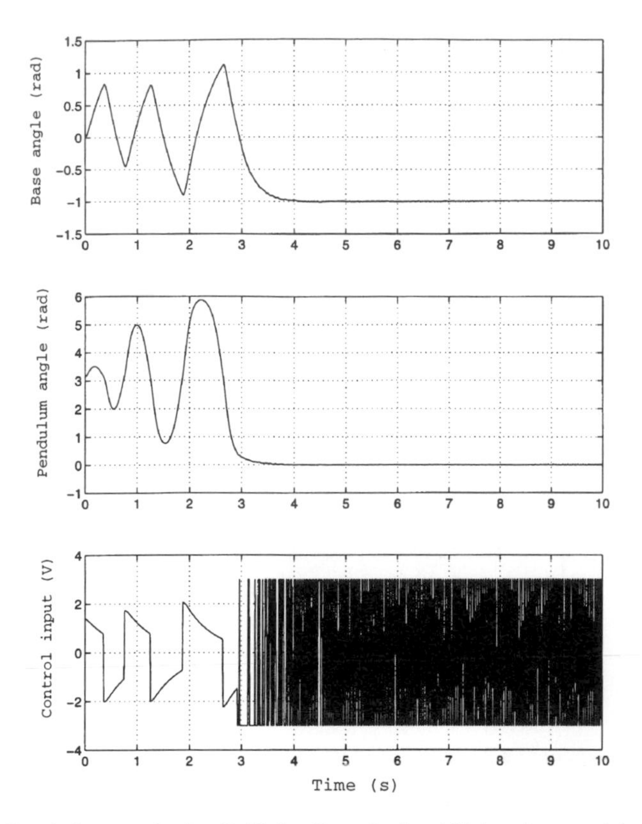

Figure 4.10 Simulation results by SMC for Case 1 of stabilizing the pendulum.

Figures 4.11 to 4.13 show the experimental results of the sliding mode control (SMC) for stabilizing the pendulum with different loads attached to the end of the pendulum. The sampling time for the control system is $\Delta t = 5\,\text{ms}$, and it is fixed for other experimental results in the later figures. The control law using a continuous approximation by a sinusoidal function is designed as

$$u = \begin{cases} u_o \sin(\pi s/2\delta) & |s| \leq \delta \\ u_o \operatorname{sign}(s) & \text{otherwise} \end{cases} \tag{4.5.1}$$

where δ is the allowable maximum width of the continuous zone from the desired ideal sliding manifold $s = 0$. It can be easily shown that the ideal discontinuous control is implemented if $\delta = 0$. The larger the value of δ, the smaller the anticipated invariance to system uncertainties and the smaller the amount of chattering in the system states. The input gains of the SMC pendulum system are selected as $\lambda_1 = 0.08$, $\lambda = 400$ and $u_o = 2.5$.

Figure 4.11 shows that, for the nominal plant, the pendulum angle is stabilized close to zero. The control force input, as we expected, swings up the pendulum from the beginning, switches to the SMC at time ~1.5 s, and then stays in the δ zone $|u| < u_o = 2.5$ after 2 s. Note that the system is stabilized at the point $(\dot{\theta}_o, \theta_1) = (0, 0)$ and it is marginally stable with respect to (θ_o, θ_1); this explains why the position of the rotating base is slowly drifting (θ_o is not constant). Similar results were obtained when the same controller was used to drive the pendulum with

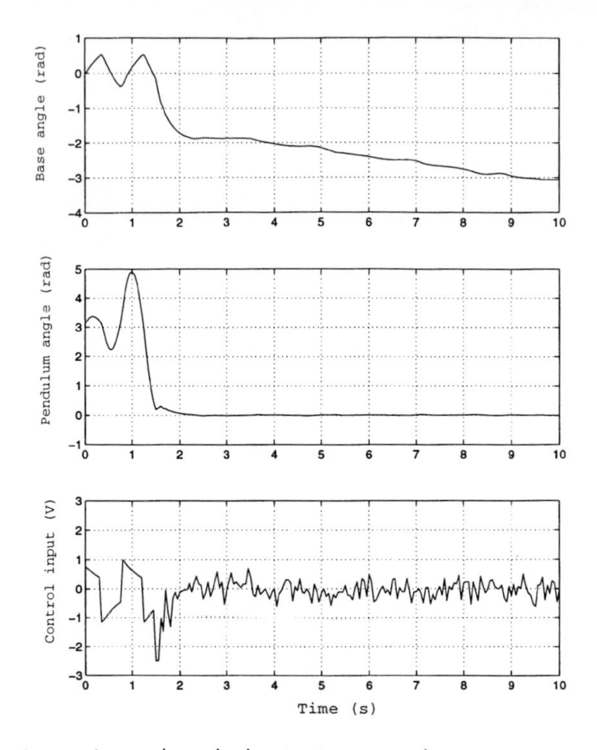

Figure 4.11 Case 1 experimental results by SMC: no weight.

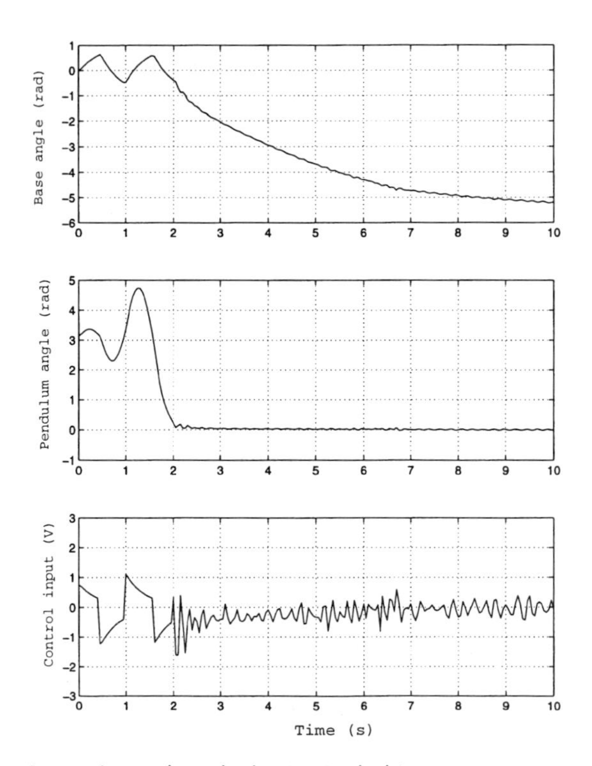

Figure 4.12 Case 1 experimental results by SMC: sloshing water.

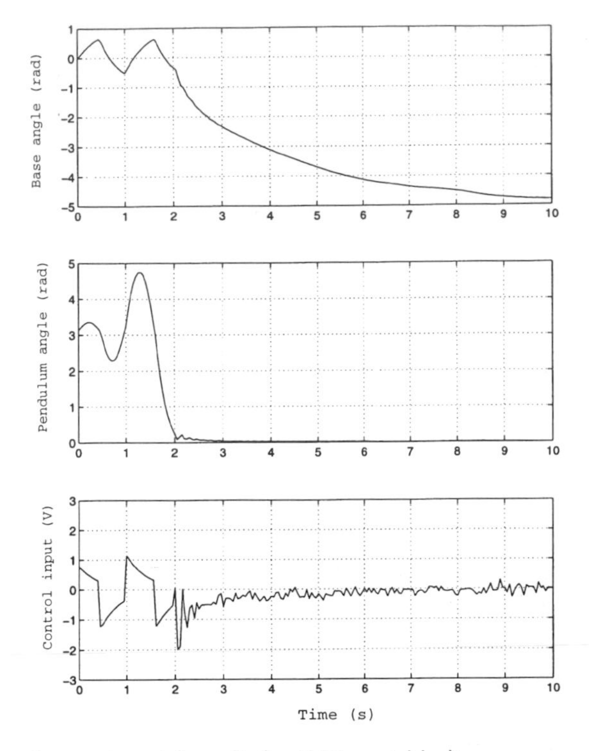

Figure 4.13 Case 1 experimental results by SMC: metal bolts.

water (Figure 4.12) and with metal bolts (Figure 4.13). We observe that the controller can still manage the balance of the inverted pendulum quite well without saturation of the control input. There are several interesting observations: small ripples are generated due to the distributed disturbance from the water in Figure 4.12; the average values of the control inputs in both cases gradually converge to zero when disturbances get settled at the final time of 10 s; and a smaller amplitude of the control input is observed at the steady state when additional weights, the metal bolts, are added to the system (Figure 4.13).

4.5.2 Stabilization of the inverted pendulum and the base

The sliding mode control for stabilizing both the pendulum and the base will be designed following (4.4.2), (4.4.3), (4.4.5) and (4.4.8) to (4.4.10). The simulation results for control (4.4.10) with the gains $\lambda_1 = 0.08$, $\lambda = 800$ and $u_o = 3$ are shown in Figure 4.14.

Figure 4.15 shows the experimental results of the SMC for the nominal pendulum using the modified controller (4.5.1) with $\lambda_1 = 0.08$, $\lambda = 800$ and $u_o = 2.5$. For a small value of δ, the control input is still similar to the discontinuous control in Figure 4.10, although its switching frequency is considerably reduced. As a result, chattering exists in both of the state responses. The results for a larger value of δ are shown in Figure 4.16. The control input is no longer saturated and varies between the extreme values ± 2.5.

The most interesting experimental results are depicted in Figures 4.17 and 4.18. The

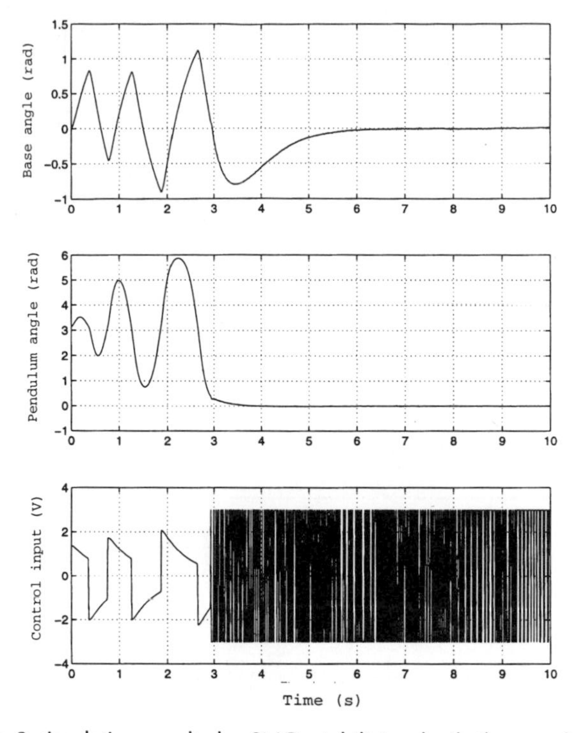

Figure 4.14 Case 2 simulation results by SMC: stabilizing both the pendulum and the base.

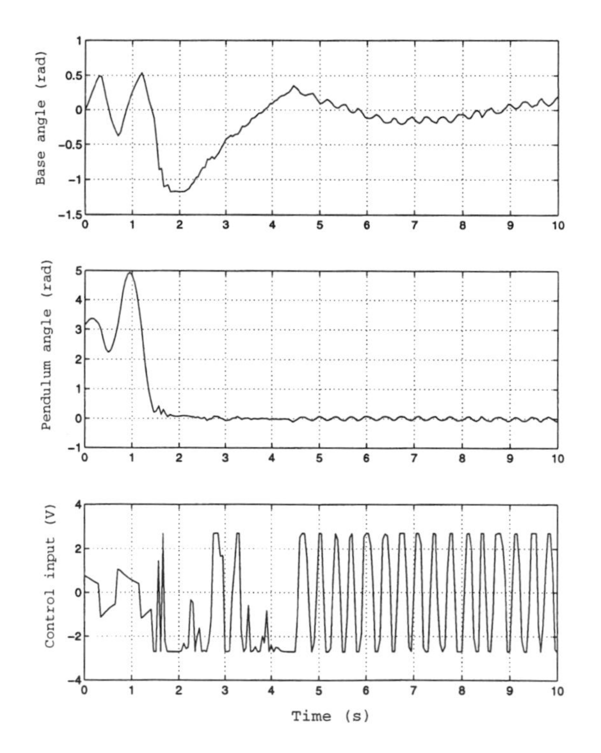

Figure 4.15 Case 2 experimental results by SMC: no weight.

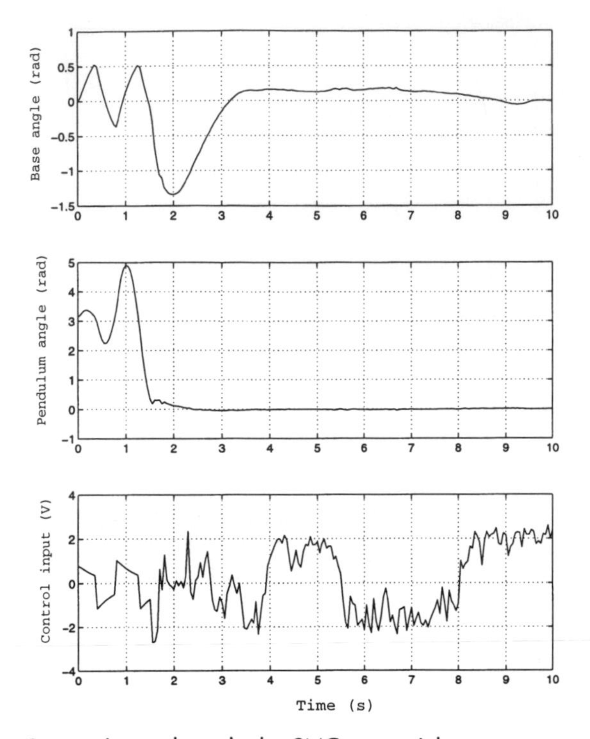

Figure 4.16 Case 2 experimental results by SMC: no weight.

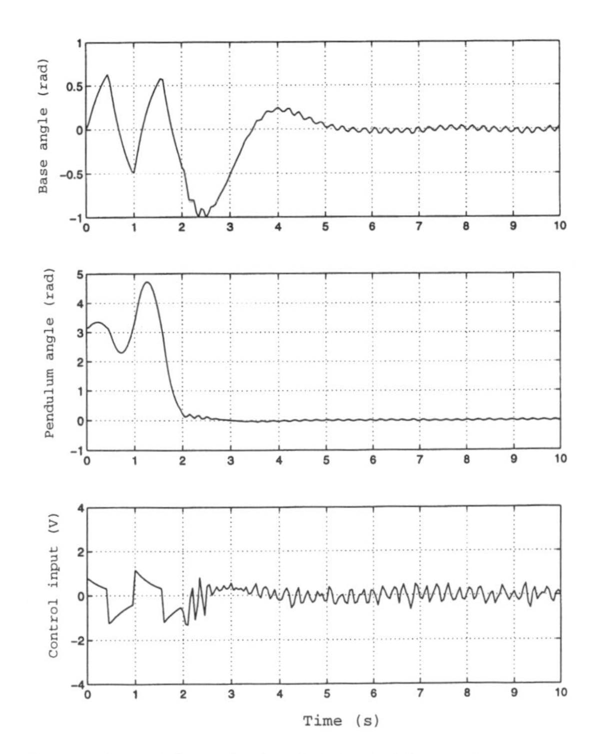

Figure 4.17 Case 2 experimental results by SMC: sloshing water.

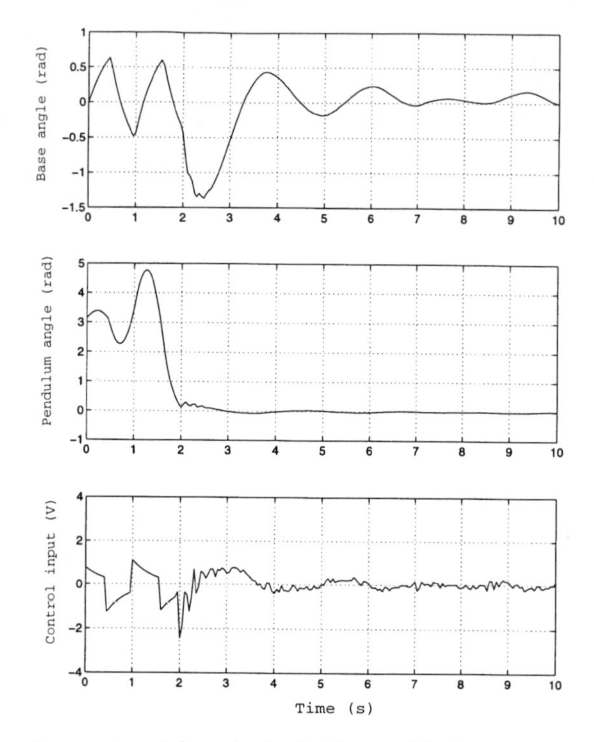

Figure 4.18 Case 2 experimental results by SMC: metal bolts.

controller is able to provide convergence of the pendulum with both metal bolts and sloshing water using the same gains for the control input. The system states are stabilized in the vicinity of the equilibrium point $(\theta_o, \theta_1) = (0, 0)$. The low-amplitude oscillations, similar to Figure 4.12 under the effect of sloshing water dynamics, are still observed in Figure 4.17, where the control input has an average value close to zero. We observe an underdamped system response in Figure 4.18 for the pendulum with metal bolts. Compared to Figure 4.13, the control input oscillations are relatively large at the beginning of the process, but they decrease to the same level after a couple of seconds, when both the pendulum and the rotating base get settled.

References

BARTOLINI, G., 1989, Chattering phenomena in discontinuous control systems, *International Journal on Systems Science*, **20**, 2471–81.

BURTON, J. A. and ZINOBER, A. S. I., 1986, Continuous approximation of variable structure control, *International Journal on Systems Science*, **17**, 875–85.

KWATNY, H. G. and SIU, T. L., 1987, Chattering in variable structure feedback systems, *Proceedings of the IFAC 10th World Congress*, **8**, 307–14.

LEDGERWOOD, T. and MISAWA E., 1992, Controllability and nonlinear control of rotational inverted pendulum, in Misawa, E. (Ed.) *Advances in Robust and Nonlinear Control Systems, ASME Journal on Dynamic Systems and Control,* **43**, 81–88.

LUK'YANOV, A. G., 1993, Optimal nonlinear block-control method, *Proceedings of the 2nd European Control Conference*, Groningen, The Netherlands, pp. 1853–55.

LUK'YANOV, A. G. and DODDS, S. J., 1996, Sliding mode block control of uncertain nonlinear plants, *Proceedings of the IFAC World Congress*, Section F-22b. pp. 241–46.

Mori, S., Nishihara, H. and Furuta, K., 1976, Control of unstable mechanical system: control of pendulum, *International Journal on Control*, **23**, 673–92.

Ordonez, R., Zumberge, J., Spooner, J. T. and Passino, K. M., 1997, Adaptive fuzzy control: experiments and comparative analyses, *IEEE Transactions on Fuzzy Systems*, **5**, 167–88.

Slotine, J. J. and Sastry, S. S., 1983, Tracking control of nonlinear systems using sliding surfaces, with application to robot manipulators, *International Journal on Control*, **38**, 465-92.

Utkin, V. I., 1992, *Sliding Modes in Control and Optimization*, Berlin: Springer-Verlag.

Widjaja, M., 1994, 'Intelligent control for swing-up and balancing of an inverted pendulum system', master's thesis, Ohio State University, Columbus OH.

CHAPTER FIVE

Control of Linear Systems

The objective of this chapter is to demonstrate the sliding mode control design methodology for linear systems. Reducing system equations to the regular form will be performed as a preliminary step in all design procedures. The core idea is to use the methods of linear control theory for reduced-order equations and to employ different methods of enforcing sliding modes with the desired dynamics.

5.1 Eigenvalue placement

We start with the conventional problem of linear control theory: eigenvalue placement in a linear time-invariant multidimensional system

$$\dot{x} = Ax + Bu \tag{5.1.1}$$

where x and u are n- and m-dimensional state and control vectors, respectively, A and B are constant matrices, $\text{rank}(B) = m$. The system is assumed to be controllable.

For any controllable system there exists a linear feedback $u = Fx$ (F being a constant matrix) such that the eigenvalues of the feedback system, i.e. of matrix $A + BF$, take the desired values and, as a result, the system exhibits the desired dynamic properties (Kwakernaak and Sivan, 1972).

Now we will show that the eigenvalue task may be solved in the framework of sliding mode control dealing with a reduced-order system. As demonstrated in Section 3.3, the design becomes simpler for systems represented in regular form. Since $\text{rank}(B) = m$, matrix B in (5.1.1) may be partitioned (after reordering the state vector components) as

$$B = \begin{bmatrix} B_1 \\ B_2 \end{bmatrix} \tag{5.1.2}$$

where $B_1 \in \Re^{(n-m)\times m}$, $B_2 \in \Re^{m\times m}$ with $\det B_2 \neq 0$. The nonsingular coordinate transformation

$$\begin{bmatrix} x_1 \\ x_2 \end{bmatrix} = Tx \qquad T = \begin{bmatrix} I_{n-m} & -B_1 B_2^{-1} \\ 0 & B_2^{-1} \end{bmatrix} \tag{5.1.3}$$

reduces the system equations (5.1.1) and (5.1.2) to regular form:

$$\begin{aligned} \dot{x}_1 &= A_{11}x_1 + A_{12}x_2 \\ \dot{x}_2 &= A_{21}x_1 + A_{22}x_2 + u \end{aligned} \tag{5.1.4}$$

where $x_1 \in \Re^{(n-m)}$, $x_2 \in \Re^m$ and A_{ij} are constant matrices for $i, j = 1, 2$.

It follows from controllability of (A, B) that the pair (A_{11}, A_{12}) is controllable as well (Utkin and Young, 1978). Handling x_2 as an m-dimensional intermediate control in the controllable $(n-m)$-dimensional first subsystem of (5.4), all $(n-m)$-eigenvalues may be assigned arbitrarily by a proper choice of matrix C in

$$x_2 = -Cx_1$$

To provide the desired dependence between components x_1 and x_2 of the state vector, sliding mode should be enforced in the manifold

$$s = x_2 + Cx_1 = 0 \tag{5.1.5}$$

where $s^T = (s_1, \ldots, s_m)$ is the difference between the desired and real values of x_2. After sliding mode starts, the motion is governed by a reduced-order system with the desired eigenvalues

$$\dot{x}_1 = (A_{11}x_1 - A_{12}C)x_1 \tag{5.1.6}$$

For a piecewise linear discontinuous control

$$u = -(\alpha|x| + \delta)\text{sign}(s) \qquad (\alpha,\ \delta \text{ positive constants}) \tag{5.1.7}$$

with $|x| = \sum_{i=1}^{n} |x_i|$, $\text{sign}(s)^T = [\text{sign}(s_1) \ldots \text{sign}(s_m)]$ calculate the time derivative of positive definite function $V = \frac{1}{2}s^T s$

$$\begin{aligned} \dot{V} &= s^T((CA_{11} + A_{21})x_1 + (CA_{12} + A_{22})x_2) - (\alpha|x| + \delta)|s| \\ &\le |s|\,|(CA_{11} + A_{21})x_1 + (CA_{12} + A_{22})x_2| - (\alpha|x| + \delta)|s| \end{aligned}$$

It is evident there exists a value of α such that, for any δ, the time derivative $\dot{V}$ is negative; this validates convergence of the state vector to manifold $s = 0$ in (5.1.5) and the existence of sliding mode with the desired dynamics. The time interval preceding the sliding motion may be decreased by increasing parameters α and δ in control (5.1.7).

A similar result may be obtained in the system with the following unit control (Section 3.5):

$$u = -(\alpha|x| + \delta)\frac{s}{\|s\|} \qquad \|s\| = (s^T s)^{1/2} \tag{5.1.8}$$

which undergoes discontinuities on manifold $s = 0$ in contrast to control (5.1.7) with discontinuity points on each surface $s_i = 0$ $(i = 1, \ldots, m)$. The time derivative of v for system (5.1.4) with control (5.1.8) is of the form

$$\begin{aligned} \dot{V} &= s^T((CA_{11} + A_{21})x_1 + (CA_{12} + A_{22})x_2) - (\alpha|x| + \delta)\|s\| \\ &\le \|s\|\,|(CA_{11} + A_{21})x_1 + (CA_{12} + A_{22})x_2| - (\alpha|x| + \delta)\|s\| \end{aligned}$$

Again, there exists α such that $\dot{V}$ is negative for any δ and states reach manifold $s = 0$ after a finite time interval.

If the system is not reduced to the regular form, the manifold $s = Cx$ (C being an $m \times (n-m)$ matrix) may be selected in terms of the original system (5.1.1) based on the approach of the equivalent control method and the conditions for sliding mode to exist (Sections 2.3 and 2.4). Assume that sliding mode in $s = 0$ has the desired dynamic properties and matrix CB is not singular. Then

$$\dot{s} = CAx + CBu$$

and the time derivative of Lyapunov function $V = \frac{1}{2}s^T s$ is of the form

$$\dot{V} = s^T CAx + s^T CBu$$

If matrix $(CB + (CB)^T)$ is positive definite, then discontinuous control

$$u = -M(x)\text{sign}(s)$$

with $M(x) = (\alpha|x| + \delta)$ and sufficiently high but finite α and an arbitrary value of δ leads to sliding mode after a finite time interval. (The proof is similar to the one for nonlinear systems in Section 3.5.)

For an arbitrary matrix CB, control should be selected in compliance with the method of Section 3.2:

$$u = -U(x)\text{sign}(s^*) \qquad s^* = (CB)^T s$$

The time derivative of Lyapunov function candidate $V = \frac{1}{2}s^T s$ for the system with $U(x) = (\alpha|x| + \delta)$,

$$\begin{aligned}\dot{V} &= (s^*)^T (CB)^{-1} CAx - |s^*|(\alpha|x| + \delta) \\ &\leq |s^*| \, |(CB)^{-1} CAx| - |s^*|(\alpha|x| + \delta)\end{aligned}$$

is negative definite for sufficiently high α and an arbitrary value of δ as well.

Finally, it is interesting to note that for systems with nonlinear unit control, a sliding mode existence condition may be derived by the algebraic stability criteria developed for linear systems. Indeed, if CB is a *Hurwitz* matrix then Lyapunov equation $(CB)P + P^T(CB) = -I_m$ has a positive definite solution $P > 0$, and the time derivative of Lyapunov function candidate $V = \frac{1}{2}s^T Ps > 0$ in the system with control $u = M(x)s/\|s\|$ may be found as

$$\begin{aligned}\dot{V} &= s^T PAx + M(x)s^T P(CA)\frac{s}{\|s\|} \\ &= s^T PAx + M(x)s^T \frac{P(CB) + (CB)^T P}{2} \frac{s}{\|s\|} \\ &\leq \|s\| \, \|PAx\| - \frac{1}{2}M(x)\|s\|\end{aligned}$$

If function $M(x) = (\alpha|x| + \delta)$ then $\dot{V}$ is negative and sliding mode occurs after a finite time interval (Section 3.5).

5.2 Invariant systems

One of the main objectives of designing feedback control systems is to reduce sensitivity with respect to disturbances and plant parameter variations. As shown in Section 3.4, sliding modes in any manifold are invariant to those factors if they act in a control space satisfying conditions (3.4.4) and (3.4.5). For linear systems

$$\dot{x} = (A + \Delta A(t))x + Bu + Qf(t) \qquad f(t) \in \Re^l \tag{5.2.1}$$

the invariance conditions were formulated in terms of system and input matrices (Drazenovic, 1969): sliding modes in any manifold are invariant with respect to parameter variations $\Delta A(t)$ and disturbance vector $f(t)$ if

$$\Delta A \in \text{range}(B) \quad \text{and} \quad Q \in \text{range}(B)$$

or there exist matrices Λ_A and Λ_Q (constant or time-varying) such that

$$\Delta A = B\Lambda_A \qquad Q = B\Lambda_Q \tag{5.2.2}$$

If the conditions (5.2.2) hold, then the regular form for (5.2.1) is similar to (5.1.4):

$$\begin{aligned} \dot{x}_1 &= A_{11}x_1 + A_{12}x_2 \\ \dot{x}_2 &= A_{21}x_1 + A_{22}x_2 + u + \Lambda_A x + \Lambda_Q f(t) \end{aligned}$$

Selecting discontinuous control in the form (5.1.7) with manifold (5.1.5) leads to the sliding mode equation (5.1.6) with desired dynamics and invariance property. Assuming that the ranges of plant parameter variations and an upper bound of the disturbance vector $\|\Lambda_Q f(t)\| \leq f_0$ are known, $s = 0$ can be made a sliding manifold in the system with control (5.1.7) under the additional condition

$$\delta \geq f_0 \tag{5.2.3}$$

A similar approach may be applied to decouple l interconnected systems

$$\dot{x}_i = A_i x_i + \sum_{\substack{j=1 \\ j \neq i}}^{l} A_{ij} x_j + B_i u_i \qquad (i = 1, \ldots, l)$$

where $x_i \in \Re^{n_i}$, $u_i \in \Re^{m_i}$ and A_i, A_{ij} and B_i are constant matrices.

Interconnection terms may be handled as disturbances and the invariance conditions may be reformulated for each subsystem $A_{ij} \in \text{range}(B_i)$. Discontinuous control in the ith system

$$u_i = -(\alpha_i + \delta_i)|x|\text{sign}(s_i) \qquad x^T = (x_i^T, \ldots, x_l^T), \;\; s_i = C_i x_i, \;\; C_i = \text{const}, \;\; s_i \in \Re^{m_i}$$

with sufficiently high but finite values α_i enforces sliding mode in manifold $s_i = 0$ governed by an $(n_i - m_i)$th order equation which does not depend on the states of the other subsystems. The dynamics of each subsystem may be designed by a proper choice of matrices C_i in the equations of the sliding manifolds.

5.3 Sliding mode dynamic compensators

The value of δ should exceed an upper estimate of a disturbance vector (5.2.3) in discontinuous control (5.1.7) designed to reject disturbances. To soften the control action, it would be desirable to reduce the control amplitude in sliding mode if the magnitude of the disturbance decreases. Usually disturbances are not accessible for measurement, which is the main obstacle for designing a control system with the above property. Nevertheless, thus formulated, the task is solvable if rather fuzzy a priori knowledge on a class of disturbances is available.

Let the disturbance in motion equation (5.2.1) with $\Delta A = 0$ satisfy the invariance condition (5.2.2) and let Q be a constant matrix. Then the system may be represented in the regular form

$$\begin{aligned} \dot{x}_1 &= A_{11}x_1 + A_{12}x_2 \\ \dot{x}_2 &= A_{21}x_1 + A_{22}x_2 + u + \Lambda_\theta f(t) \end{aligned} \tag{5.3.1}$$

and the equation of sliding mode in manifold $s = Cx_1 + x_2 = 0$ (5.1.6) does not depend on the disturbance vector. Assume that the components of vector $f(t)$ cannot be measured and a 'disturbance model' is taken in the form of a time-varying linear dynamic system

$$f^{(k)} + \sum_{i=1}^{k-1} \theta_i(t) f^{(i)} = 0 \tag{5.3.2}$$

The scalar coefficients $\theta_i(t)$ can vary arbitrarily over bounded intervals

$$|\theta_i(t)| \leq \vartheta_{i0} \tag{5.3.3}$$

It is assumed that neither initial conditions nor functions $\theta_i(t)$ are measured, and only ranges ϑ_{i0} are known. The equation (5.3.2) embraces a rather wide class of disturbances. For $k = 2$ it includes exponential and harmonic functions, polynomials of any finite power (beginning from a certain time), all kinds of products of these functions, etc.

The controller is designed as a dynamic system with control u as an output

$$u^{(k)} + \sum_{i=0}^{k-1} d_i u^{(i)} = v \tag{5.3.4}$$

where d_i are constant scalar coefficients whose choice is dictated by convenience of implementation only.

The input v will be selected as a piecewise linear function of the controller and system states. Each of m control channels of the system has a kth order dynamic element, the total order of the system being equal to $n + mk$. The state coordinate of the additional dynamic system can be measured.

Let us write the motion equations of the extended system in the space consisting of $x_1, x_2, \ldots, x_{k+2}$, if

$$\dot{x}_i = x_{i+1} \qquad (i = 2, \ldots, k+1) \tag{5.3.5}$$

Since $\dot{x}_2 = x_3$, it follows from the second equation of (5.3.1) that

$$u = x_3 - A_{21}x_1 - A_{22}x_2 - \Lambda_Q f \tag{5.3.6}$$

Differentiating (5.3.6) k times and substituting the right-hand sides of (5.3.1), (5.3.5) and (5.3.6) for the time derivatives of x_i and u, we obtain

$$u^{(i)} = x_{i+3} + \sum_{j=1}^{i+2} A_j^i x_j - \Lambda_Q f^{(i)} \qquad (i = 1, \dots, k-1) \tag{5.3.7}$$

$$u^{(k)} = \dot{x}_{k+2} + \sum_{j=1}^{k+2} A_j^k x_j - \Lambda_Q f^{(k)} \tag{5.3.8}$$

where A_j^i and A_j^k are constant matrices. By substituting the values of derivatives $u^{(i)}$ from (5.3.7) and (5.3.8) into (5.3.4) and replacing the kth derivative of the disturbance vector in accordance with (5.3.2) by a linear combination of vectors $f, \dots, f^{k-1}$, we obtain

$$\dot{x}_{k+2} = \sum_{i=1}^{k+2} A_i x_i + \sum_{i=0}^{k-1} (d_i - \theta_i(t)) \Lambda_Q f^{(i)} + v \tag{5.3.9}$$

where A_i are constant matrices.

Bearing in mind that vectors $\Lambda_Q f^{(i)} (i = 0, \dots, k-1)$ may be computed from (5.3.7), equation (5.3.9) may be represented as

$$\dot{x}_{k+2} = \sum_{i=1}^{k+2} \bar{A}(t)_i x_i + \sum_{i=0}^{k-1} (d_i - \theta_i(t)) u^{(i)} + v \tag{5.3.10}$$

where $\bar{A}_i(t)$ are matrices depending on $\theta_i(t)$, and consequently, on time.

Introduce notation

$$\bar{x}^T = [\bar{x}_1^T \ \bar{x}_2^T], \ \bar{x}_1^T = [x_1^T \dots x_{k+1}^T], \ \bar{x}_2 = x_{k+2}, \ \bar{u}^T = [u^T \ \dot{u}^T \dots (u^{(k-1)})^T]$$

$$\sum_{i=0}^{k-1} (d_i - \theta_i(t)) u^{(i)} = \theta(t)\bar{u} \qquad \theta = [(d_1 - \theta_i(t)) \dots (d_i - \theta_{k-1}(t))$$

and rewrite the first equation in (5.3.1), equation (5.3.5) and equation (5.3.10) as

$$\begin{aligned} \frac{d\bar{x}_1}{dt} &= \bar{A}_{11}\bar{x}_1 + \bar{A}_{12}\bar{x}_2 \\ \frac{d\bar{x}_2}{dt} &= \bar{A}_{21}(t)\bar{x}_1 + \bar{A}_{22}(t)\bar{x}_2 + \theta(t)\bar{u} + v \end{aligned} \tag{5.3.11}$$

where $\bar{A}_{11}$ and $\bar{A}_{12}$ are constant matrices; $\bar{A}_{21}(t)$, $\bar{A}_{22}(t)$ and $\theta(t)$ are time-varying matrices with bounded elements since the coefficients $\theta_i(t)$ in equation (5.3.3) were assumed to be bounded.

The system (5.3.11) is in regular form and by handling $\bar{x}_2$ as an intermediate control in the first block, the desired dynamics may be assigned by a proper choice of matrix C

in $\bar{x}_2 = -C\bar{x}_1$. Then, following the design methodology of Section 3.3, sliding mode is enforced in the manifold $s = \bar{x}_2 + C\bar{x}_1 = 0$ by control

$$v = -(\alpha(|\bar{x}_1| + |\bar{x}_2| + |\bar{u}|) + \delta)\text{sign}(s) \tag{5.3.12}$$

Indeed the time derivative of Lyapunov function candidate $V = \frac{1}{2}s^T s$,

$$\dot{V} = s^T((C\bar{A}_{11} + \bar{A}_{21})\bar{x}_1 + (C\bar{A}_{12} + \bar{A}_{22})\bar{x}_{21} + \theta\bar{u}) - (\alpha(|\bar{x}_1| + |\bar{x}_2| + |\bar{u}|) + \delta)|s|$$

is negative for a sufficiently high, but finite value of α and any δ. After a finite time interval, sliding mode governed by

$$\frac{d\bar{x}_1}{dt} = (\bar{A}_{11} - \bar{A}_{12}C)\bar{x}_1 \tag{5.3.13}$$

will occur with the desired dynamics and invariance properties with respect to disturbances.

The objective of the design was to decrease the magnitude of control with decreasing disturbances without measurement of the disturbances. This is the case for our system: in the solution to (5.3.13), $\bar{x}_1(t)$ and $\bar{x}_2(t)$ tend to zero, which means that functions $u^{(i)}$ tend to $-\Lambda_Q f^{(i)}$ (5.3.7); since $u^{(i)}$ $i = 1, \ldots, (k-1)$ are components of vector $\bar{u}$, the control v (5.3.12) decreases with the disturbances. The output of the additional dynamic system u is a continuous function and tends to $-\Lambda_Q f$, which leads to disturbance rejection. In real systems, often there is no need to introduce an additional dynamic system; its part may be played by actuators with outputs usually accessible for measurement. Then, an actuator input is the control to be designed and its magnitude depends on disturbances and their derivatives.

Example 5.1

The disturbance rejection method will be illustrated with a second-order system where the plant and actuator are integrators (Figure 5.1). An external disturbance $f(t)$ is applied to the plant and is not accessible for measurement. The control u is designed as a piecewise linear function of the controlled value $x = x_1$ that should be reduced to zero but also of the actuator output y. Then the behaviour of the plant and the system is governed by systems of the first and second order, respectively:

$$\begin{aligned} u &= x_2 - f \\ \dot{x}_1 &= x_2 \\ \dot{x}_2 &= v - \dot{f} \end{aligned} \tag{5.3.14}$$

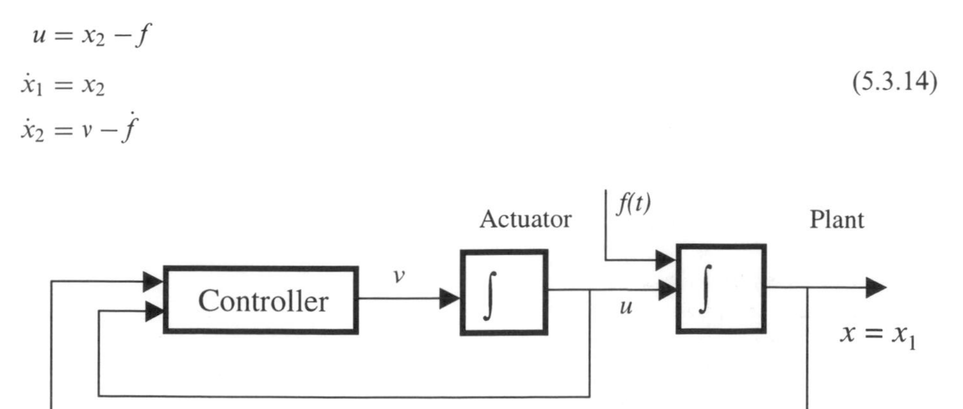

Figure 5.1 Sliding mode dynamic compensator in a second-order system.

Select control for system (5.3.14) in the form similar to (5.3.12) $v = -(\alpha|x_1| + \beta|u| + \delta)\text{sign}(s)$, $s = x_2 + Cx_1$ with α, β and $c > 0$ being constant coefficients. The control low may be also written as

$$v = -\Psi x_1 - \Psi_u u - \delta \text{ sign}(s)$$

where

$$\Psi = \alpha \text{ sign}(x_1 s) \quad \text{and} \quad \Psi_u = \beta \text{ sign}(us)$$

System (5.3.14) is governed by piecewise linear differential equations

$$\dot{x}_1 = x_2$$

$$\dot{x}_2 = -\Psi x_1 - \Psi_u x_2 - \delta \text{ sign}(s) + \Psi_u f + \dot{f}$$

As it was shown for a similar second-order system with no disturbance in Section 2.1 (with $f(t) \equiv 0$), the coefficients of control can be selected such that the state reaches the switching line, and sliding mode is enforced at each point of this line. After some finite time interval, the state tends to zero in sliding mode with motion equation $\dot{x} + cx = 0$. The state planes of the two unstable linear structures of the system are shown in Figure 1.9 and the state plane of the asymptotically stable system with variable structure is shown in Figure 1.10. If $f(t) \neq 0$, singular points ($\dot{x}_1 = \dot{x}_2 = 0$) of each of the four linear structures, corresponding to the four combinations of $\pm\alpha$ and $\pm\beta$, are shifted from the origin. The magnitudes and signs of the shifts depend on coefficients α, β and disturbance $f(t)$. Figure 5.2 separately shows right and left semiplanes of the system state plane for the area $|x_2| \leq f(t)$, where $\text{sign}(u) = -\text{sign}(f(t))$ by virtue of the plant equation. The first-order actuator takes part of an additional dynamic system (5.3.4), therefore the disturbance $f(t)$ is assumed to satisfy conditions (5.3.2) for $k = 1$:

$$\dot{f} = \theta(t) f \qquad |\theta(t)| \leq \beta$$

The state plane shows that the singular points are shifted such that the state trajectories in the vicinity of the switching line $s = 0$ are directed towards it and sliding mode

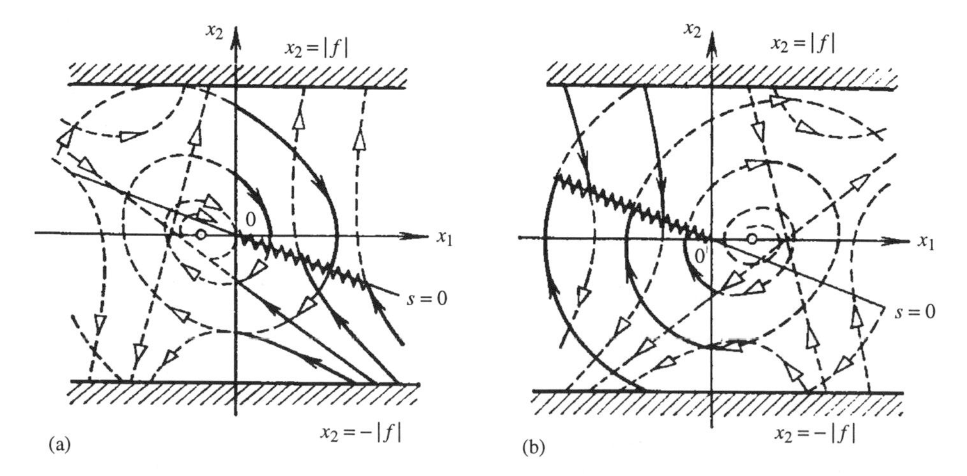

Figure 5.2 The state plane: (a) right semiplane and (b) left semiplane.

occurs in the system. The same conclusion can be made analytically. Since $\text{sign}(u) = -\text{sign}(f(t))$, the term $\Psi_u f$ in the motion equation is equal to $-\beta|f|\ \text{sign}(s)$ and for the points $x_2 = -cx_1$ on the switching line

$$\dot{s} = (-c^2 + \Psi_u - \Psi)x_1 - \delta\ \text{sign}(s) - \beta|f|\ \text{sign}(s) + \dot{f}$$

The values of s and $\dot{s}$ have opposite signs, or the conditions (2.4.1) for sliding mode to exist hold for $\alpha > c^2 + \beta$ and an arbitrary positive δ. Thus the change of signs in the main and local feedback enables one to reject unmeasured disturbances. Strictly speaking, the curves of Figure 5.2 are not state trajectories because the disturbance makes the system time-varying. Each curve should be regarded as the locus of points for which, at each fixed time, the tangent to the curve coincides with the state velocity vector.

5.4 Ackermann's formula

Ackermann's formula for linear state feedback enables us to determine a scalar control law in explicit form so we can obtain a system with desired eigenvalues (Ackermann, 1992). A similar task arises when designing sliding mode control in linear systems with a linear discontinuity surface since the corresponding sliding mode equation is linear and depends on the coefficients of the surface equation. The traditional approach to sliding mode control design implies transforming the linear system into the regular form (Section 5.1) consisting of two blocks and handling the state vector of one of the blocks as a fictitious control.

In this section the design method of scalar sliding mode control rests upon Ackermann's formula and is oriented towards obtaining a discontinuity plane equation in explicit form as well as in terms of the original system, without transforming it into regular form (Ackermann and Utkin, 1998). For a linear plant operating under uncertainty conditions, the control enforces sliding motions governed by a linear equation with the desired eigenvalue placement and independent of disturbance.

Consider a controllable system described by a differential equation

$$\dot{x} = Ax + b(u + f(x, t)) \tag{5.4.1}$$

where x is an n-dimensional state vector, u is a scalar control, A and b are known matrix and vector, and $f(x, t)$ is a nonlinear disturbance with known upper bound $|f(x, t)| < f_o(x, t)$. It follows from (5.4.1) that the control and disturbance vectors (bu and bf) are collinear, so the invariance condition (5.2.2) is satisfied; consequently, the sliding mode in any plane is invariant with respect to the disturbance.

The design of sliding mode control in (5.4.1) implies the selection of a plane $s = c^T x = 0$ (c^T is an n-dimensional constant row vector) and consequently design of the control enforcing the sliding mode in $s = 0$. The sliding mode equation is of the $(n-1)$th order and does not depend on the disturbance. The desired dynamic properties may be provided by a proper choice of the vector c. Traditionally, the sliding mode equation is derived first and then the conventional methods of the linear theory are applied.

The aim of this section is to show how, using Ackermann's formula, the vector c may be found in an explicit form without the sliding motion equation, as far as the eigenvalue placement task is concerned. The desired eigenvalues $\lambda_1, \lambda_2, \ldots, \lambda_n$ of linear system $\dot{x} = Ax + bu_a$ may be assigned using Ackermann's formula (Ackermann, 1992):

$$u_a = -k^T x \quad k^T = e^T P(A) \tag{5.4.2}$$

where

$$e^T = [0 \ldots 0 \quad 1][b \quad Ab \ldots A^{n-1}b]^{-1}$$
$$P(\lambda) = (\lambda - \lambda_1)(\lambda - \lambda_2)\ldots(\lambda - \lambda_{n-1})(\lambda - \lambda_n).$$

Suppose now that the real or pairwise conjugate complex values $\lambda_1, \lambda_2, \ldots, \lambda_{n-1}$ are the desired eigenvalues of the sliding mode.

Theorem 5.1

If

$$c^T = e^T P_1(A) \tag{5.4.3}$$

with $P_1(\lambda) = (\lambda - \lambda_1)(\lambda - \lambda_2)\ldots(\lambda - \lambda_{n-1}) = p_1 + p_2\lambda + \ldots + p_{n-1}\lambda^{n-2} + \lambda^{n-1}$

then $\lambda_1, \lambda_2, \ldots, \lambda_{n-1}$ are the eigenvalues of the sliding mode dynamics in the plane $s = c^T x = 0$. □

Proof

According to Ackermann's formula (5.4.2), $\lambda_1, \lambda_2, \ldots, \lambda_n$ are eigenvalues of the matrix $A^* = A - bk^T$, with λ_n being an arbitrary value. Vector c^T is a left eigenvector of A^* corresponding to λ_n. Indeed, as follows from (5.4.2) and (5.4.3),

$$c^T A^* = c^T A - c^T b e^T P(A)$$

Since

$$\begin{aligned} c^T b &= e^T P_1(A) b \\ &= [0 \ 0 \ \ldots \ 0 \ 1][b \ Ab \ \ldots \ A^{n-1}b]^{-1}[b \ Ab \ \ldots \ A^{n-1}b][p_1 \ p_2 \ \ldots \ 1]^T \\ &= 1 \end{aligned} \tag{5.4.4}$$

and $P(A) = P_1(A)(A - \lambda_n I)$, thus

$$c^T A^* = c^T A - e^T P_1(A)(A - \lambda_n I)$$

which reduces with (5.4.3) to

$$c^T A^* = \lambda_n c^T \tag{5.4.5}$$

The system $\dot{x} = (A - bk^T)x + b(u - u_a + f(x, t))$ is now transformed such that $s = c^T x$ becomes the last state and the first $n - 1$ states $x^* = [x_1 \quad x_2 \quad \ldots \quad x_{n-1}]^T$ remain unchanged, i.e.

$$\begin{bmatrix} x^* \\ s \end{bmatrix} = \begin{bmatrix} I & 0 \\ & c^T \end{bmatrix} x = Tx$$

For T to be invertible, the last component of c^T must be nonzero. Since this vector is nonzero, the condition can always be satisfied by reordering the components of the state vector x. Under conditions (5.4.4) and (5.4.5), the transformed system is

$$\dot{x}^* = A_1 x^* + a_s s + b^*(u - u_a + f(x, t)) \tag{5.4.6}$$

$$\dot{s} = \lambda_n s + u - u_a + f(x, t) \tag{5.4.7}$$

where

$$\begin{bmatrix} A_1 & a_s \\ 0 & \lambda_n \end{bmatrix} = T(A - bk^T)T^{-1} \qquad \begin{bmatrix} b^* \\ 1 \end{bmatrix} = Tb$$

The spectrum of matrix A_1 consists of desired eigenvalues $\lambda_1, \lambda_2, \ldots, \lambda_{n-1}$.

To derive the sliding mode equation in the plane $s = 0$, the solution to the algebraic equation $\dot{s} = 0$ with respect to u should be substituted into (5.4.6). It results in

$$\dot{x}^* = A_1 x^* \tag{5.4.8}$$

with the desired dynamics, independent of the unknown disturbance $f(x, t)$. ■

The result has a transparent geometric interpretation. Vector c^T is a left eigenvector of the matrix A^* corresponding to the eigenvalue λ_n. This means that the plane $s = c^T x = 0$ is an invariant subspace of A^* with the motion determined by the previously selected set of $(n - 1)$ eigenvalues $\lambda_1, \lambda_2, \ldots, \lambda_{n-1}$. If sliding mode is enforced in the plane $s = c^T x = 0$, then it exhibits the desired dynamics. Note that the design of the plane $s = c^T x = 0$ does not imply assigning the eigenvalue λ_n; it only appears in the proof of the theorem and may take an arbitrary value.

The discontinuous control u is designed to enforce sliding mode in the plane $s = 0$. This implies that conditions (2.4.1) should be satisfied, i.e. the values of s and $\dot{s}$ should have different signs in some vicinity of the plane:

$$\begin{aligned} \dot{s} &= c^T A x + u + f(x, t) \\ u &= -M(x, t)\operatorname{sign}(s) \end{aligned} \tag{5.4.9}$$

where $M(x, t)$ is chosen such that

$$M(x, t) > |C^T A x| + f_0(x, t)$$

If the control may take only two extreme values $+M_0$ or $-M_0$ (which is common in applications), then (5.4.9) with $M(x, t) = M_0$ enforces a sliding mode in the plane $s = 0$ governed by (5.4.8) as well. Of course, the domain of initial conditions and the disturbance should be bounded.

Example 5.2

Let $\lambda = -1$ be the desired eigenvalue of sliding motion for the second-order system

$$\dot{x} = Ax + b(u + f(x, t))$$

where

$$A = \begin{bmatrix} 0 & 0 \\ 0 & 1 \end{bmatrix} \qquad b = \begin{bmatrix} 1 \\ 1 \end{bmatrix} \qquad x = \begin{bmatrix} x_1 \\ x_2 \end{bmatrix}$$

According to (5.4.3),

$$c^T = [0 \quad 1][b \quad Ab]^{-1}P_1(A) \qquad P_1(A) = A + I$$
$$c^T = [-1 \quad 2]$$

and the sliding surface equation is of the form (note that $c^T b = 1$)

$$s = -x_1 + 2x_2 = 0$$

By the equivalent control method, the solution to the system $s = 0$, $\dot{s} = 0$ with respect to x_2 and u,

$$u_{eq} = -x_1 - f(x_1, t), \qquad x_2 = \frac{1}{2}x_1$$

should be substituted into the original system to derive the sliding motion equation:

$$\dot{x}_1 = -x_1$$

The sliding mode is determined by the eigenvalue $\lambda = -1$ and does not depend on the disturbance $f(x, t)$.

The design procedure based on Ackermann's formula is summarized as follows:

1. The desired spectrum of the sliding motion $\lambda_1, \lambda_2, \ldots, \lambda_{n-1}$ is selected.
2. The equation of the discontinuity plane $s = c^T x = 0$ is found as

$$c^T = e^T(A - \lambda_1 I)(A - \lambda_2 I)\ldots(A - \lambda_{n-1} I).$$

3. The discontinuous control (5.4.9) is designed.

Remark 5.1

It follows from (5.4.9) that sliding mode may be enforced in an unperturbed system by $u = -(\alpha\|x\| + \delta)\text{sign}(s)$ with some finite positive number α and any positive δ. The control is decreasing in the system with asymptotically stable sliding modes. □

5.4.1 Simulation results

The design procedure will be demonstrated for sliding mode stabilization of an inverted pendulum (Figure 5.3) subjected to a bounded unknown disturbance force. The linearized motion equations have the following form (Kortüm and Lugner, 1994):

$$\dot{x} = Ax + b(u + f(t))$$

where

$$A = \begin{bmatrix} 0 & 0 & 1 & 0 \\ 0 & 0 & 0 & 1 \\ 0 & a_{32} & 0 & 0 \\ 0 & a_{42} & 0 & 0 \end{bmatrix} \qquad b = \begin{bmatrix} 0 \\ 0 \\ b_3 \\ b_4 \end{bmatrix} \qquad x = \begin{bmatrix} x \\ \alpha \\ \dot{x} \\ \dot{\alpha} \end{bmatrix}$$

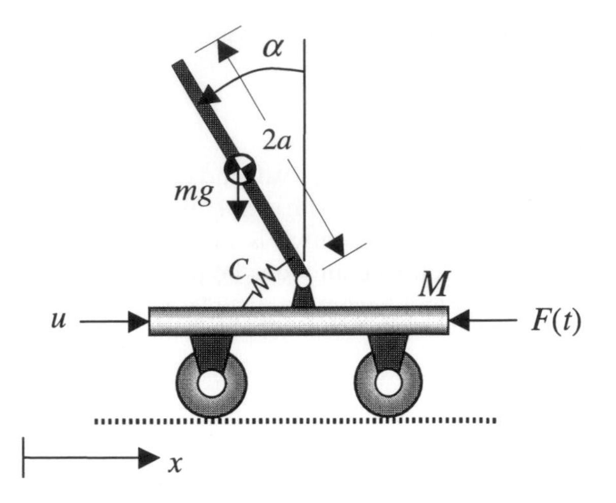

Figure 5.3 Inverted pendulum with trolley system.

and

$$
\begin{aligned}
a_{32} &= -3(C - mga)/a(4M_t + m)\\
a_{42} &= -3(M_t + m)(C - mga)/a^2m(4M_t + m)\\
b_3 &= 4/(4M_t + m)\\
b_4 &= 3/(4M_t + m)
\end{aligned}
$$

M_t and m are the masses of the trolley and the pendulum, $2a$ is the pendulum length, C is the spring stiffness, $g = 9.81\,\mathrm{m\,s^{-2}}$, u and $f(t)$ are control and disturbance forces, $|f(t)| \le f_0 = \text{const}$ and f_0 is assumed to be known.

Sliding mode control is designed for $M_t = 5, m = 1, a = 1, C = 1$. Let $\lambda_1 = -1, \lambda_2 = -2, \lambda_3 = -3$ be the desired eigenvalues of sliding motion. According to (5.4.3) the discontinuity plane equation

$$
\begin{aligned}
&s = c^Tx = 0\\
&c^T = [0 \quad 0 \quad 0 \quad 1][b \quad Ab \quad A^2b \quad A^3b]^{-1}(A + I)(A + 2I)(A + 3I),
\end{aligned}
$$

I is an identity matrix.

For the given values of the parameters in matrices A and B

$$c^T = [-4.77 \quad 48.4 \quad -8.75 \quad 18.7]$$

The control is assumed to take two extreme values only:

$$u = -M_0\text{sign}(s) \qquad M_0 = \text{const}$$

As follows from the above studies, for any $M_0 > f_0$, there exists a domain of initial conditions such that sliding mode is enforced in the plane $s = 0$. The simulation examples are given for sliding mode control with

$$M_0 = 40 \quad \text{and} \quad f(t) = f_0\sin(3t), \quad f_0 = 0.5$$

Figure 5.4 shows the stabilization process for initial conditions

$$x(0) = 0.5, \alpha(0) = 0.2, \dot{x}(0) = 0, \dot{\alpha}(0) = 0$$

Sliding mode occurs after a finite time interval, and thereafter both coordinates x and α do not depend on the time-varying disturbance and they tend to zero. The system without feedback is unstable and control is bounded, therefore the motion may become unstable should the initial conditions be increased. The system is still stable for $\alpha(0) = 0.38$ (Figure 5.5) and becomes unstable with $\alpha(0) = 0.39$ (Figure 5.6).

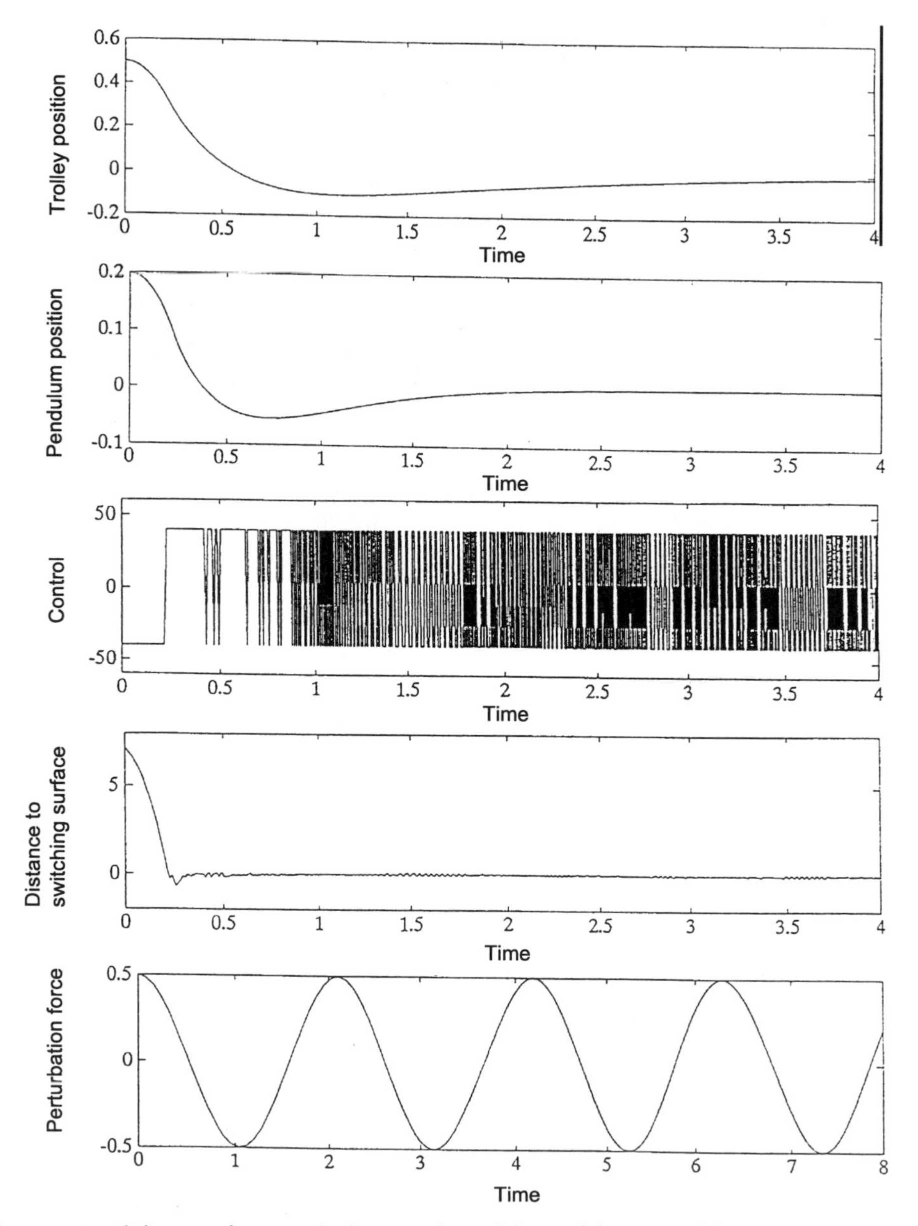

Figure 5.4 Sliding mode control of inverted pendulum, $x(0) = 0.5$, $\alpha(0) = 0.2$.

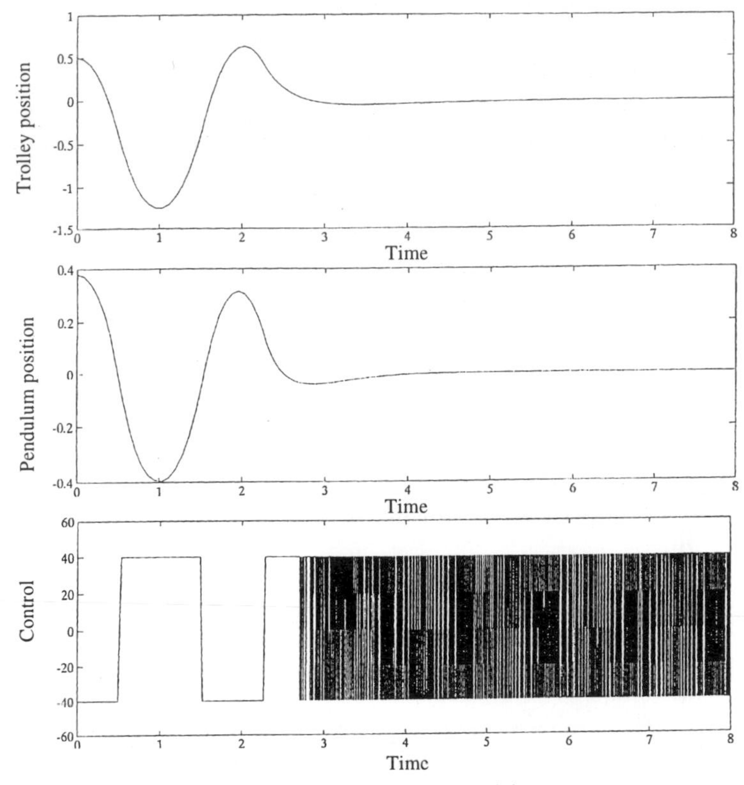

Figure 5.5 Sliding mode control of inverted pendulum, $\alpha(0) = 0.38$.

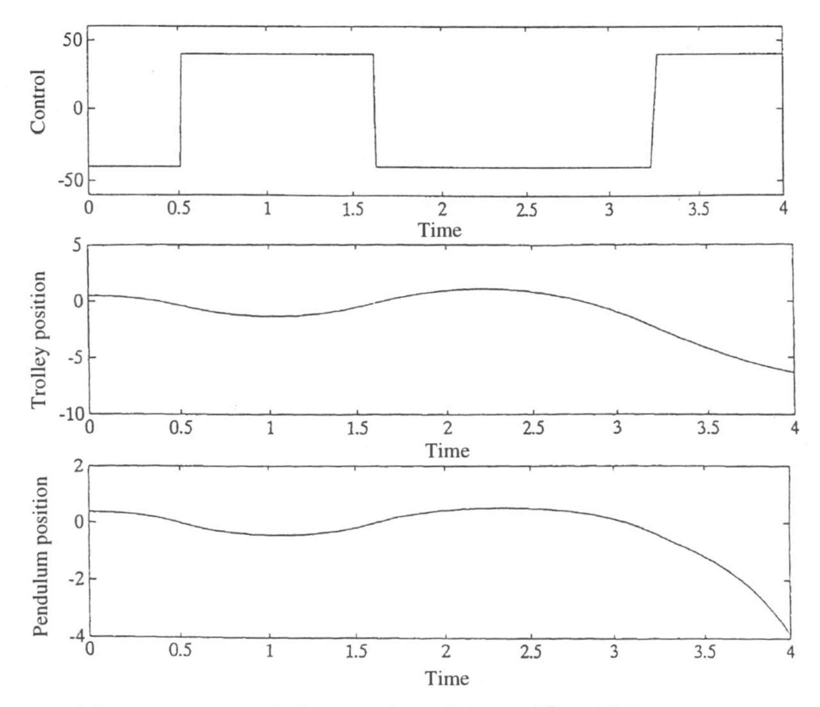

Figure 5.6 Sliding mode control of inverted pendulum, $\alpha(0) = 0.39$.

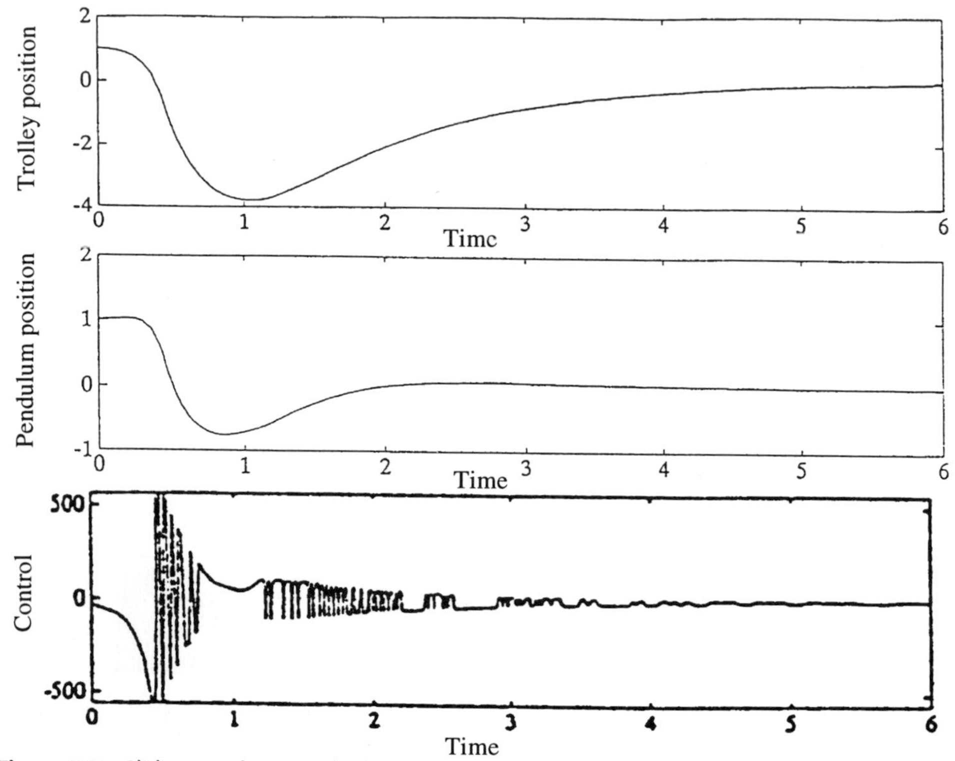

Figure 5.7 Sliding mode control of inverted pendulum: unperturbed system with state-dependent control magnitude.

As explained in Remark 5.1, sliding mode may be enforced in an unperturbed system ($f(t) = 0$) for arbitrary initial conditions with piecewise linear control

$$u = -\beta(|x| + |\dot{x}| + |\alpha| + |\dot{\alpha}| + \delta)\text{sign}(s)$$

where β, δ are positive values and some finite β and arbitrary δ.

The simulation results with $\beta = 30$, $\delta = 0$, $x(0) = 1$, $\dot{x}(0) = 0$, $\alpha(0) = 1$, $\dot{\alpha}(0) = 0$ are shown in Figure 5.7. Both the state vector and the control tend to zero in sliding mode.

5.5 Output feedback sliding mode control

Implementation of the design methods developed in the previous sections implies that all components of a state vector have to be accessible for measurement. However, this is not the case for many practical situations. Two approaches may be studied for these cases. The first method is the design of state observers to restore the state vector using available measurements of some states, and the second method is to derive a class of systems such that the control task may be solved by designing a static output feedback controller. The second approach to sliding mode control design under incomplete information on the system states is studied in this section. The observer design methods are studied in Chapter 6.

It is assumed that for the system

$$\dot{x} = Ax + Bu \tag{5.5.1}$$

$$y = Cx \tag{5.5.2}$$

with l-dimensional output vector y,

(i) the pair (A, B) is controllable and the pair (A, C) is observable
(ii) rank $B = m$ and rank $C = l$
(iii) $l > m$

The system (5.5.1), (5.5.2) is described as output pole-assignable if the eigenvalues of matrix $A + BLC$, or of a feedback system with linear control $u = Ly$ (L is a constant $m \times l$ matrix) take any desired values. The class of such linear systems may be found based on the well-known result by Kimura (1975).

Theorem 5.2

If the system (5.5.1), (5.5.2) is controllable and observable and satisfies the relation

$$n \le l + m - 1 \tag{5.5.3}$$

then it is pole-assignable by gain output feedback. The Theorem will be used further for pole assignment in systems with sliding mode control. □

There exists a nonsingular coordinate transformation reducing the system to regular form (see also Section 3.3):

$$\begin{bmatrix} \dot{x}_1 \\ \dot{x}_2 \end{bmatrix} = \begin{bmatrix} A_{11} & A_{12} \\ A_{21} & A_{22} \end{bmatrix} \begin{bmatrix} x_1 \\ x_2 \end{bmatrix} + \begin{bmatrix} 0 \\ I_m \end{bmatrix} u \tag{5.5.4}$$

where the pair (A_{11}, A_{12}) is controllable. Equation (5.5.2) is transformed into

$$y = [C_1 \quad C_2] \begin{bmatrix} x_1 \\ x_2 \end{bmatrix} = C_1 x_1 + C_2 x_2 \qquad (C_1 \in \Re^{l \times (n-m)},\ C_2 \in \Re^{l \times m})$$

If C_2 is a matrix with full rank, then it may be represented in the following form (reordering the components of vector x_2 if needed):

$$C_2 = \begin{bmatrix} C_{21}^* \\ C_{22}^* \end{bmatrix} \qquad (C_{12}^* \in \Re^{(l-m) \times m},\ C_{22}^* \in \Re^{m \times m},\ \det(C_{22}^*) \neq 0)$$

and for a nonsingular matrix

$$P = \begin{bmatrix} I_{l-m} & -C_{21}^*(C_{22}^*)^{-1} \\ 0 & (C_{22}^*)^{-1} \end{bmatrix}$$

$$y^* = Py = [PC_1 \quad PC_2]x = \begin{bmatrix} C_{11} & 0 \\ C_{21} & I_m \end{bmatrix} \begin{bmatrix} x_1 \\ x_2 \end{bmatrix},$$ C_{11} and C_{21} are constant matrices

The switching manifold $s = 0$ in sliding mode control is defined as

$$s = Fy^* = [F_1 \quad I_m]\begin{bmatrix} C_{11}x_1 \\ C_{21}x_1 + x_2 \end{bmatrix} = (F_1C_{11} + C_{21})x_1 + x_2$$

where $F \in R^{m\times l}$, $F_1 \in R^{m\times(l-m)}$.

The control input is selected as a discontinuous function of the output

$$u = -K \cdot \text{sign}(s) \qquad K = \text{diag}[k_1 \quad k_2 \quad \dots \quad k_m]$$

The control gains k_i are chosen such that

$$k_i > |f_i| \qquad f^T = [f_1, \dots, f_m]$$

$$f = ((F_1C_{11} + C_{12})A_{11} + A_{21})x_1 + ((F_1C_{11} + C_{21})A_{12} + A_{22})x_2$$

which means that the time derivative of Lyapunov function candidate $V = \frac{1}{2}s^Ts$,

$$\dot{V} = s^Tf - s^TK \cdot \text{sign}(s) < 0$$

is negative definite and sliding mode will be enforced after a finite time interval.

In sliding mode, $s = 0$ and

$$x_2 = -(F_1C_{11} + C_{21})x_1 \tag{5.5.5}$$

Substituting (5.5.5) into (5.5.4) yields

$$\dot{x}_1 = (A_{11} - A_{12}C_{21})x_1 + A_{12}\upsilon \tag{5.5.6}$$

where $\upsilon = -F_1C_{11}x_1 = -F_1y_1 \in R^m$ is handled as a control input.

The original system (5.5.1), (5.5.2) in sliding mode is replaced by reduced order system (5.5.6) and $s = 0$:

$$\dot{x}_1 = A^*x_1 + A_{12}\upsilon \tag{5.5.7}$$

$$y_1 = C_{11}x_1 \tag{5.5.8}$$

where $A^* = (A_{11} - A_{12}C_{12})$; pair (A^*, A_{12}) is controllable due to controllability of pair (A_{11}, A_{12}) (Utkin and Young, 1978).

Now we deal with a pole placement task for a reduced-order controllable system with an $(l - m)$-dimensional output vector. The condition of Kimura's (Theorem 5.2) may be reformulated for eigenvalue placement of system (5.5.7), (5.5.8).

Theorem 5.3

If the original system (5.5.1), (5.5.2) is controllable, the system (5.5.7), (5.5.8) is observable and satisfies the relation

$$(n - m) \le (l - m) + m^* - 1 \qquad m^* = \text{rank}\ (A_{12}) \tag{5.5.9}$$

then it is pole-assignable by gain output feedback. □

Under condition (5.5.9) there exists matrix F_1 such that eigenvalues of sliding mode equation

$$\dot{x}_1 = (A^* - A_{12}F_1C_{11})x_1 \tag{5.5.10}$$

take the desired values. Since $m^* \leq m$, condition (5.5.9) $n \leq l + m^* - 1$ means that the original system (5.5.1), (5.5.2) is pole-assignable by gain output feedback if the reduced-order system (5.5.7), (5.5.8) is pole-assignable.

If the pair (C_{11}, A^*) is not observable, the same procedure is applicable to the observable part of x_1. The unobservable system (5.5.7), (5.5.8) may be presented in the following form (Kwakernaak, 1972):

$$\begin{bmatrix} \dot{x}_1' \\ \dot{x}_1'' \end{bmatrix} = \begin{bmatrix} A_{11}^* & 0 \\ A_{21}^* & A_{22}^* \end{bmatrix} \begin{bmatrix} x_1' \\ x_1'' \end{bmatrix} + \begin{bmatrix} A_{12}' \\ A_{12}'' \end{bmatrix} v$$

$$y_1 = C_{11}' x_1'$$

with an observable pair (C_{11}', A_{11}^*). Then the upper subsystem is pole-assignable if it satisfies condition (5.5.9), whereas the eigenvalues of the lower subsystem are those of A_{21}^* and can be changed under no conditions.

Remark 5.2

The above eigenvalue placement method is applicable for systems with plant parameter variations and disturbances (5.2.1):

$$\dot{x} = (A + \Delta A)x + Bu + Qf$$
$$y = Cx$$

If invariance conditions (5.2.2) hold then the motion in sliding mode depends on neither ΔA nor $f(t)$ and is governed by equation (5.5.10) as well. □

Example 5.3

Assumption The original system is pole-assignable, but the sliding mode control system is not pole-assignable.

$$\begin{bmatrix} \dot{x}_1 \\ \dot{x}_2 \\ \dot{x}_3 \\ \dot{x}_4 \end{bmatrix} = \begin{bmatrix} 0 & 1 & 1 & 0 \\ 1 & 0 & 2 & 0 \\ 0 & 0 & 0 & 0 \\ 0 & 0 & 0 & 0 \end{bmatrix} \begin{bmatrix} 0 & 0 \\ 0 & 0 \\ 1 & 0 \\ 0 & 1 \end{bmatrix} + \begin{bmatrix} u_1 \\ u_2 \end{bmatrix}$$

$$\begin{bmatrix} y_1 \\ y_2 \\ y_3 \end{bmatrix} = \begin{bmatrix} 1 & 0 & 0 & 0 \\ 0 & 0 & 1 & 0 \\ 0 & 0 & 0 & 1 \end{bmatrix} \begin{bmatrix} x_1 \\ x_2 \\ x_3 \\ x_4 \end{bmatrix} \qquad (n = 4, m = 2, l = 3)$$

This system is controllable and observable. Condition (5.5.9) is satisfied. The switching manifold is designed as

$$s = \begin{bmatrix} f_1 & 1 & 0 \\ f_2 & 0 & 1 \end{bmatrix} \begin{bmatrix} y_1 \\ y_2 \\ y_3 \end{bmatrix} = 0$$

with parameters f_1 and f_2 to be selected.

The reduced-order system is

$$\begin{aligned} \begin{bmatrix} \dot{x}_1 \\ \dot{x}_2 \end{bmatrix} &= \begin{bmatrix} 0 & 1 \\ 1 & 0 \end{bmatrix} \begin{bmatrix} x_1 \\ x_2 \end{bmatrix} + \begin{bmatrix} 1 & 0 \\ 2 & 0 \end{bmatrix} \upsilon \\ y_1 &= [1 \quad 0] \begin{bmatrix} x_1 \\ x_2 \end{bmatrix} \end{aligned} \tag{5.5.11}$$

This system is controllable and observable as well, but condition (5.5.9) is not satisfied, because

$$\text{rank} \begin{bmatrix} 1 & 0 \\ 2 & 0 \end{bmatrix} = m^* = 1 \quad \text{and} \quad n > l + m^* - 1$$

It is impossible to locate poles arbitrarily. Substituting $\upsilon = F_1 y = -(f_1, f_2)^T x_1$ into (5.5.11) yields

$$\begin{bmatrix} \dot{x}_1 \\ \dot{x}_2 \end{bmatrix} = \begin{bmatrix} -f_1 & 1 \\ 1 - 2f_1 & 0 \end{bmatrix} \begin{bmatrix} x_1 \\ x_2 \end{bmatrix}$$

Since one parameter is free, only one pole may be located arbitrarily.

Example 5.4

Assumption Both the original system and the system with the sliding mode control are pole-assignable.

$$\begin{bmatrix} \dot{x}_1 \\ \dot{x}_2 \\ \dot{x}_3 \\ \dot{x}_4 \end{bmatrix} = \begin{bmatrix} 0 & 1 & 1 & 0 \\ 1 & 0 & 0 & 1 \\ 0 & 1 & 0 & 0 \\ 0 & 0 & 0 & 0 \end{bmatrix} \begin{bmatrix} 0 & 0 \\ 0 & 0 \\ 1 & 0 \\ 0 & 1 \end{bmatrix} + \begin{bmatrix} u_1 \\ u_2 \end{bmatrix}$$

$$\begin{bmatrix} y_1 \\ y_2 \\ y_3 \end{bmatrix} = \begin{bmatrix} 1 & 0 & 0 & 0 \\ 0 & 2 & 1 & 0 \\ 0 & 0 & 0 & 1 \end{bmatrix} \begin{bmatrix} x_1 \\ x_2 \\ x_3 \\ x_4 \end{bmatrix} \qquad (n = 4, m = 2, l = 3)$$

This system is controllable and observable, and satisfies condition (5.5.9). The switching manifold is

$$s = \begin{bmatrix} f_1 & 1 & 0 \\ f_2 & 0 & 1 \end{bmatrix} \begin{bmatrix} y_1 \\ y_2 \\ y_3 \end{bmatrix} = 0$$

The reduced-order system

$$\begin{bmatrix}\dot{x}_1\\ \dot{x}_2\end{bmatrix}=\begin{bmatrix}0 & -1\\ 1 & 0\end{bmatrix}\begin{bmatrix}x_1\\ x_2\end{bmatrix}+\begin{bmatrix}1 & 0\\ 0 & 1\end{bmatrix}\upsilon$$

$$y_1=[1 \quad 0]\begin{bmatrix}x_1\\ x_2\end{bmatrix}$$

is controllable and observable, and satisfies condition (5.5.9) therefore it is pole-assignable. Indeed any eigenvalues of the system with control

$$\upsilon=-F_1y_1=-(f_1,f_2)^Tx_1$$

$$\begin{bmatrix}\dot{x}_1\\ \dot{x}_2\end{bmatrix}=\begin{bmatrix}f_1 & -1\\ 1+f_2 & 0\end{bmatrix}\begin{bmatrix}x_1\\ x_2\end{bmatrix}$$

may be assigned by a proper choice of coefficients f_1 and f_2.

5.6 Control of time-varying systems

Designing the desired dynamics of time-varying control systems is a considerably more difficult task than control design for systems with constant parameters, since the properties of time-varying systems cannot be interpreted in terms of their eigenvalues. Decoupling system motions into motions of lower dimension enables one to simplify the control design. Therefore it is of interest to apply sliding mode control methodology to reduce the order of the motion equations. We will follow the so-called *block control principle* (Drakunov *et al.*, 1990) for reducing the original design problem to a set of independent problems with lower dimensions. In time-varying system

$$\dot{x}=A(t)x+B(t)u \qquad x\in\Re^n,\quad u\in\Re^m,\quad \text{rank}(B)=m \tag{5.6.1}$$

let the elements of $A(t)$ and $B(t)$ be bounded with time derivatives of proper orders. Similar to (5.1.1) and (5.1.2), it is assumed that matrix $B(t)$ may be represented in the form

$$B(t)=\begin{bmatrix}B_1(t)\\ B_2(t)\end{bmatrix} \quad \text{and} \quad \det(B_2(t))\neq 0 \quad \text{for any } t$$

The nonsingular coordinate transformation

$$\begin{aligned}\begin{bmatrix}\dot{x}_1\\ \dot{x}_2\end{bmatrix}&=TAT^{-1}\begin{bmatrix}x_1\\ x_2\end{bmatrix}+T\begin{bmatrix}B_1\\ B_2\end{bmatrix}u+\dot{T}T^{-1}\begin{bmatrix}x_1\\ x_2\end{bmatrix}\\ \begin{bmatrix}x_1\\ x_2\end{bmatrix}&=Tx, \quad T=\begin{bmatrix}I_{n-m} & -B_1B_2^{-1}\\ 0 & B_2^{-1}\end{bmatrix}\end{aligned} \tag{5.6.2}$$

reduces the system equations (5.6.1) to the regular form consisting of two blocks:

$$\begin{aligned}\dot{x}_1&=A_{11}x_1+A_{12}x_2\\ \dot{x}_2&=A_{21}x_1+A_{22}x_2+u\end{aligned} \tag{5.6.3}$$

where $x_1\in\Re^{(n-m)}$, $x_2\in\Re^m$ and A_{ij} are time-varying matrices $(i,j=1,2)$.

The state vector x_2 of the second block in (5.6.3) is handled as control for the first block; x_2 should be selected as a function of the state vector x_1 to shape the desired dynamics of the first block. We will confine our choice to linear functions with time-varying gains

$$x_2 = -C(t)x_1 \qquad C(t) \in \Re^{m\times(n-m)} \tag{5.6.4}$$

Shaping the desired dynamics of the system

$$\dot{x}_1 = (A_{11} - A_{12}C)x_1$$

in accordance with some performance criterion, this is reduced to designing the linear time-varying system of the $(n - m)$th order with the feedback matrix $C(t)$. This task is easier than the original task of nth order with control u being a linear function of the full state vector $x^T = [x_1^T \; x_2^T]$.

If the matrix $C(t)$ is found, then discontinuous control enforcing sliding mode in the manifold

$$s = x_2 + C(t)x_1 = 0 \qquad s^T = (s_1, s_2, \ldots, s_m) \tag{5.6.5}$$

should be designed. The equation of the motion projection on subspace s in the system with control

$$u = -(\alpha(|x_1| + |x_2|) + \delta)\text{sign}(s)$$

is of form

$$\dot{s} = (CA_{11} + A_{21} + \dot{C})x_1 + (CA_{12} + A_{22})x_2 - (\alpha(|x_1| + |x_2|) + \delta)\text{sign}(s)$$

where α and δ are positive constants. The coefficient α may be found such that for any positive δ, the conditions (2.4.1) hold for any function s_i. Then sliding mode exists on each plane $s_i = 0$ and their intersection $s = 0$.

Although design of desired dynamics in sliding mode is easier than for the original system, we deal with a time-varying system and special methods should be employed for stabilization. We will discuss one of them resulting in exponentially stable time-varying systems.

The idea of the design procedure is to take the first system in (5.6.3) with intermediate control x_2 and to decouple it into two subsystems, similar to applying transformation (5.6.2) to the original system. If $m \leq n - m$ and $\text{rank}(A_{12}) = m$ then the first system in (5.6.3) may be transformed into

$$\begin{aligned} \dot{x}_1' &= A_{11}'x_1' + A_{12}'x_1'' \\ \dot{x}_1'' &= A_{21}'x_1' + A_{22}'x_1'' + x_2 \qquad x_1' \in \Re^{n-2m}, \;\; x_1'' \in \Re^m \end{aligned} \tag{5.6.6}$$

where

$$\begin{bmatrix} x_1' \\ x_1'' \end{bmatrix} = T_1 x_1 \quad T_1 = \begin{bmatrix} I_{n-2m} & -B_1'(B_2')^{-1} \\ 0 & (B_2')^{-1} \end{bmatrix} \quad A_{12} = \begin{bmatrix} B_1' \\ B_2' \end{bmatrix} \quad \det(B_2') \neq 0 \tag{5.6.7}$$

Vector x_1'' in the first subsystem of (5.6.7) is handled as control and it is assumed that $\operatorname{rank}(A_{12}') = n - 2m$. Then for any matrix $A_1(t)$

$$\dot{x}_1' = A_1 x_1' \tag{5.6.8}$$

$$\text{if} \quad x_1'' = -C_1(t) x_1' \tag{5.6.9}$$

where $C_1 = (A_{12}')^+(A_{11}' - A_1)$ with $(A_{12}')^+$ being the pseudoinverse to A_{12}', i.e. $A_{12}'(A_{12}')^+ = I_{n-2m}$.

Condition (5.6.9) holds if

$$s_1 = x_1'' + C_1(t) x_1' \tag{5.6.10}$$

tends to zero. The equation for s_1 is derived from (5.6.6), (5.6.9):

$$\dot{s}_1 = S' x_1' + S'' x_1'' + x_2$$

with matrices S' and S'' depending on the elements of (5.6.6), (5.6.9) and their time derivatives.

For any matrix $A_2(t)$

$$\dot{s}_1 = A_2 s_1 \tag{5.6.11}$$

$$\text{if} \quad x_2 = A_2 s_1 - S' x_1' - S'' x_1'' \tag{5.6.12}$$

Since vectors x_1, x_1', x_1'' and s_1 are correlated through nonsingular transformations (5.6.7) and (5.6.9), equation (5.6.12) may be presented as

$$x_2 = -C(t) x_1 \tag{5.6.13}$$

with $C(t)$ depending on the matrices A and B in the original system (5.6.1), their time derivatives and matrices A_1 and A_2.

Condition (5.6.13) holds if control u in (5.6.1) is designed as a discontinuous function of the state enforcing sliding mode in the manifold

$$s = x_2 + C(t) x_1 = 0$$

The above design procedure with control

$$u = -(\alpha(|x_1| + |x_2|) + \delta)\operatorname{sign}(s)$$

is applicable for this task. After sliding mode in $s = 0$ occurs, the conditions (5.6.12) and (5.6.13) hold and s_1 will be governed by autonomous equation (5.6.11). As follows from the first equation of (5.6.6) and (5.6.8) to (5.6.10),

$$\dot{x}_1' = A_1 x_1' + A_{12}' s \tag{5.6.14}$$

The system dynamics in sliding mode are determined by differential equations (5.6.11) and (5.6.14), or by matrices A_1 and A_2 which may be selected by the designer. For example, they may be assigned constant with spectra such that exponential convergence at the desired rate is provided.

Generalization of the design method for the cases $\mathrm{rank}(A_{12}) < m$ and $\mathrm{rank}(A'_{12}) < n - 2m$ may be found in Drakunov *et al.* (1990). Drakunov *et al.* show that exponential stability of sliding mode may be provided for controllable time-varying systems.

References

ACKERMANN, J., 1992, *Sampled-Data Control Systems*, Berlin: Springer-Verlag.

ACKERMANN, J. and UTKIN, V., 1998, Sliding mode control design based on Ackermann's formula, *IEEE Transactions on Automatic Control*, **43**, 234–237.

DRAKUNOV, S. *et al.*, 1990, Block control principle I, *Automation and Remote Control*, **51**, 601–9.

DRAZENOVIC, B., 1969, The invariance conditions in variable structure systems, *Automatica*, **5**, 287-95.

KIMURA, H., 1975, Pole assignment by gain output feedback, *IEEE Transactions on Automatic Control*, **20**, 509–16.

KORTÜM, W. and LUGNER, P., 1994, *Systemdynamik und Regelung von Fahrzeugen. Einfuehrung und Beispiele* (in German), Berlin: Springer-Verlag.

KWAKERNAAK, H. and SIVAN, R., 1972, *Linear Optimal Control Systems*, New York: Interscience.

UTKIN, V. and YOUNG, K.-K.D., 1978, Methods for constructing discontinuity planes in multidimensional variable structure systems, *Automation and Remote Control*, **39**, 1466–70.

CHAPTER SIX

Sliding Mode Observers

All design methods of the previous chapters, except for Section 5.5, were developed under the assumption that the state vector is available. In practice, however, only a part of its components may be measured directly. The output feedback sliding mode control method of Section 5.5 is applicable to rather limited types of systems. An alternative approach is designing asymptotic observers which are dynamic systems for estimating all the components of the state vector using the measured components directly. First, we will study the conventional full-order and reduced-order observers intended for linear time-invariant systems. Next we present sliding mode modifications for state observation of time-invariant (Utkin, 1992) and time-varying systems with disturbance estimation (Hashimoto *et al.*, 1990).

6.1 Linear asymptotic observers

The idea underlying observer design methods may be illustrated for a linear time-invariant system (5.1.1):

$$\dot{x} = Ax + Bu \tag{6.1.1}$$

with output vector

$$y = Cx \quad y \in \Re^{l}, \ \ C = \text{const}, \ \ \text{rank}(C) = l \tag{6.1.2}$$

The pair (C, A) is assumed to be observable.

A linear asymptotic observer is designed in the same form as the original system (6.1.1) with an additional input depending on the mismatch between the real values (6.1.2) and the estimated values of the output vector:

$$\dot{\hat{x}} = A\hat{x} + Bu + L(C\hat{x} - y) \tag{6.1.3}$$

where $\hat{x}$ is an estimate of the system state vector and $L \in \Re^{n \times l}$ is an input matrix. Of course, the state vector of the observer $\hat{x}$ is available since the auxiliary dynamic system is implemented in a controller.

The motion equation with respect to mismatch $\bar{x} = \hat{x} - x$ is of form

$$\dot{\bar{x}} = (A + LC)\bar{x} \tag{6.1.4}$$

The behaviour of the mismatch governed by homogeneous equation (6.1.4) is determined by eigenvalues of matrix $A + CL$. For observable systems they may be assigned arbitrarily by a proper choice of input matrix L (Kwakernaak and Sivan, 1972). It means that any desired rate of convergence of the mismatch to zero or estimate $\hat{x}(t)$ to state vector $x(t)$ may be provided. Then any full-state control algorithms with vector $\hat{x}(t)$ are applicable.

The order of the observer may be reduced due to the fact that $\text{rank}(C) = l$ and the observed vector may be represented as

$$y = C_1x_1 + C_2x_2 \quad x^T = [x_1^T \ \ x_2^T], \ x_1 \in \Re^{n-l}, \ \ x_2 \in \Re^l, \ \ \det(C_2) \neq 0$$

It is sufficient to design an observer only for vector x_1, then the components of vector x_2 are calculated as

$$x_2 = C_2^{-1}(y - C_1x_1) \tag{6.1.5}$$

Write the equation of the system (6.1.1), (6.1.2) in space (x_1, y) as

$$\begin{aligned} \dot{x}_1 &= A_{11}x_1 + A_{12}y + B_1u \\ \dot{y} &= A_{21}x_1 + A_{22}y + B_2u \end{aligned} \tag{6.1.6}$$

$$\text{where} \quad TAT^{-1} = \begin{bmatrix} A_{11} & A_{12} \\ A_{21} & A_{22} \end{bmatrix} \quad TB = \begin{bmatrix} B_1 \\ B_2 \end{bmatrix} \quad T = \begin{bmatrix} I_{n-l} & 0 \\ C_1 & C_2 \end{bmatrix}$$

(The coordinate transformation is nonsingular, $\det(T) \neq 0$.)

The design of a reduced-order observer rests upon coordinate transformation

$$x' = x_1 + L_1y \tag{6.1.7}$$

and the system behaviour is considered in the space (x', y). The coordinate transformation is obviously nonsingular for any L_1.

The equation with respect to x' is obtained from (6.1.5) to (6.1.7):

$$\begin{aligned} \dot{x}' &= (A_{11} + L_1A_{21})x' + A'_{12}y + (B_1 + L_1B_2)u \\ A'_{12} &= A_{12} + L_1A_{22} - (A_{11} + L_1A_{21})L_1 \end{aligned}$$

The observer is designed in the form of a dynamic system of the $(n - l)$th order system

$$\dot{\hat{x}}' = (A_{11} + L_1A_{21})\hat{x}' + A'_{12}y + (B_1 + L_1B_2)u \tag{6.1.8}$$

with $\hat{x}'$ as an estimate of the state vector x'. The mismatch $\bar{x}' = \hat{x}' - x'$ is governed by

$$\dot{\bar{x}}' = (A_{11} + L_1A_{21})\bar{x}' \tag{6.1.9}$$

Again if the original system is observable, the eigenvalues of matrix $A_{11} + L_1A_{21}$ may be assigned arbitrarily (Kwakernaak and Sivan, 1972). It means that $\bar{x}'$ tends to zero and $\hat{x}'$ tends to x' at any desired rate. The components of the state vector x_1, and x_2, are thus found from (6.1.5) and (6.1.7).

6.2 Observers for linear time-invariant systems

Let us proceed to the design of a state observer with inputs as discontinuous functions of mismatches where motion preceding sliding mode and motion in the intersection of discontinuity surfaces may be handled independently. The observer is described by differential equations

$$\begin{aligned} \dot{\hat{x}}_1 &= A_{11}\hat{x}_1 + A_{12}\hat{y} + B_1 u + L_1 v \\ \dot{\hat{y}} &= A_{21}\hat{x}_1 + A_{22}\hat{y} + B_2 u - v \end{aligned} \tag{6.2.1}$$

where $\hat{x}_1$ and $\hat{y}$ are the estimates of the system state,

$$v = M\,\mathrm{sign}(\hat{y} - y) \qquad M > 0, \quad M = \mathrm{const}$$

The vector y is measured, hence $\hat{y} - y$ is available.

The discontinuous vector function $v \in \Re^l$ is chosen such that sliding mode is enforced in the manifold $\bar{y} = \hat{y} - y = 0$ and the mismatch between the output vector y and its estimate $\hat{y}$ is reduced to zero. A matrix L_1 must be found such that the mismatch $\bar{x}_1 = \hat{x}_1 - x_1$ between x_1 and its estimate $\hat{x}_1$ decays at the desired rate. Equations with respect to $\bar{x}_1$ and $\bar{y}$ are obtained from (6.1.6) and (6.2.1):

$$\begin{aligned} \dot{\bar{x}}_1 &= A_{11}\bar{x}_1 + A_{12}\bar{y} + L_1 v \\ \dot{\bar{y}} &= A_{21}\bar{x}_1 + A_{22}\bar{y} - v \\ v &= M\,\mathrm{sign}(\bar{y}) \end{aligned} \tag{6.2.2}$$

As shown in Section 2.4, the sliding mode is enforced in the manifold $\bar{y} = 0$ if the matrix multiplying v in the second equation of (6.2.2) is negative definite and M takes a high but finite value. This is the case for our system, since v is multiplied by a negative identity matrix. Hence for bounded initial conditions, sliding mode can be enforced in manifold $\bar{y} = 0$. It follows from the equivalent control methods that the solution v_{eq} to equation $\dot{\bar{y}} = 0$ should be substituted into the first equation of (6.2.2) with $\bar{y} = 0$ to derive the sliding mode equation

$$\begin{aligned} v_{eq} &= A_{21}\bar{x}_1 \\ \dot{\bar{x}}_1 &= (A_{11} + L_1 A_{21})\bar{x}_1 \end{aligned} \tag{6.2.3}$$

which coincides with (6.1.9). Hence the desired rate of convergence of $\bar{x}_1$ to zero and convergence of $\hat{x}_1$ to x_1 can be provided by a proper choice of matrix L_1 and then x_2 is found from (6.1.5).

The observer with the input as a discontinuous function of the mismatch (6.2.2) in sliding mode is equivalent to the reduced-order observer (6.1.8). However, if the plant and observed signal are affected by noise, the nonlinear observer may turn out to be preferable due to its filtering properties, which coincide with those of a Kalman filter (Drakunov, 1983).

6.3 Observers for linear time-varying systems

6.3.1 Block-observable form

For the time-varying system

$$\dot{x} = A(t)x + B(t)u \tag{6.3.1}$$

$$y = C(t)x \tag{6.3.2}$$

where $x \in \Re^n$, $u \in \Re^m$, $y \in \Re^l$, the output vector $y(t)$ and matrices $A(t)$, $B(t)$ and $C(t)$ are assumed to be known. An observer is to be designed to estimate the state vector $x(t)$.

For any nonsingular transformation of the state x into (y_o^T, x_1^T), $y_o \in \Re^{l_0}$, $x_1 \in \Re^{n-l_0}$, equation (6.3.1) is represented as follows:

$$\dot{y}_0 = A_{00}(t)y_o + A_{01}^*(t)x_1 + B_0(t)u \tag{6.3.3}$$

$$\dot{x}_1 = A_{10}^*(t)y_o + A_{11}^*(t)x_1 + B_1^*(t)u \tag{6.3.4}$$

The system (6.3.3), (6.3.4) with known vector y_0 is called the *block-observable form*. Superscripts and subscripts in (6.3.3), (6.3.4) denote block matrices of the transformed system matrices A, B in (6.3.1). The system (6.3.1), (6.3.2) can be represented in block-observable form if the rank l_0 and the principal minor position of the time-varying matrix $C(t)$ do not vary in time. In this case, after reordering vectors x and y, there exists $(l - l_0) \times l_0$ matrix $\Lambda_0(t)$ such that

$$C(t) = \begin{bmatrix} C_0(t) \\ \Lambda_0(t)C_0(t) \end{bmatrix} \qquad (\text{rank}C_0(t) = l_0)$$

and $l_0 \times n$ matrix C_0 is of the form

$$C_o(t) = [C_0'(t) \quad C_0''(t)]$$

with nonsingular $l_0 \times l_0$ matrix $C_0''(t)$. Vector y_0 is found as

$$y_0 = C_o(t)x = C_0'(t)x_1 + C_0''(t)x_1^*$$

where $x^T = [x_1^T \quad x_1^*]$ is transformed into $[y_o^T \quad x_1^T]$

$$\begin{bmatrix} y_0 \\ x_1 \end{bmatrix} = T_0 \begin{bmatrix} x_1 \\ x_1^* \end{bmatrix} = \begin{bmatrix} C_0' & C_0'' \\ I_{n-l_0} & 0 \end{bmatrix} \begin{bmatrix} x_1 \\ x_1^* \end{bmatrix} \qquad (\det(T_0) \neq 0) \tag{6.3.5}$$

The output equation (6.3.2) is written as

$$y = \begin{bmatrix} C_0 \\ \Lambda_0 C_0 \end{bmatrix} x = \begin{bmatrix} C_0 \\ \Lambda_0 C_0 \end{bmatrix} \begin{bmatrix} x_1 \\ x_1^* \end{bmatrix} = \begin{bmatrix} y_0 \\ \Lambda_0 y_0 \end{bmatrix} = \begin{bmatrix} I_{l_0} \\ \Lambda_0 \end{bmatrix} y_0$$

where I_{l_0} is an identity matrix. Vector y_0 consists of linearly independent components of vector y and the system can be transformed into the block-observable form using (6.3.5).

This procedure leads to block-observable forms for $0 < l_0 < n$. The $l_0 = 0$ case implies that the original system is unobservable (under the assumption that the position of the principal minor does not vary in time). For $l_0 = n$ the state vector may be obtained directly as a solution of the equation $y = C(t)x$.

Treating vector $A^*_{01}(t)x_1$ in block (6.3.3) as an output vector of the subsystem (6.3.4) and assuming that the rank and position of the principal minor of the matrix $A^*_{01}(t)$ do not vary in time, subsystem (6.3.3) can be represented in block-observable form. The rank of $A^*_{01}(t)$ is equal to $l_1 (0 \le l_1 \le l_0)$. There exists $(l_0 - l_1) \times l_1$ matrix $\Lambda_1(t)$ such that

$$A^*_{01}(t) = \begin{bmatrix} C_1(t) \\ \Lambda_1(t)C_1(t) \end{bmatrix}$$

$$\text{rank} A^*_{01}(t) = l_1 \qquad \text{rank } C_1(t) = l_1$$

$C_1(t)$ is represented as

$$C_1(t) = [C_1'(t) \quad C_1''(t)] \qquad (\det C_1''(t) \neq 0)$$

where $C_1'(t) \in \Re^{l_1 \times (n-l_0-l_1)}$, $C_1''(t) \in \Re^{l_1 \times l_1}$. Then $y_1(t)$ is given by

$$y_1(t) = C_1(t)x_1 = C_1'(t)x_2 + C_1''(t)x_2^*$$

where $y_1 \in \Re^{l_1}$, $x_2 \in \Re^{n-l_0-l_1}$ and $x_2^* \in \Re^{l_1}$.

The transformation matrix of $x_1^T = (x_2^T, x_2^{*T})$ into (y_1^T, x_2^T) is

$$\begin{bmatrix} y_1 \\ x_2 \end{bmatrix} = \begin{bmatrix} C_1' & C_1'' \\ I_{n-l_1-l_0} & 0 \end{bmatrix} \begin{bmatrix} x_2 \\ x_2^* \end{bmatrix} = T_1 \begin{bmatrix} x_2 \\ x_2^* \end{bmatrix} \qquad (\det(T_1) \neq 0)$$

Applying this transformation to system (6.3.4), the following equation is obtained:

$$\begin{bmatrix} \dot{y}_1 \\ \dot{x}_2 \end{bmatrix} = T_1 A^*_{10} y_0 + (T_1 A^*_{11} T_1^{-1} + T_1 \dot{T}_1^{-1}) \begin{bmatrix} y_1 \\ x_2 \end{bmatrix} + T_1 B_1^* u \tag{6.3.6}$$

where

$$T_1 A^*_{10} = \begin{bmatrix} A_{10} \\ A^*_{20} \end{bmatrix} \qquad (T_1 A^*_{11} T_1^{-1} + T_1 \dot{T}_1^{-1}) = \begin{bmatrix} A_{11} & A^*_{12} \\ A^*_{21} & A^*_{22} \end{bmatrix}$$

$$T_1 B_1^* = \begin{bmatrix} B_1 \\ B_2^* \end{bmatrix}$$

Equation (6.3.6) is rewritten as

$$\dot{y}_1 = A_{10}(t)y_o + A_{11}(t)y_1 + A^*_{12}(t)x_2 + B_1(t)u$$
$$\dot{x}_2 = A^*_{20}(t)y_o + A^*_{21}(t)y_1 + A^*_{22}(t)x_2 + B_2^*(t)u$$

Block (6.3.3) is rewritten as

$$\dot{y}_0 = A_{00}(t)y_o + A_{01}(t)y_1 + B_0(t)u \tag{6.3.7}$$

where $A_{01}(t) = \begin{bmatrix} I_{l_1} \\ \Lambda_1(t) \end{bmatrix}$ (rank $A_{01}(t) = l_1$)

For $l_1 < n - l_0$ the subsystem (6.3.7) can be represented in block-observable form again ($l_1 > 0$, otherwise the system (6.3.1), (6.3.2) is unobservable). After each step the dimension of x_i is less than the dimension of x_{i-1}, so the procedure terminates after a finite number of steps.

Recall that the procedure is implementable if in the block-observable form of the ith subsystem

$$\begin{aligned} \dot{y}_i &= \sum_{j=0}^{i} A_{i,j}(t)y_j + A^*_{i,j+1}(t)x_{i+1} + B_i(t)u \\ \dot{x}_{i+1} &= \sum_{j=0}^{i} A^*_{i,j}(t)y_j + A^*_{i+1,i+1}(t)x_{i+1} + B^*_{i+1}(t)u \end{aligned} \tag{6.3.8}$$

the rank and the position of the principal minor of the matrix $A^*_{i,i+1}(t)$ do not vary in time.

Finally, this procedure terminates after r steps and $y_r = C_r x_r = C''_r x^*_{r+1}$ $(x_{r+1} = 0)$. The rth subsystem is written as

$$\begin{aligned} \dot{y}_r &= A_{r0}(t)y_o + A_{r1}(t)y_1 + \ldots + A_{rr}(t)y_r + A^*_{r,r+1}(t)x_{r+1} + B_r u \\ &= \sum_{j=0}^{r} A_{r,j}(t)y_j + B_r(t)u \end{aligned}$$

Similar to (6.3.3) and (6.3.7), $A^*_{i,i+1}(t)x_{i+1}$ in (6.3.8) may be replaced by $A_{i,i+1}(t)y_{i+1}$ as follows:

$$\dot{y}_i = A_i(t)y^*_i + A_{i,i+1}(t)y_{i+1} + B_i(t)u \quad (i = 1, \ldots, r-1) \tag{6.3.9}$$

$$\dot{y}_r = A_r(t)y^*_r + B_r(t)u \tag{6.3.10}$$

where $A_i(t) = (A_{i1}, A_{i2}, \ldots, A_{ii})$, $y^{*T}_i = (y^T_0, y^T_1, \ldots, y^T_i)$, $y_i \in \Re^{l_i}$, rank $A_{i,i+1}(t) = l_{i+1}$ and $A_{i,i+1}$ is a full-rank matrix. Then the system (6.3.9), (6.3.10) is represented as a set of block observable forms

$$\frac{d}{dt}\begin{bmatrix} y_0 \\ y_1 \\ y_2 \\ \vdots \\ \vdots \\ y_r \end{bmatrix} = \begin{bmatrix} A_{00} & A_{01} & 0 & 0 & \cdots & 0 \\ A_{10} & A_{11} & A_{12} & 0 & \cdots & 0 \\ A_{20} & A_{21} & A_{22} & A_{23} & \cdots & 0 \\ \vdots & \vdots & \vdots & \vdots & \ddots & \vdots \\ \vdots & \vdots & \vdots & \vdots & \cdots & A_{r-1,r} \\ A_{r0} & A_{r1} & A_{r2} & A_{r3} & \cdots & A_{rr} \end{bmatrix} \begin{bmatrix} y_0 \\ y_1 \\ y_2 \\ \vdots \\ \vdots \\ y_r \end{bmatrix} + \begin{bmatrix} B_0 \\ B_1 \\ B_2 \\ \vdots \\ \vdots \\ B_r \end{bmatrix} u$$

with full-rank matrices $A_{i,i=1}$.

6.3.2 Observer design

Using the sliding mode approach, the design procedure for an observer for a time-varying system may be decoupled into r trivial and independent stabilization subproblems.

Let the observer equation be of the form

$$\dot{\hat{y}}_i = A_i(t)\hat{y}_i^* + A_{i,\,i+1}(t)\hat{y}_{i+1} + B_i(t)u + v_i \qquad (i = 0, \ldots, r-1) \tag{6.3.11}$$

$$\dot{\hat{y}}_r = A_r(t)\hat{y}_r^* + B_r(t)u + v_r \tag{6.3.12}$$

Observer inputs v_i are designed as

$$v_0 = M_0 \operatorname{sign}(y_0 - \hat{y}_0) \tag{6.3.13}$$

$$v_i = M_i \operatorname{sign}(A_{i-1,\,i}^{+} z_i) \qquad \tau_i \dot{z}_i + z_i = v_{i-1} \qquad (i = 1, \ldots, r) \tag{6.3.14}$$

where the left pseudoinverse matrix $A_{i-1,\,i}^{+}$ $(A_{i-1,i}^{+} A_{i-1,i} = I_{l_i})$ exists, since $A_{i-1,i}$ is a full-rank matrix. The equations for the mismatches $\bar{y}_i = y_i - \hat{y}_i$ are

$$\frac{d}{dt}\begin{bmatrix} \bar{y}_0 \\ \bar{y}_1 \\ \bar{y}_2 \\ \vdots \\ \vdots \\ \bar{y}_r \end{bmatrix} = \begin{bmatrix} A_{00} & A_{00} & 0 & 0 & \cdots & 0 \\ A_{10} & A_{11} & A_{12} & 0 & \cdots & 0 \\ A_{20} & A_{21} & A_{22} & A_{23} & \cdots & 0 \\ \vdots & \vdots & \vdots & \vdots & \ddots & \vdots \\ \vdots & \vdots & \vdots & \vdots & \cdots & A_{r-1,\,r} \\ A_{r0} & A_{r1} & A_{r2} & A_{r3} & \cdots & A_{rr} \end{bmatrix} \begin{bmatrix} \bar{y}_0 \\ \bar{y}_1 \\ \bar{y}_2 \\ \vdots \\ \vdots \\ \bar{y}_r \end{bmatrix} + \begin{bmatrix} v_0 \\ v_1 \\ v_2 \\ \vdots \\ \vdots \\ v_r \end{bmatrix}.$$

For bounded initial conditions and any finite numbers $M_1, \ldots, M_r$ there exists a number (M_0) such that after a finite time interval, sliding mode occurs in the manifold $\bar{y}_0 = 0$ since each component of $\bar{y}_0$ and its time derivative have different signs (2.4.1). According to the equivalent control method (Section 2.3), the solution of $\dot{\bar{y}}_0 = 0$ with $\bar{y}_0 = 0$,

$$(v_0)_{\text{eq}} = (M_0 \operatorname{sign}(\bar{y}_0))_{eq} = A_{01}\bar{y}_1$$

should be substituted into (6.3.14) to find the sliding mode equation. Then the output of the first-order filter z_1 approaches the equivalent control input v_{0eq}:

$$\lim_{\tau_1 \to 0} z_1 = v_{0_{eq}} = A_{01}\bar{y}_1 \quad \text{and} \quad \bar{y}_1 = \lim_{\tau_1 \to 0} A_{01}^{+} z_1$$

Similarly, $\bar{y}_2$ can be found from the second block (subsystem with respect to $\bar{y}_1$)

$$\bar{y}_2 = \lim_{\tau_2 \to 0} A_{12}^{+} z_2$$

Consequently, sliding mode will occur at each block, and then all the subvectors $\bar{y}_1, \ldots, \bar{y}_r$ will converge to zero. Since $y_i = \hat{y}_i + \bar{y}_i$ $(i = 1, \ldots, r)$ all subvectors of the state vector $(y_1, \ldots, y_r)$ and correspondingly x will be found.

Remarks 6.1

1. The procedure is invariant with respect to $l_i \times (l_i + \ldots + l_i)$ matrices $A_i(t)$ for $i = 0, \ldots, r$.
2. There is no need to enforce sliding mode in the last block since

   $$v_{r-1,eq} = A_{r-1,r}(t)\bar{y}_r$$

   and the last subvectors y_r may be found as

   $$y_r = \hat{y}_r + A^{+}_{r-1,r}(t)z_r$$

3. When there is an unknown disturbance vector $f(t)$ in the last block,

   $$\dot{y}_r = A_r(t)y_r + B_r(t)u + f(t) \quad f(t) \in \Re^{l_r}$$

   the mismatch equation is of the form

   $$\dot{\bar{y}}_r = A_r(t)\bar{y}_r + f(t) - v_r$$

   After sliding mode occurs in this subsystem, $\bar{y}_r = 0$ and

   $$v_{req} = f(t) \quad \text{and} \quad \lim_{\tau_f \to 0} z_f = v_{req} = f(t)$$

   if $\tau_f \dot{z}_f + z_f = v_r$.

As a result, the last block enables us to find the equivalent disturbance vector which includes external disturbances, parameter variations and nonlinear state functions. This approach is developed in Chapter 8 as one of the methods for chattering suppression. □

6.3.3 Simulation results

As an example of observer design, let us consider the following time-varying system:

$$\dot{x} = A(t)x + B(t)u + f(t)$$
$$y = C(t)x$$

where

$$A(t) = \begin{bmatrix} a_{00} & a_{01} & 0 & a_{03} \\ 0 & a_{11} & a_{12} & 0 \\ a_{20} & 0 & 0 & a_{23} \\ 0 & a_{31} & 0 & a_{33} \end{bmatrix} \qquad B(t) = \begin{bmatrix} b_{00} & 0 \\ 0 & b_{11} \\ 0 & 0 \\ 0 & 0 \end{bmatrix}$$

$$C(t) = \begin{bmatrix} c_{00} & c_{01} & 0 & 0 \\ 0 & c_{11} & 0 & 0 \end{bmatrix} \qquad f(t) = [0 \quad 0 \quad f_2 \quad f_3]$$

The elements of the coefficient matrices, the inputs and the disturbances are as follows:

$$a_{00} = -4e^{-t} \qquad a_{01} = -2 + \sin(t/2)$$
$$a_{11} = -5\cos(0.3t) \qquad a_{23} = -2e^{-t}$$

$$a_{31} = -1 - \frac{\sqrt{t+2}}{t+1} \qquad a_{33} = -2$$
$$a_{03} = a_{12} = a_{20} = -1$$

$$b_{00} = -3 \quad b_{11} = -5e^{-t/5}$$

$$c_{00} = 4 + \sin(t/2) \qquad c_{10} = 2 - e^{-t/5}$$
$$c_{11} = -5 + \tfrac{1}{2}\sin(-0.3t)$$

$$u_0 = 5\cos t \quad u_1 = -3$$

For output matrix $C(t)$

$$\operatorname{rank}\begin{bmatrix} c_{00} & c_{01} \\ 0 & c_{11} \end{bmatrix} = 2 \qquad t \in [0, \infty)$$

The rank and the principal minor position of the time-varying matrix $C(t)$ do not vary in time.

The block-observable form can be easily obtained following the method of this chapter:

$$\frac{d}{dt}\begin{bmatrix} y_0 \\ y_1 \end{bmatrix} = \begin{bmatrix} A_{00} & A_{01} \\ A_{10} & A_{11} \end{bmatrix}\begin{bmatrix} y_0 \\ y_1 \end{bmatrix} + \begin{bmatrix} B_0 \\ B_1 \end{bmatrix} u + \begin{bmatrix} 0 \\ f \end{bmatrix}$$

where

$$y_0 = \begin{bmatrix} c_{00} & c_{01} \\ 0 & c_{11} \end{bmatrix}\begin{bmatrix} x_1 \\ x_2 \end{bmatrix}$$

$$y_1 = \begin{bmatrix} x_3 \\ x_4 \end{bmatrix}$$

The observer, (6.3.11) to (6.3.14), for our system has the form

$$\frac{d}{dt}\begin{bmatrix} \hat{y}_0 \\ \hat{y}_1 \end{bmatrix} = \begin{bmatrix} A_{00} & A_{01} \\ A_{10} & A_{11} \end{bmatrix}\begin{bmatrix} \hat{y}_0 \\ \hat{y}_1 \end{bmatrix} + \begin{bmatrix} B_0 \\ B_1 \end{bmatrix} u + \begin{bmatrix} v_0 \\ v_1 \end{bmatrix}$$

where v_0 and v_1 are observer inputs:

$$
\begin{aligned}
v_0 &= M_0 \operatorname{sign}(y_0 - \hat{y}_0) \\
v_1 &= M_1 \operatorname{sign}(A_{01}^{-1} z_1) \\
\tau_1 \dot{z}_1 + z_1(t) &= v_0(t) \\
z_1(0) &= 0
\end{aligned}
$$

Simulations were performed for two cases.

System with zero disturbance

The initial values of the system states are given by

$$x_0(0) = x_1(0) = x_2(0) = x_3(0) = 5$$

Since the disturbance is assumed to be zero, only the state estimation is considered. The initial values of the observer state are equal to zero.

Because the sampling interval T_s confines the switching frequency, the filter constant τ_1, which must theoretically approach zero, is selected as $4T_s$. The sampling interval is 100 μs. Figure 6.1 depicts the system states $x_2(t)$, $x_3(t)$ and the outputs $\hat{x}_2(t)$, $\hat{x}_3(t)$ of the sliding mode observer. Notice how the estimated states rapidly converge to the real states. The gain matrices M_0 and M_1 are

$$M_0 = \begin{bmatrix} 5 & 0 \\ 0 & 5 \end{bmatrix} \qquad M_1 = \begin{bmatrix} 10 & 0 \\ 0 & 10 \end{bmatrix}$$

System with disturbances

Since the unknown disturbances f_2, f_3 are to be estimated, an additional filter is introduced (item 3 of Remark 6.1):

$$
\begin{aligned}
&\tau_f \dot{z}_f + z_f(t) = v_1(t) \\
&z_f(0) = [0 \quad 0]^T
\end{aligned}
$$

The estimated disturbance is the output of this filter:

$$\hat{f}(t) = z_f(t) \qquad \hat{f}(t)^T = [\hat{f}_2(t) \quad \hat{f}_3(t)]$$

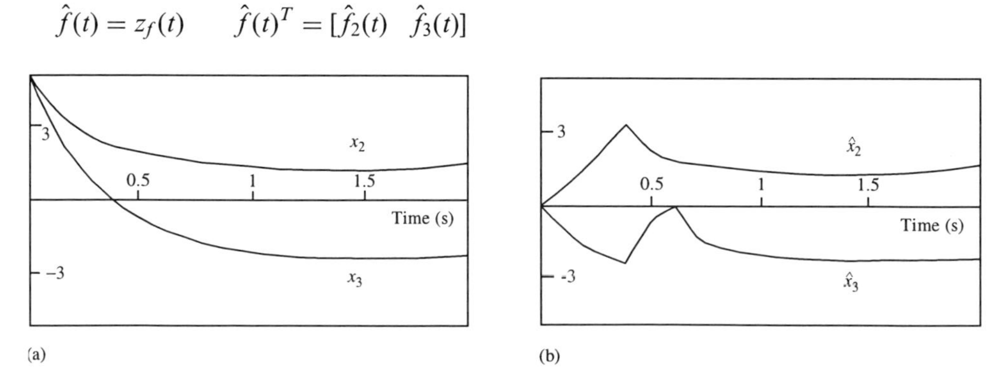

Figure 6.1 Values for (a) the real states $x_2(t)$ and $x_3(t)$, and (b) the outputs of the sliding mode observer $\hat{x}_2(t)$ and $\hat{x}_3(t)$.

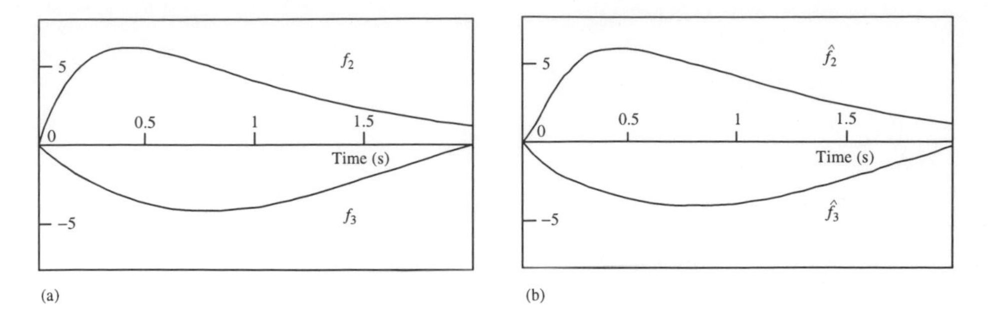

Figure 6.2 Disturbances: (a) real, and (b) estimated.

The time constant τ_1 is chosen to be 500 μs. Other conditions for simulation are the same as in Case 1. The estimation responses in Figure 6.2 demonstrate that the unknown disturbances can be found by the sliding mode observer.

References

DRAKUNOV, S., 1983, On adaptive quasioptimal filter with discontinuous parameters, *Automation and Remote Control*, **44**, 1167–75.

HASHIMOTO, H. *et al.*, 1990, VSS observer for linear time-varying system, *Proceedings of IECON'90*, Pacific Grove CA, pp. 34–39.

KWAKERNAAK, H. and SIVAN, R., 1972, *Linear Optimal Control Systems*, New York: Interscience.

UTKIN, V., 1992, *Sliding Modes in Control and Optimization*, Berlin: Springer-Verlag.

CHAPTER SEVEN

Integral Sliding Mode

The robustness property of conventional sliding mode control with respect to variations of system parameters and external disturbances can only be achieved after the occurrence of sliding mode. During the reaching phase, however, there is no guarantee of robustness. Integral sliding mode seeks to eliminate the reaching phase by enforcing sliding mode throughout the entire system response. Different from the conventional design approach, the order of the motion equation in integral sliding mode is equal to the order of the original system, rather than reduced by the dimension of the control input. As a result, robustness of the system can be guaranteed starting from the initial time instance. Uniform formulations of this new sliding mode design principle will be developed in this chapter. It is shown with examples that this generalized scheme of integral sliding mode enables a wide scope of application areas including control in robotics and electric drives. The concept of integral sliding mode can also be extended to construct a new type of perturbation estimator which solves the chattering problem without loss of robustness and control accuracy (Section 8.5). Further details on integral sliding mode can be found in Utkin and Shi (1996).

7.1 Motivation

Sliding mode plays a dominant role in variable-structure systems (VSS) theory. The core idea of designing VSS control algorithms consists of enforcing sliding mode in some manifold of the system space. Traditionally, these manifolds are constructed as the intersection of hypersurfaces in the state space. This intersection domain is normally called a switching manifold. Once the system reaches the switching plane, the structure of the feedback loop is adaptively altered to *slide* the system state along the switching plane; the system response depends thereafter on the gradient of the switching plane and remains insensitive to variations of system parameters and external disturbances under so-called *matching conditions* (Section 3.4). The order of the motion equation in sliding mode is equal to $(n - m)$, with n being the dimension of the state space and m the dimension of the control input. However, during the reaching phase, before sliding mode occurs, the system possesses no such insensitivity property; therefore, insensitivity cannot be ensured throughout an entire response. The robustness during the reaching phase is normally improved by high-gain feedback control. Stability problems inevitably limit the application of such high-gain feedback control schemes.

As an extension of traditional sliding mode schemes, the concept of integral sliding mode concentrates on robustness during the entire response. The order of the motion equation in this new type of sliding mode is equal to the dimension of the plant model. Therefore, sliding mode is established without a reaching phase, implying that the invariance of the system to parametric uncertainty and external disturbances is guaranteed starting from the initial time instant. This chapter generalizes the sliding mode control concept and emphasizes the background philosophy used for developing such new variable-structure systems.

We assume there already exists an ideal system consisting of a nominal plant model and a properly designed feedback control. To this existing controller, a discontinuous term is added based on integral sliding mode to ensure the desired performance despite parametric uncertainty and external disturbances. Design examples in some application areas are given to illustrate the methodology of this design philosophy. The basic idea of integral sliding mode for linear systems can be found in Ackermann and Utkin (1994).

Integral sliding mode may also be employed to prevent *chattering* in a control loop, often caused by controller discontinuities exciting unmodelled dynamics. For *chattering* prevention, the discontinuous control term is lowpass filtered before being fed to the plant, thus moving the discontinuity to an auxiliary control loop without unmodelled dynamics to be excited. The filtered control acts as a perturbation compensator and preserves the invariance property of sliding mode. The *chattering* prevention aspect of integral sliding mode is discussed in detail in Section 8.5.

7.2 Problem statement

For a given dynamic system represented by the state-space equation

$$\dot{x} = f(x) + B(x)u \tag{7.2.1}$$

with $x \in \Re^n$ being the state vector and $u \in \Re^m$ being the control input vector (rank $B(x) = m$), suppose there exists a feedback control law $u = u_0(x)$, which may be continuous or discontinuous, such that system (7.2.1) can be stabilized in a desired way (e.g. its state trajectory follows a reference trajectory with a given accuracy). We denote this ideal closed loop system as

$$\dot{x}_0 = f(x_0) + B(x_0)u_0 \tag{7.2.2}$$

where x_0 represents the state trajectory of the ideal system under control u_0. However, in practical applications, system (7.2.1) operates under uncertainty conditions that may be generated by parameter variations, unmodelled dynamics and external disturbances. Under this consideration, the real trajectory of the closed-loop control system may be summarized by

$$\dot{x} = f(x) + B(x)u + h(x, t) \tag{7.2.3}$$

in which vector $h(x, t)$ comprises the perturbation due to parameter variations, unmodelled dynamics and external disturbances and is assumed to fulfill the following *matching condition* (Section 3.4):

$$h(x, t) \in \text{span}\{B(x)\} \tag{7.2.4}$$

or equivalently,

$$h(x, t) = B(x)u_h \quad \text{with} \quad u_h \in \Re^m \tag{7.2.5}$$

In other words, control u is assumed to be able to influence all components of vector $h(x, t)$ via control matrix $B(x)$.

Assume that $h(x, t)$ is bounded and that an upper bound can be found as

$$|h_i(x, t)| \leq h_i^+(x, t) \qquad (i = 1, \ldots, n) \tag{7.2.6}$$

with $h_i^+(x, t)$ being known positive scalar functions. The control design challenge thus becomes: Find a control low $u(x, t)$ such that the trajectories of system (7.2.3) satisfy $x(t) \equiv x_0(t)$ starting from the initial time instant, i.e. $x(0) = x_0(0)$.

7.3 Design principles

For system (7.2.3), first redesign the control law to be

$$u = u_0 + u_1 \tag{7.3.1}$$

where $u_0 \in \Re^m$ is the *ideal control* defined in (7.2.2) and $u_1 \in \Re^m$ is designed to reject the perturbation term $h(x, t)$. Substitution of control law (7.3.1) into (7.2.3) yields

$$\dot{x} = f(x) + B(x)u_0 + B(x)u_1 + h(x, t) \tag{7.3.2}$$

Now define a sliding manifold as

$$s = s_0(x) + z, \quad \text{with} \quad s, s_0(x), z \in \Re^m \tag{7.3.3}$$

which consists of two parts: the first part $s_0(x)$ may be designed as a linear combination of the system states, similar to the conventional sliding mode design; the second part z introduces the integral term and will be determined below.

The philosophy of integral sliding mode is: In order to achieve $x(t) \equiv x_0(t)$ at all times $t > 0$, the equivalent control of u_1, denoted by u_{1eq}, should fulfill

$$B(x)u_{1eq} = -h(x, t) \tag{7.3.4}$$

or, in terms of (7.2.5),

$$u_{1eq} = -u_h \tag{7.3.5}$$

The equivalent control u_{1eq} accurately describes the system trajectories when 'sliding' along the manifold $s(x) = 0$ in (7.3.3). See also Section 2.3 or Utkin (1992) for details of the mathematical derivation of equivalent control.

In order to adequately define auxiliary variable $z(x, t)$ in (7.3.3) to achieve (7.3.5), set the time derivative $\dot{s}$ equal to zero,

$$\dot{s} = \dot{s}_0(x) + \dot{z} = \frac{\partial s_0}{\partial x}\{f(x) + B(x)u_0(x) + B(x)u_{1eq}(x) + B(x)u_h\} + \dot{z} = 0 \tag{7.3.6}$$

To ensure requirement (7.3.5), define

$$\dot{z} = -\frac{\partial s_0}{\partial x}\{f(x) + B(x)u_0(x)\}, \qquad z(0) = -s_0(x(0)) \tag{7.3.7}$$

where initial condition $z(0)$ is determined based on the requirement $s(0) = 0$. In other words, sliding mode is to occur starting from the initial time instance. Since equation (7.3.5) is satisfied, the motion equation of the system in sliding mode will be

$$\dot{x} = f(x) + B(x)u_0(x) \tag{7.3.8}$$

similar to the ideal system trajectories (7.2.2).

Definition 7.1: Integral sliding mode

A sliding mode is said to be an *integral sliding mode* if its motion equation is of the same order as the original system, (i.e. the order of sliding motion is equal to n).

□

The control u_1 in (7.3.1) is defined to enforce sliding mode along the manifold (7.3.3) via discontinuous function

$$u_1 = -M(x)\,\text{sign}(s) \tag{7.3.9}$$

where $M(x)$ is a positive scalar function for control gain. Substitution of (7.3.9) and (7.3.7) into (7.3.6) yields

$$\dot{s} = \frac{\partial s_0}{\partial x}B(x)u_h - \frac{\partial s_0}{\partial x}B(x)M(x)\,\text{sign}(s) \tag{7.3.10}$$

In equation (7.3.10) s_0 should be selected such that matrix $(\partial s_0/\partial x)B(x)$ is *nonsingular* during the entire system response. Then the scalar function $M(x)$ may be selected depending on the property of $(\partial s_0/\partial x)B(x)$ such that sliding mode is enforced in the manifold $s = 0$ (Sections 2.4 and 3.2).

7.4 Perturbation and uncertainty estimation

A crucial part of the nature of sliding mode control schemes is the control discontinuity. In closed loop, the ‘switching’ in the control action often results in high-frequency oscillations in practical implementations. Fast dynamics, like those of actuators and sensors, which were neglected in the control design process, are excited by the sliding mode control switches, occurring at high but finite frequencies. This phenomenon, which is common to all high-gain control systems, is known as *chattering*.

Several methods have been presented in the literature to alleviate *chattering*. The key idea is to limit the controller gain or the controller bandwidth. A detailed discussion of the causes of chattering and the various tools to prevent this phenomenon can be found in Chapter 8. The remainder of this section is devoted to a brief description of using integral sliding mode to design disturbance estimators and rejecting disturbance without causing chattering.

High-gain controllers are frequently limited by loop dynamics, especially by actuator dynamics preventing direct implementation of sliding mode schemes. On the other hand, discontinuous control inputs are often dictated by the nature of the system, e.g. by conventional pulse-width modulation (PWM) units in power electronics. To resolve such seemingly contradictory specifications, recall that the actual effect of a discontinuous controller on a given plant is equal to the average of the control action, the so-called equivalent control (Section 2.3).

With this in mind, reformulate the integral sliding mode principle in Section 7.3 in terms of a perturbation estimator. Instead of equation (7.3.1), change the control input to

$$u = u_0 + u_{1eq} \tag{7.4.1}$$

However, control (7.4.1) cannot be implemented directly since the equivalent value of discontinuous control u_1 depends on unknown disturbance $h(x, t)$ in (7.2.3). It was shown in Section 2.3 that the equivalent value of a discontinuous control is equal to the average value measured by a first-order linear filter, with the discontinuous control as its input. The time constant of the filter should be sufficiently fast such that the plant and disturbance dynamics are allowed to pass through the filtering without significant phase lag. Therefore, substitute $u_{1eq} = u_{1av}$ with u_{1av} defined by

$$\mu \dot{u}_{1av} + u_{1av} = u_1 \tag{7.4.2}$$

where the time constant μ should be made small enough not to distort the slow component of the switched action, equal to u_{1eq}. In most practical applications the spectrum of the perturbation does not overlap with the high-frequency components of the switching unit. But one may be tempted to ask: If the discontinuity in the real control path is smoothed, how can sliding mode be generated? Furthermore, does u_{1av} $(= u_{1eq})$ still cancel the perturbation term u_h? Yes, sliding mode can be generated; and yes, the perturbation term is still cancelled, if the discontinuous control u_1 is shifted from the plant input to the input of an auxiliary dynamic system. Here is the explanation.

Similar to equations (7.3.3) and (7.3.7), redesign the switching function

$$s = s_0(x) + z \tag{7.4.3}$$

with z defined in

$$\dot{z} = -\frac{\partial s_0}{\partial x}\{f(x) + B(x)u - B(x)u_1\} \qquad z(0) = -s_0(x(0)) \tag{7.4.4}$$

The time derivative of sliding variable s in (7.4.3) can be calculated as

$$\begin{aligned} \dot{s} &= \frac{\partial s_0}{\partial x}\{f(x) + B(x)u + B(x)u_h\} - \frac{\partial s_0}{\partial x}\{f(x) + B(x)u - B(x)u_1\} \\ &= \frac{\partial s_0}{\partial x}B(x)u_h + \frac{\partial s_0}{\partial x}B(x)u_1 \end{aligned} \tag{7.4.5}$$

Design the same discontinuous control for u_1 as shown in (7.3.9) and assume that matrix $(\partial s_0/\partial x)B(x)$ is *nonsingular* during the entire system response, the sliding mode may be enforced in the system using methods given in Sections 2.4 and 3.2. Solving for u_1 after setting $\dot{s} = 0$ in (7.4.5) reveals that $u_{1eq} = -u_h$ holds as well, implying that u_{1av} $(= u_{1eq})$ is indeed an estimate of the perturbation term u_h.

In this case, equation (7.4.4) can be interpreted as an internal process for generating sliding mode defined by (7.4.3); discontinuity appears only in the internal process, thus no chattering is excited in the real control path. Moreover, since u_{1av} cancels the perturbation u_{1h} without precise knowledge of the system model and associated parameters, a high degree of robustness is maintained. The information needed for this control scheme is the upper bound of the perturbation. From a conceptual viewpoint, integral sliding mode is used here only for estimating the system perturbation rather than for the purpose of control. The control action to the real plant will be continuous and is significantly enhanced by the perturbation compensator.

7.5 Application examples

Table 7.1 lists the four application examples in this section and explains their significance.

Table 7.1 Four application examples and their significance

Application	Significance
Linear time-invariant systems	Linear time-invariant systems are a special case of the proposed general design principle
Control of robot manipulators	Robust control of rigid-body robots under parameter uncertainties
Implementation of pulse-width modulation for electric drives	Design philosophy of integral sliding mode can be directly applied to practical systems
Robust current control for permanent-magnet synchronous motors	Version 1: Robust current control of permanent-magnet synchronous motors Version 2: Use of proposed perturbation estimator to achieve advanced performance

7.5.1 Linear time-invariant systems

Consider a controllable linear time-invariant system with scalar control

$$\dot{x} = Ax + B(u + d(x, t)) \tag{7.5.1}$$

with state vector $x \in \Re^n$, scalar control $u \in \Re$, known system matrix $A \in \Re^{n \times n}$, $B \in \Re^n$ and $d(x, t)$ being a nonlinear perturbation with known upper bound

$$|d(x, t)| < d^+(x, t) \tag{7.5.2}$$

Design control u, expressed in equation (7.3.1) as $u = u_0 + u_1$, where u_0 is predetermined such that system $\dot{x} = Ax + Bu_0$ follows a given trajectory with desired accuracy.

For example, u_0 may be designed as linear static feedback control $u_0 = -k^T x, k \in \Re^n$, in which gain vector k can be determined by *pole-placement* or *Linear Quadratic Gauss* methods.

For pole-placement design, *Ackermann's formula* (Section 5.4) may be used. Assuming that the desired eigenvalues for system $\dot{x} = Ax + Bu_0$ are $\lambda_1, \lambda_2, \ldots, \lambda_n$, the control gain k^T can be determined explicitly depending on Ackermann's formula

$$k^T = e^T P(A), \tag{7.5.3}$$

where $e^T = [0, \ldots, 0, 1][B, AB, \ldots, A^{n-1}B]^{-1}$ and $P(\cdot)$ is the characteristic polynomial of the system, defined by $P(\lambda) = (\lambda - \lambda_1)(\lambda - \lambda_2)\ldots(\lambda - \lambda_{n-1})(\lambda - \lambda_n)$.

According to (7.3.3) and (7.3.7), design the sliding manifold as

$$s = C^T x + z = 0 \qquad (C \in \Re^n) \tag{7.5.4}$$

with

$$\dot{z} = -C^T(Ax + Bu_0) \qquad z(0) = -C^T x(0) \tag{7.5.5}$$

In particular, vector $C \in \Re^n$ may be selected to be equal to $B \in \Re^n$, resulting in

$$s = B^T x + z \tag{7.5.6}$$

and

$$\dot{z} = -B^T Ax - (B^T B)u_0 \qquad z(0) = -B^T x(0) \tag{7.5.7}$$

The time derivative of s can be calculated as

$$\dot{s} = (B^T B)(u_1 + d(x, t)) \tag{7.5.8}$$

Solving for u_1 by formally setting $\dot{s} = 0$ shows that $u_{1eq} = -d(x, t)$. Thus the motion equation in sliding mode coincides with the motion of the ideal system $\dot{x} = Ax + Bu_0$ without perturbation $d(x, t)$. Furthermore, since $s(0) = B^T x(0) + z(0) = 0$, sliding mode will occur from the initial time instant $t = 0$.

For a controllable linear time-invariant system ($B \neq 0$), $C^T B = B^T B > 0$ holds and the second part of the control, u_1, can be designed as

$$u_1 = -m_0(x)\,\mathrm{sign}(s) \tag{7.5.9}$$

where $m_0(x) > d^+(x, t)$ should be satisfied such that the functions s and $\dot{s}$ have different signs, implying that sliding mode can be enforced.

For systems where only discontinuous control inputs are allowed, e.g. for switching controlled devices like power converters, the control input should be designed as

$$u = -m_0\,\mathrm{sign}(s) \tag{7.5.10}$$

instead of $u = u_0 + u_1$. To enforce sliding mode, control gain m_0 should satisfy

$$m_0 > |u_0| + d^+(x, t) \tag{7.5.11}$$

Integral sliding mode may also be called *full-order sliding mode* (Ackermann and Utkin, 1994).

7.5.2 Control of robot manipulators

The model of a rigid-body robot manipulator (Section 12.1) with n degrees of freedom can be written as

$$M(q)\ddot{q} + N(q, \dot{q}) = \tau \tag{7.5.12}$$

where $M \in \Re^{n\times n}$ is the mass matrix; $N \in \Re^{n\times 1}$ is the vector including centrifugal, Coriolis and gravity forces; $q \in \Re^{n\times 1}$ represents the joint angle vector and $\tau \in \Re^{n\times 1}$ denotes the joint torque vector.

Using the *computed torque method* based on the nominal model without perturbations, the required nominal joint torque for the tracking control of the joint position is defined as

$$\tau_0 = M_0(q)(\ddot{q}_d - K_D\dot{q}_e - K_P q_e) + N_0(q, \dot{q}) \tag{7.5.13}$$

where M_0, N_0 are the nominal values of M, N; $K_P \in \Re^{n\times n}, K_D \in \Re^{n\times n}$ are positive definite, diagonal gain matrices determining the closed-loop performance; and the tracking error is defined as $q_e(t) = q(t) - q_d(t)$ with $[q_d(t) \quad \dot{q}_d(t) \quad \ddot{q}_d(t)]$ being the desired trajectory and its time derivatives.

Substituting (7.5.13) into (7.5.12) under the assumption of exact knowledge of the model parameters, i.e. $M = M_0$ and $N = N_0$, the resulting closed-loop error dynamics are given by

$$\ddot{q}_e + K_D\dot{q}_e + K_P q_e = 0 \tag{7.5.14}$$

implying that tracking error $q_e(t)$ tends to zero asymptotically.

However, for a real robot with uncertain parameters $M \neq M_0$ and $N \neq N_0$, the error dynamics are perturbed as

$$\ddot{q}_e + K_D\dot{q}_e + K_P q_e = M_0^{-1}\tau_p \tag{7.5.15}$$

where τ_p is given by

$$\tau_p = (\bar{M}\ddot{q} + \bar{N}) \tag{7.5.16}$$

with $\bar{M} = M_0 - M$ being the parameter error for matrix $M(q)$ and $\bar{N} = N_0 - N$ being the parameter error for vector $N(q, \dot{q})$. Thus, no matter how the constant matrices K_P and K_D are chosen, the tracking error $q_e(t)$ will not tend to zero.

In order to suppress the perturbation caused by modelling uncertainty, design a robust controller based on the proposed integral sliding mode principle and show that the ideal closed-loop error dynamics as given by equation (7.5.14) can still be achieved.

The disturbance torque τ_p contains $\ddot{q}$, which is a function of the control input $\tau = \tau_0 + \tau_1$, so in order to design the control input τ_1 rejecting the system disturbance, it is necessary to reformulate the system model such that the resulting disturbance term is not a function of the control input τ_1. The ideal robot dynamics with $M = M_0, N = N_0$ and $\tau = \tau_0$ can be rewritten in terms of the error dynamics as

$$\ddot{q}_{e0} = -M_0^{-1}N_0 + M_0^{-1}\tau_0 - \ddot{q}_d \tag{7.5.17}$$

For the real system (7.5.12) under control $\tau = \tau_0 + \tau_1$, error dynamics similar to the ideal tracking error (7.5.17) can be derived as

$$\ddot{q}_e = -M^{-1}N + M^{-1}\tau - \ddot{q}_d \tag{7.5.18}$$

According to the proposed integral sliding mode design method in equations (7.3.3) and (7.3.7), let the switching function be $s = s_0 + z$ with

$$s_0 = [C \quad I]\begin{bmatrix} q_e \\ \dot{q}_e \end{bmatrix} \tag{7.5.19}$$

and

$$\dot{z} = -[C \quad I]\begin{bmatrix} \dot{q}_e \\ -M_0^{-1}N_0 + M_0^{-1}\tau_0 - \ddot{q}_d \end{bmatrix} \qquad z(0) = -Cq_e(0) - \dot{q}_e(0) \tag{7.5.20}$$

where $C \in \Re^{n\times n}$ is a positive definite gain matrix and $I \in \Re^{n\times n}$ is an $n \times n$ unit matrix.

The time derivative of the sliding variable $s(t)$ can then be obtained by differentiation of equation (7.5.19) with substitution of error dynamics (7.5.18) and auxiliary variable z as defined in (7.5.20):

$$\dot{s} = \dot{s}_0 + \dot{z} = \zeta_1 + \zeta_2\tau_0 + M^{-1}\tau_1 \tag{7.5.21}$$

where $\zeta_1 = (M_0^{-1}N_0 - M^{-1}N)$ and $\zeta_2 = (M^{-1} - M_0^{-1})$ are the mismatches between the nominal parameters M_0 and N_0, and the real system parameters $M(q)$ and $N(q, \dot{q})$, viewed as system perturbations similar to (7.5.16) in equation (7.5.15). Note that for the derivation of (7.5.21), the joint torque is composed of two additive parts:

$$\tau = \tau_0 + \tau_1 \tag{7.5.22}$$

where τ_0 is defined in (7.5.13) and τ_1 is the discontinuous part to reject the system perturbations. Define

$$\tau_1 = -\Gamma_0 \operatorname{sign}(s) \tag{7.5.23}$$

where Γ_0 is a positive constant and design a Lyapunov function candidate $V = \frac{1}{2}s^T s$. The time derivative of V along the solutions of (7.5.21) is given by

$$\dot{V} = s^T\dot{s} = s^T(\zeta_1 + \zeta_2\tau_0) - s^T M^{-1}\Gamma_0 \operatorname{sign}(s) \tag{7.5.24}$$

Since the kinetic energy of a robot, i.e. $\frac{1}{2}\dot{q}^T M(q)\dot{q}$, is always positive for $\|\dot{q}\| \neq 0$, matrix $M(q)$ is positive definite, and the inverse matrix $M(q)^{-1}$ is positive definite as well. The control gain Γ_0 can be chosen as

$$\Gamma_0 > \frac{\|\zeta_1 + \zeta_2\tau_0\|}{\|M^{-1}\|} \tag{7.5.25}$$

which guarantees $\dot{V} < 0$, implying that sliding mode will be enforced in finite time. Definition of the initial conditions in (7.5.20) as $z(0) = -Cq_e(0) - \dot{q}_e(0)$ eliminates the reaching time to the sliding manifold. Once sliding mode occurs and the system is confined to the manifold $s(t) = 0$, the equivalent control of τ_1 can be used to examine the system behaviour. The equivalent control is obtained by formally setting $\dot{s} = 0$, yielding

$$\tau_{1eq} = -M(\zeta_1 + \zeta_2\tau_0) \tag{7.5.26}$$

Substitution of $\tau = \tau_0 + \tau_{1eq}$ in equation (7.5.12) with equivalent control (7.5.26) leads to the motion equation during sliding mode which can be simplified as

$$M_0(q)\ddot{q} + N_0(q, \dot{q}) = \tau_0 \tag{7.5.27}$$

Control τ_0 in (7.5.13) thus achieves the ideal closed-loop error dynamics (7.5.14) as if perturbations caused by the parametric uncertainty did not exist.

7.5.3 Pulse-width modulation for electric drives

In contrast to the examples in Sections 7.5.1 and 7.5.2, the integral sliding mode design philosophy is directly exploited to implement *pulse-width modulation* (PWM) in an electric drive system instead of applying equations (7.3.3) and (7.3.7). Without loss of generality, an electric drive supplied by a power converter can be described by the affine dynamic system

$$\dot{x} = f(x) + B(x)u \tag{7.5.28}$$

where $x \in R^n$ represents the current and flux components and $u \in R^m$ represents the control voltages taking only two values, $-u_0$ and $+u_0$, with u_0 being the DC bus (also called the DC link or link) voltage. For field-oriented control design (Chapter 10), equation (7.5.28) is often transformed into a rotating coordinate system aligned with one of the flux vectors (rotor flux or stator flux). Using a transformation matrix T, a nonlinear projector with sinusoidal entries, the system equation (7.5.28) may be transformed into the new coordinate system, denoted as (d, q):

$$\dot{x}_{dq} = f_{dq}(x_{dq}) + B_{dq}(x_{dq})u_{dq} \tag{7.5.29}$$

where u_{dq} is the new control input in coordinate system (d, q). Suppose that the control u_{dq} has been determined to satisfy the given specifications. The task then is to transform the control u_{dq} back to the original coordinate system using the inverse transformation T^{-1}. Denote this transformed control as u^* and let

$$u^* = T^{-1}u_{dq} \tag{7.5.30}$$

Now the question arises: How to obtain the actual control u for system (7.5.28) which may only take two discontinuous values $-u_0$ and $+u_0$ and should be exactly equivalent to u^*? The solution is to make the equivalent value of the control u to be equal to the equivalent value of u^*, i.e. $u_{eq} = u^*_{eq}$. Design a sliding mode manifold

$$s = \int_0^t (u^*(\zeta) - u(\zeta))\, d\zeta = 0 \tag{7.5.31}$$

with associated control

$$u = u_0 \ \mathrm{sign}(s) \tag{7.5.32}$$

For a Lyapunov function candidate $V = \frac{1}{2}s^T s$, the time derivative of V is given by

$$\dot{V} = s^T \dot{s} = s^T u^* - s^T u_0 \operatorname{sign}(s) \qquad (7.5.33)$$

It is obvious that sliding mode can be enforced if the DC bus voltage satisfies $u_0 > \|u^*\|$, or in other words, the DC bus voltage should be high enough to enforce the desired motion.

An example of applying sliding mode PWM to the current control of permanent-magnet synchronous motors (PMSMs) can be found in Section 10.2.3.

7.5.4 Robust current control for PMSMs

For high-performance operation of permanent-magnet synchronous motors (PMSMs), current control may be implemented using the so-called *field-oriented control* (FOC) approach (Chapter 10 has more details). From a control viewpoint, this approach uses a state transformation after which the decoupling and linearization tasks can be performed easily. However, FOC needs precise knowledge of the motor parameters. Practically, those parameters cannot be known exactly since (1) the model used for FOC is a simplified motor model; (2) the motor parameters are normally obtained by an identification procedure in which errors are always present; and (3) these parameters may vary with the rotor position and ambient temperature. As a result, the motor torque and the motor flux cannot be controlled independently, resulting in torque pulsation and lower efficiency.

Two solutions exist for solving this problem: adaptive control and robust control. Adaptive control recursively calculates the motor parameters depending on the state measurements, whereas robust control tries to suppress the parameter uncertainty using high control gains. Adaptive control involves a high computational overhead and an additional convergence problem; robust control may result in a low control efficiency and may excite high-frequency unmodelled dynamics. In the following, we propose two versions of control approaches based on integral sliding mode. These control strategies belong to the category of robust control. However, as one can see from the second version of the control design based on the perturbation estimation described in Section 7.4, the control approach is actually an adaptation to the system perturbation, resulting in continuous control actions suitable for interfacing with a PWM unit.

In the (d, q) synchronously rotating reference frame (Section 10.2.2 gives details of modelling a PMSM in different coordinate frames), the voltage equations of a PMSM motor are expressed by the following nonlinear differential equations:

$$\begin{aligned} \frac{di_d}{dt} &= \frac{1}{L}u_d - \frac{R}{L}i_d + \omega_e i_q \\ \frac{di_q}{dt} &= \frac{1}{L}u_q - \frac{R}{L}i_q - \omega_e i_d - \lambda\omega_e \end{aligned} \qquad (7.5.34)$$

where

i_d = d-axis stator current

i_q = q-axis stator current

u_d = d-axis stator voltage

u_q = q-axis stator voltage
L = armature inductance
R = armature resistance
ω_e = electrical angular velocity
λ = flux linkage of permanent magnet

In practice, parameters L, R and λ are not known exactly. For the control design, the nominal values of these parameters, denoted as L_0, R_0 and λ_0, are used. In the ideal case with $L = L_0$, $R = R_0$ and $\lambda = \lambda_0$, we may exploit FOC to design the current controller. For the motor currents $i_d(t)$ and $i_q(t)$ to track desired current references $i_d^*(t)$ and $i_q^*(t)$, control voltages $u_d = u_{d0}$ and $u_q = u_{q0}$ can be designed to achieve the desired performance

$$|i_d(t) - i_d^*(t)| < \varepsilon_d \qquad |i_q(t) - i_q^*(t)| < \varepsilon_q \qquad (\forall t > t_0) \tag{7.5.35}$$

where $\varepsilon_d, \varepsilon_q$ and t_0 are specified by the control designer. However, in practice, $L \neq L_0$, $R \neq R_0$ and $\lambda \neq \lambda_0$, and the FOC design may result in an unacceptable control error. To suppress the parameter uncertainty, the control voltages are augmented to

$$\begin{aligned} u_d &= u_{d0} + u_{d1} \\ u_q &= u_{q0} + u_{q1} \end{aligned} \tag{7.5.36}$$

For the first version of the control design, u_{d1} and u_{q1} are selected to be discontinuous so they suppress the perturbation caused by discrepancies between the true motor parameters and the nominal motor parameters (the nominal parameters were used for FOC design of $u_d = u_{d0}$ and $u_q = u_{q0}$):

$$\begin{aligned} u_{d1} &= -M_d \operatorname{sign}(s_d) \\ u_{q1} &= -M_q \operatorname{sign}(s_q) \end{aligned} \tag{7.5.37}$$

where M_d and M_q are the control gains to be determined later. The sliding mode variables s_d and s_q serve as switching functions and are of integral form

$$\begin{aligned} s_d &= s_{d0} + z_d \\ s_q &= s_{q0} + z_q \end{aligned} \tag{7.5.38}$$

in which s_{d0} and s_{q0} are defined as $s_{d0} = i_d - i_d^*$, $s_{q0} = i_q - i_q^*$; z_d and z_q are given as follows:

$$\begin{aligned} \dot{z}_d &= -\left(\frac{1}{L_0}u_{d0} - \frac{R_0}{L_0}i_d + \omega_e i_q\right) + \frac{di_d^*}{dt} \\ z_d(0) &= -(i_d(0) - i_d^*(0)) \\ \dot{z}_q &= -\left(\frac{1}{L_0}u_{q0} - \frac{R_0}{L_0}i_q - \omega_e i_d - \lambda_0\omega_e\right) + \frac{di_q^*}{dt} \\ z_q(0) &= -(i_q(0) - i_q^*(0)) \end{aligned} \tag{7.5.39}$$

with di_d^*/dt and di_q^*/dt being provided by an outer control loop, e.g. a speed control loop. Now let us analyze the system stability and determine the control gains M_d

and M_q. First we deal with the d-component. Taking the time derivative of s_d yields

$$\dot{s}_d = \dot{s}_{d0} + \dot{z}_d = \frac{1}{L}u_{d1} + \varepsilon_1 u_{d0} - \varepsilon_2 i_d \tag{7.5.40}$$

where $\varepsilon_1 = (1/L - 1/L_0)$ and $\varepsilon_2 = (R/L - R_0/L_0)$. Substitute (7.5.37) into (7.5.40) to obtain

$$\dot{s}_d = -\frac{M_d}{L}\operatorname{sign}(s_d) + (\varepsilon_1 u_{d0} - \varepsilon_2 i_d) \tag{7.5.41}$$

To enforce sliding mode in equation (7.5.41), the discontinuous control gains M_d should be selected as

$$M_d > \max\{L|\varepsilon_1 u_{d0} - \varepsilon_2 i_d|\} \tag{7.5.42}$$

The right-hand side of this inequality is assumed to be bounded. Once sliding mode is achieved, $s_d = 0$ holds; the equivalent control of u_{d1} compensates exactly for the perturbation in terms of u_{hd} (see equations (7.2.5) and (7.3.5)):

$$(u_{d1})_{eq} = -u_{hd} = -L(\varepsilon_1 u_{d0} - \varepsilon_2 i_d) \tag{7.5.43}$$

Similar derivations hold for the q-component, where $\dot{s}_q$ can be given as

$$\dot{s}_q = \dot{s}_{q0} + \dot{z}_q = \frac{1}{L}u_{q1} + \varepsilon_1 u_{q0} - \varepsilon_2 i_q - \varepsilon_3 \omega_e \tag{7.5.44}$$

in which ε_1 and ε_2 are the same as for the d-component in equation (7.5.40) and $\varepsilon_3 = (\lambda - \lambda_0)$. Substitution of (7.5.37) into (7.5.44) yields

$$\dot{s}_q = -\frac{M_q}{L}\operatorname{sign}(s_q) + (\varepsilon_1 u_{q0} - \varepsilon_2 i_q - \varepsilon_3 \omega_e) \tag{7.5.45}$$

Enforcing sliding mode in equation (7.5.45) requires

$$M_q > \max\{L|\varepsilon_1 u_{q0} - \varepsilon_2 i_q - \varepsilon_3 \omega_e|\} \tag{7.5.46}$$

Again, the equivalent control of u_{q1} compensates exactly the perturbation in terms of u_{hq}:

$$(u_{q1})_{eq} = -u_{hq} = -L(\varepsilon_1 u_{q0} - \varepsilon_2 i_q - \varepsilon_3 \omega_e) \tag{7.5.47}$$

Note that, unlike M_d, control gain M_q depends on the electrical rotor speed ω_e if the flux linkage λ is not known precisely.

Since the DC bus voltage of a drive system is always limited, the amplitude of the control voltages, i.e. $\sqrt{u_d^2 + u_q^2}$, is also limited, implying that M_d and M_q cannot be selected arbitrarily: increasing M_q leads to a decrease of M_d. As long as inequalities (7.5.42) and (7.5.46) hold, sliding mode can be enforced. Otherwise, stability of sliding mode is not guaranteed, which means the parameter uncertainties cannot be fully compensated. This is one of the drawbacks of the first version of control design.

Another drawback of the first version lies in the fact that the resulting controllers u_d and u_q, are difficult to implement by a conventional PWM unit, since they contain a discontinuous part. Note that the standard FOC relies on a PWM unit to adopt the continuous controllers u_{d0} and u_{q0} for the discontinuous control voltages of PMSM motors. However, u_d and u_q as defined in (7.5.36) are a mixture of continuous and discontinuous parts. After the coordinate transformation, i.e. from the (d, q) frame to the stator (α, β) frame resulting in u_α and u_β, and further transformation to the phase coordinate frame resulting in u_a, u_b and u_c, they should be applied to the PWM unit. If these controls contain a discontinuous part, the associated sudden jumps in the PWM duty ratio may be harmful to the employed power converter. This problem can be overcome by the sliding mode PWM approach due to introduced integral action (Section 7.5.3). Now let us modify the control design to solve these drawbacks and obtain the second version, which is based on the perturbation estimator of Section 7.4.

The controls u_d and u_q are of similar form to equation (7.5.36):

$$\begin{aligned} u_d &= u_{d0} + \tilde{u}_{d1} \\ u_q &= u_{q0} + \tilde{u}_{q1} \end{aligned} \tag{7.5.48}$$

where the continuous functions $\tilde{u}_{d1}$ and $\tilde{u}_{q1}$ will be determined later. Design discontinuous controls u_{d1} and u_{q1} to have the same form as given by equation (7.5.37):

$$\begin{aligned} u_{d1} &= -M_d \operatorname{sign}(s_d) \\ u_{q1} &= -M_q \operatorname{sign}(s_q) \end{aligned} \tag{7.5.49}$$

The switching functions s_d and s_q are also identical with equation (7.5.38):

$$\begin{aligned} s_d &= s_{d0} + z_d \\ s_q &= s_{q0} + z_q \end{aligned} \tag{7.5.50}$$

where the integral terms z_d and z_q, which implement the integral sliding mode control are of the form

$$\begin{aligned} \dot{z}_d &= -\left(\frac{1}{L_0}u_d - \frac{R_0}{L_0}i_d + \omega_e i_q - \frac{u_{d1}}{L_0}\right) + \frac{di_d^*}{dt} \\ z_d(0) &= -(i_d(0) - i_d^*(0)) \\ \dot{z}_q &= -\left(\frac{1}{L_0}u_q - \frac{R_0}{L_0}i_q - \omega_e i_d - \lambda_0\omega_e - \frac{u_{q1}}{L_0}\right) + \frac{di_q^*}{dt} \\ z_q(0) &= -(i_q(0) - i_q^*(0)) \end{aligned} \tag{7.5.51}$$

The time derivatives $\dot{s}_d$ and $\dot{s}_q$ are

$$\begin{aligned} \dot{s}_d &= -\frac{M_d}{L_0}\operatorname{sign}(s_d) + (\varepsilon_1(u_{d0} + \tilde{u}_{d1}) - \varepsilon_2 i_d) \\ \dot{s}_q &= -\frac{M_q}{L_0}\operatorname{sign}(s_q) + (\varepsilon_1(u_{q0} + \tilde{u}_{q1}) - \varepsilon_2 i_q - \varepsilon_3\omega_e) \end{aligned} \tag{7.5.52}$$

The conditions for enforcing sliding mode in (7.5.52) are

$$\begin{aligned} M_d &> \max\{L_0|\varepsilon_1(u_{d0} + \tilde{u}_{d1}) - \varepsilon_2 i_d|\} \\ M_q &> \max\{L_0|\varepsilon_1(u_{q0} + u_{q1}) - \varepsilon_2 i_q - \varepsilon_3\omega_e|\} \end{aligned} \tag{7.5.53}$$

Now, the discontinuous controls u_d and u_q appear only in the control algorithms, i.e. in the control computer rather than in the real control path, so M_d and M_q can be selected as high as required to enforce sliding mode. After the occurrence of sliding mode, the equivalent control can be written as

$$\begin{aligned}(u_{d1})_{eq} &= L_0(\varepsilon_1(u_{d0} + \tilde{u}_{d1}) - \varepsilon_2 i_d) \\ (u_{q1})_{eq} &= L_0(\varepsilon_1(u_{q0} + \tilde{u}_{q1}) - \varepsilon_2 i_q - \varepsilon_3 \omega_e)\end{aligned} \tag{7.5.54}$$

These equations show that the equivalent values of u_{d1} and u_{q1} are indeed equal to the disturbances to be compensated. And the equivalent values can be obtained by applying lowpass filters with u_{d1} and u_{q1} as the filter inputs (Section 2.3).

Finally, the disturbance compensation terms $\tilde{u}_{d1}$ and $\tilde{u}_{q1}$ can be obtained as

$$\begin{aligned}\tilde{u}_{d1} &= (u_{d1})_{eq} \\ \tilde{u}_{q1} &= (u_{q1})_{eq}\end{aligned} \tag{7.5.55}$$

Now $\tilde{u}_{d1}$ and $\tilde{u}_{q1}$ are continuous and thus acceptable for the PWM unit.

Remark 7.1

If the resulting $\tilde{u}_{d1}$ and $\tilde{u}_{q1}$ are too high such that $\sqrt{u_d^2 + u_q^2}$ exceeds the DC bus voltage, the system disturbances cannot be compensated completely. However, the perturbation estimation given by equation (7.5.54) remains true. This case indicates that the system disturbances are too large and cannot be fully compensated by any control algorithm. □

Another example of using the perturbation estimation concept will be given in Chapter 12, showing the effectiveness of this control scheme in the torque control of a flexible robot joint. In this application the moment of inertia of the motor rotor and the link, the joint stiffness and the fractions at both motor and link sides are assumed to be unknown.

7.6 Summary

This chapter has developed a new sliding mode design concept – integral sliding mode. The proposed uniform formulation of integral sliding mode allows a wide range of application. The main advantage of this new design principle is that the robustness provided by sliding mode can be guaranteed over an entire system response starting from the initial time instant. We have emphasized the basic idea and the background philosophy used to develop such a new approach. And we gave detailed examples of practical applications. The chattering problem, which was the major drawback of sliding mode control, can be solved using the proposed algorithms, while preserving the robustness provided by sliding mode design and the accuracy of the control system.

References

ACKERMANN, J. and UTKIN, V. I., 1994, Sliding mode control design based on Ackermann formula, *Proceedings of the IEEE Conference on Decision and Control*, Orlando FL, pp. 3622–27.

UTKIN, V. I., 1992, *Sliding Modes in Control and Optimization*, London: Springer-Verlag.
UTKIN, V. I. and SHI, J., 1996, Integral sliding mode in systems operating under uncertainty conditions, *Proceedings of the IEEE Conference on Decision and Control*, Kobe, Japan, pp. 4591–96.

CHAPTER EIGHT

The Chattering Problem

Almost ever since sliding mode ideas have been put forward, the audible noise some sliding mode controllers exhibit has irritated control engineers and has often led to resentment, even rejection of the technique. The phenomenon is best known as chattering. Two main causes have been identified. First, fast dynamics in the control loop which were neglected in the system model, are often excited by the fast switching of sliding mode controllers. Second, digital implementations in microcontrollers with fixed sampling rates may lead to discretization chatter. This chapter concentrates on the first cause, the unmodelled dynamics in the control loop and introduces four methods to reliably prevent chattering. Sliding mode in discrete-time systems without discretization chatter is discussed in Chapter 9.

8.1 Problem analysis

The term 'chattering' describes the phenomenon of finite-frequency, finite-amplitude oscillations appearing in many sliding mode implementations. These oscillations are caused by the high-frequency switching of a sliding mode controller exciting unmodelled dynamics in the closed loop. 'Unmodelled dynamics' may refer to sensors and actuators neglected in the principal modelling process since they are generally significantly faster than the main system dynamics. However, since *ideal* sliding mode systems are infinitely fast, all system dynamics should be considered in the control design.

Fortunately, preventing chattering usually does not require a detailed model of all system components. Rather, a sliding mode controller may first be designed under idealized assumptions of no unmodelled dynamics. In a second design step, possible chattering is to be prevented by one of the methods discussed in this chapter. The solution of the chattering problem is of great importance when exploiting the benefits of a sliding mode controller in a real-life system. Without proper treatment in the control design, chattering may be a major obstacle to the implementation of sliding mode in a wide range of applications. Note that the switching action itself, as the core of a continuous-time sliding mode system, is *not* called chattering since, in the ideal case, the switching is intended and its frequency tends to infinity; this book uses 'chattering' to describe undesired system oscillations with *finite* frequency caused by system imperfections.

This section seeks to provide an in-depth analysis of the chattering problem. Both analytical and numerical studies are used to examine how unmodelled dynamics in a closed system with a controller discontinuity are excited, leading to oscillations in the system trajectory.

8.1.1 Example system: model

A simple first-order plant with second-order 'unmodelled' actuator dynamics is used as an example for illustration purposes throughout this section. For clarity of presentation, we refrain from using a more complex system like the inverted pendulum. The model of the first-order system with state and output $x(t)$ is given by

$$\dot{x}(t) = ax(t) + d(x, t) + bw(t) \tag{8.1.1}$$

where $a^- \leq a \leq a^+$ and $0 < b^- \leq b \leq b^+$ are unknown parameters within known bounds, $w(t)$ is the control variable and disturbance $d(t)$ is assumed to be uniformly bounded for all operating conditions (x, t) as $|d(x, t)| \leq d^+$. Control variable $w(t)$ is the output of an 'unmodelled' actuator with stable dynamics dominated by second-order function

$$w(t) = \frac{\omega^2}{p^2 + 2\omega p + \omega^2} u(t) = \frac{1}{(\mu p + 1)^2} u(t) \tag{8.1.2}$$

where $u(t)$ is the actual control input to plant (8.1.1) and p denotes the Laplace variable. In equation (8.1.2) and in the sequel, a mixed representation of time domain and frequency (Laplace) domain functions is used for ease of presentation, although it is not formally correct. For example, it is understood in (8.1.2) that time-domain control variable $w(t)$ is the output of the lowpass filter described by the inverse of the Laplace transfer function in p with time-domain input $u(t)$.

In equation (8.1.2) $\omega > 0$ is the unknown actuator bandwidth with $\omega \gg a$ in equation (8.1.1). The small time constant $\mu = 1/\omega > 0$ was substituted to symbolize that the actuator dynamics are assumed to be significantly faster than the system dynamics (8.1.1).

The goal of control is to make state and output $x(t)$ of system (8.1.1) track a desired trajectory $x_d(t)$ with a known amplitude bound as $|x_d(t)| \leq x_d^+$ and a known bound on the rate of change $|\dot{x}_d(t)| \leq v_d^+$. The parameters for the simulation examples in this section are $a = 0.5$, $b = 1$, $d(t) = 0.2\sin(10t) + 0.3\cos(20t) \leq 0.5$, $\omega = 50$, thus $\mu = 0.02$, with a limit on available control resources of $|u(t)| \leq 2.01$ and a desired trajectory $x_d(t) = \sin t$, i.e. $x_d^+ = 1$ and $v_d^+ = 1$. Note that with $a > 0$, the example plant (8.1.1) is *unstable*.

8.1.2 Example system: ideal sliding mode

Standard sliding mode control design for the ideal plant (8.1.1), i.e. neglecting actuator dynamics (8.1.2) by setting $w(t) = u(t)$, defines the sliding variable as

$$s(t) = x_d(t) - x(t) \tag{8.1.3}$$

and the associated sliding mode controller as

$$w(t) = M \text{ sign } s(t) \tag{8.1.4}$$

Stability of the closed-loop system and tracking of desired $x_d(t)$ are manifested by examination of the Lyapunov function candidate. Although stability analysis in this simple example does not require detailed derivation via a Lyapunov function, the method is used here to maintain a uniform approach throughout the chapter. The candidate is

$$V(t) = \frac{1}{2b} s^2(t) \tag{8.1.5}$$

Differentiation of (8.1.5) along the system trajectories (8.1.1) under control (8.1.4) and without actuator dynamics (8.1.2) yields

$$\begin{aligned} \dot{V}(t) &= \frac{1}{b} s(t)\dot{s}(t) \\ &= g(x, x_d, t)s(t) - M|s(t)| \end{aligned} \tag{8.1.6}$$

where the term

$$g(x, x_d, t) = \frac{\dot{x}_d(t) - ax(t) - d(t)}{b}$$

has an upper bound

$$|g(x, x_d, t)| \leq g^+ = \frac{v_d^+ + a^+x_d^+ + d^+}{b^-} \tag{8.1.7}$$

under the assumption that $x(t) \approx x_d(t)$. For $M \geq g^+ + \xi/\sqrt{2b^-}$ with scalar $\xi > 0$, substitution of the control law (8.1.4) into (8.1.6) leads to

$$\dot{V}(t) \leq -\xi V^{1/2}(t) \tag{8.1.8}$$

which testifies to convergence to $s(t) = 0$ within finite time (Section 3.5 gives more details). Solution of (8.1.8) for an arbitrary initial condition $V(0) > 0$ yields

$$V(t) = \left(-\frac{\xi}{2}t + V^{1/2}(0)\right)^2 \tag{8.1.9}$$

which implies that $V(t)$ is identical to zero after finite time $t_{sm} \geq (2/\xi)V^{1/2}(0)$. Reaching time t_{sm} is a conservative estimate of the maximum time necessary to reach $s(t) = 0$. In practice, sliding mode often occurs earlier.

Subsequently, the system is invariantly confined to the manifold $s(t) = 0$ in (8.1.3) despite parametric uncertainty in a and b and unknown disturbance $d(x, t)$. A block diagram of the ideal sliding mode system is shown in Figure 8.1.

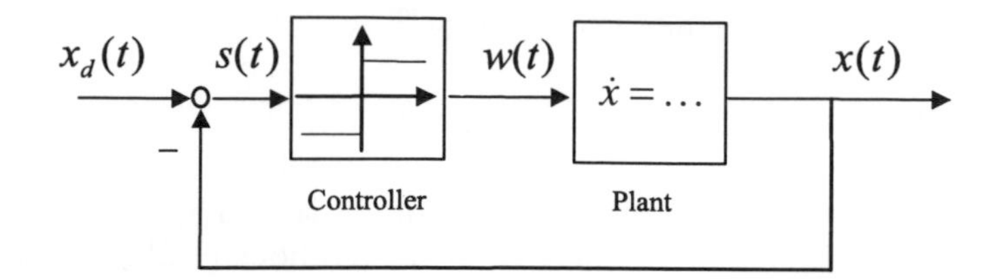

Figure 8.1 Block diagram of ideal sliding mode control loop. A discontinuous controller forces the output $x(t)$ of the plant to exactly track the desired trajectory $x_d(t)$. No chattering occurs since the control loop is free of unmodelled dynamics.

The behaviour of plant (8.1.1) in sliding mode under control (8.1.4) can be examined using the equivalent control method (Section 2.3). Since $s(t)$ is invariantly identical to zero after reaching the sliding manifold, $\dot{s}(t)$ can be formally set to zero. Solving

$$\begin{aligned}\dot{s}(t) &= \dot{x}_d(t) - \dot{x}(t) \\ &= b(g(x, t) - w(t)) \\ &\equiv 0\end{aligned} \tag{8.1.10}$$

for the continuous equivalent control yields

$$w_{eq}(t) = g(x, t) \tag{8.1.11}$$

which can be viewed as an average of the discontinuous control $w(t)$ in (8.1.4). Applying equivalent control $w_{eq}(t)$ to plant (8.1.1) would result in exactly the same motion trajectory as applying discontinuous control $w(t)$ (8.1.4), but this is not possible since $g(x, t)$ contains unknown terms. Substitution of $w_{\text{eq}}(t)$ into (8.1.1) validates the exact tracking performance in sliding mode with $x_d(t) = x(t)$.

For the simulation of (8.1.1) under control (8.1.4) with $M = 2.01$ in Figure 8.2, initial condition $x(0) = 1$ was chosen. After reaching the sliding manifold $s(t) = 0$ at $t \approx 0.45$ s, system trajectory $x(t)$ coincides exactly with desired $x_d(t)$, and control $w(t)$ is switched at very high frequency, creating a solidly black area. For illustration, Figure 8.2(b) shows equivalent control $w_{eq}(t)$ in (8.1.11) as a white dashed line in this black area. Setting the parameter bounds to $a^- = a^+ = a = 0.5$ and $b^- = b^+ = b = 1.0$ results in $g^+ = 2$, which leads to slow convergence to $s(t) = 0$ due to small $\xi \approx 0.014$. This slow convergence was chosen to illustrate the reaching process.

8.1.3 Example system: causes of chattering

In a practical application, unmodelled dynamics in the closed-loop like actuator (8.1.2) often prevent ideal sliding mode from occurring and cause fast, finite-amplitude oscillations. Figurc 8.3 shows a block diagram of the closed control loop including the previously neglected actuator dynamics.

In order to study the causes of these oscillations, first revisit the differences between *continuous* and *discontinuous* systems. In accordance with singular perturbation theory – Kokotovic (1984) gives a comprehensive survey – for systems with *continuous* motion equations, fast motion components like those of actuators for large ω in (8.1.2)

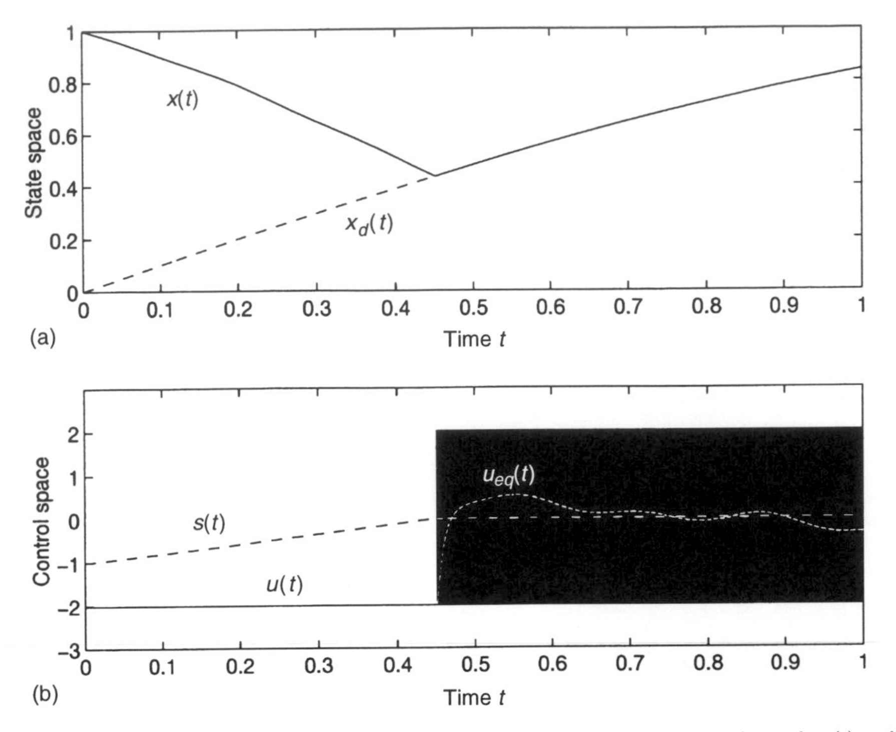

Figure 8.2 Ideal sliding mode in first-order system. State $x(t)$ converges to desired $x_d(t)$ in finite time, i.e. $s(t) = 0$ after $t \approx 0.45$ s. Thereafter, control $u(t)$ switches with infinite frequency and shows as a black area. Equivalent control $u_{eq}(t)$ is drawn as a dashed line. (a) output and desired output, (b) control inputs and sliding variable.

decay rapidly provided they are stable (as is the case for $\omega > 0$). The slow motion component of plant (8.1.1) thus continuously depends on the steady-state solution of (8.1.2). In other words, the algebraic solution of (8.1.2) for $\mu \to 0$ may be substituted into (8.1.1) as an approximation, and continuous control design may very well neglect the actuator dynamics. In the case of (8.1.2) $w = u$ leads to Figures 8.1 and 8.2, as predicted.

In systems with *discontinuities* the solution to the motion equation depends on the small time constants of fast components as well. But unlike in systems with continuous control, discontinuities in the control excite the unmodelled dynamics, leading to oscillations in the state vector. This phenomenon is also known as 'chattering' in the control literature. These oscillations are known to result in low control accuracy, high heat losses in electrical power circuits and high wear of moving mechanical parts.

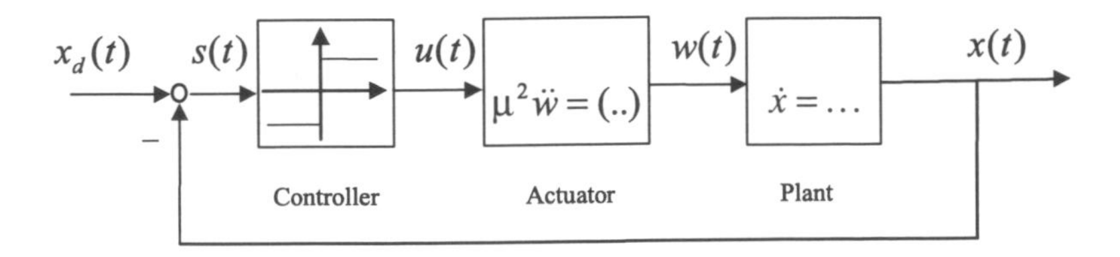

Figure 8.3 Control loop with actuator dynamics neglected in ideal control design. Sliding mode does not occur since the actuator dynamics are excited by the fast switching of the discontinuous controller, leading to chattering in the loop.

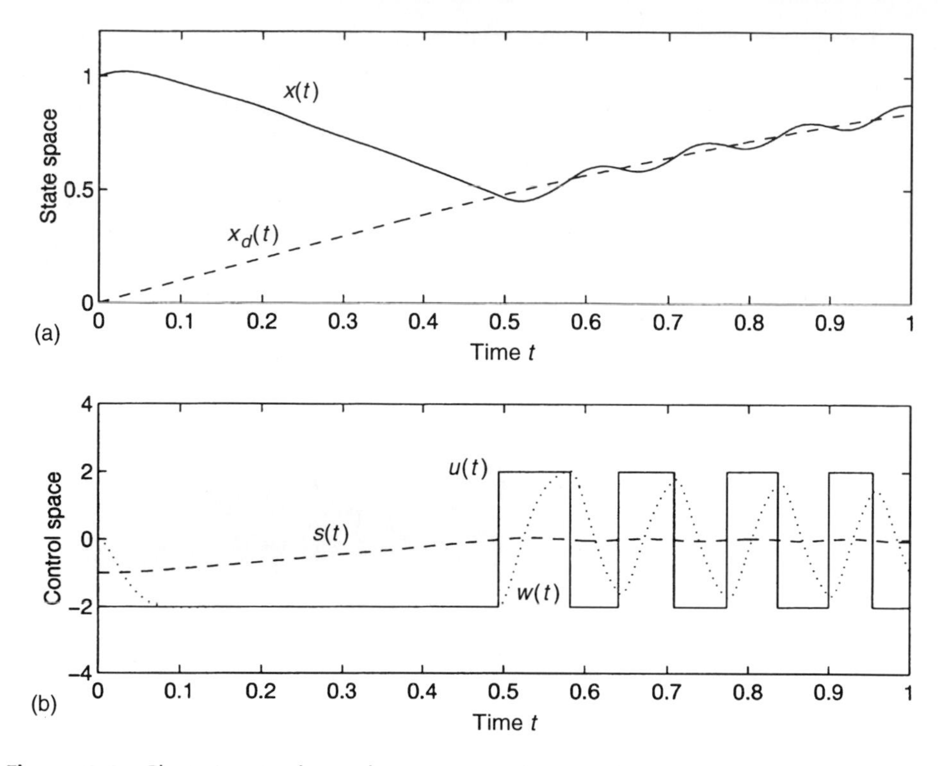

Figure 8.4 Chattering in first-order system with second-order actuator dynamics under discontinuous control. After switches in control $u(t)$, actuator output $w(t)$ lags behind, leading to oscillatory system trajectories. (a) output and desired output, (b) control inputs and sliding variable.

Figure 8.4 shows the chattering behaviour of system (8.1.1) under control (8.1.4), but with actuator dynamics (8.1.2) in the loop of Figure 8.3. Figure 8.4(a) depicts output $x(t)$ oscillating around the desired $x_d(t)$ after $t \approx 0.5$ s. In Figure 8.4(b), control $u(t)$ switches with finite frequency (solid line), whereas output $w(t)$ of the actuator (dotted line) clearly cannot follow the steps in control command $u(t)$. Note that an increase of the actuator bandwidth would increase the frequency of the square-wave behaviour of $u(t)$, but would not be able to eliminate the oscillations. In fact, when a mechanical device is performing oscillations similar to Figure 8.4, audible noise often results at high frequency, which led to the name 'chattering'. Chattering is extremely harmful to mechanical system components.

For further study of the nature of the oscillations, consider the situation immediately after a switch in control $u(t)$ from $u(t_{sw}^-) = -M$ to $u(t_{sw}^+) = M$ at time t_{sw}. Immediately after the switch, input $u(t_{sw}^+)$ and output $w(t_{sw}^+)$ of the actuator differ by $2M$. Although the discrepancy between $u(t)$ and $w(t)$ decreases after the switch at a rate faster than the motion of system (8.1.1), $u(t)$ and $w(t)$ are not 'close' in the sense of singular perturbation theory (Utkin, 1993). Consequently, the small time constants cannot be neglected when examining the behaviour of a system with discontinuities in the motion equation.

Consider the system trajectories in Figure 8.4. Initially, $x(t)$ converges to $x_d(t)$ until $t \approx 0.5$ s. Thereafter, instead of tracking $x_d(t)$ exactly as in the ideal case shown in Figure 8.2, $x(t)$ goes through cycles of divergence and convergence. This suggests that the manifold $s(t) = x_d(t) - x(t) = 0$ is attractive for large deviations, but the trajectories might diverge in some small vicinity $\varepsilon(\mu)$ of $s(t) = 0$, where scalar ε depends on the bandwidth of the unmodelled actuator dynamics (8.1.2). The motion trajectory is ultimately confined to this vicinity, i.e. $|s(t)| \leq \varepsilon$; however, inside the ε-vicinity, oscillations of finite frequency and finite amplitude occur.

Stability for large deviations, i.e. for $|s(t)| > \varepsilon$, can be illustrated using the fact that $u(t)$ is constant for $|s(t)| > \varepsilon$. The actuator dynamics (8.1.2) decay rapidly since they are stable and $w(t) \approx u(t)$ after some short time interval. In the example of Figure 8.4, the actuator dynamics have decayed after less than 0.1 s. Hence the stability analysis of (8.1.5) to (8.1.8) can be used to establish convergence of the system trajectories to $s(t) = 0$ until the first switching of $u(t)$ takes place (at $t \approx 0.5$ s in Figure 8.4).

In order to examine the subsequent system behaviour, assume steady-state conditions with $u(t) = w(t) = -M$ for $|s(t)| > \varepsilon$. The step response of the actuator for the first switch at t_{sw} from $u(t) = -M$ to $u(t) = +M$ at $s(t) = 0$ is given by

$$w(t) = M\left(1 - 2\left(\frac{t - t_{sw}}{\mu} + 1\right)e^{-(t-t_{sw})/\mu}\right) \tag{8.1.12}$$

For some initial time interval $\Delta t = t - t_{sw}$, actuator output $w(t) < u(t) = M$ and $\dot{V}(t) > 0$ in (8.1.6) results for the case $g(x, x_d, t)s(t) > 0$. It is only after the decay of the exponential term in (8.1.12), i.e. after $\Delta t(\mu)$, that $|w(t)| > g^+$ is established once more and $\dot{V}(t) < 0$ indicates convergence to sliding manifold $s(t) = 0$. During the time interval Δt, the maximum deviation from ideal tracking can be approximated by

$$|\Delta s| \leq \varepsilon(\mu) = (g^+ + M)\Delta t(\mu) \tag{8.1.13}$$

Similar derivations hold for the next switch from $u(t) = +M$ to $u(t) = -M$.

Summarizing the above shows that, in the non-ideal system, $s(t)$ is converging towards zero for $|s(t)| > \varepsilon$. Hence, for large deviations from the sliding manifold, system (8.1.1) with unmodelled dynamics (8.1.2) under control (8.1.4) behaves similar to the ideal system, converging to the sliding manifold. Therefore its motion is ultimately confined to $|s(t)| \leq \varepsilon$ after some finite time interval. Inside the $\varepsilon(\mu)$-vicinity, stability cannot be guaranteed. In fact, temporary *divergence* can be shown for $|s(t)| < \varepsilon$.

To qualitatively illustrate the influence of unmodelled dynamics on the system behaviour, consider the simplest case $a = 0$, $d(x, t) = 0$, $b = 1$, $x_d(t) = 0$ in (8.1.1) and (8.1.3) as shown in Figure 8.5.

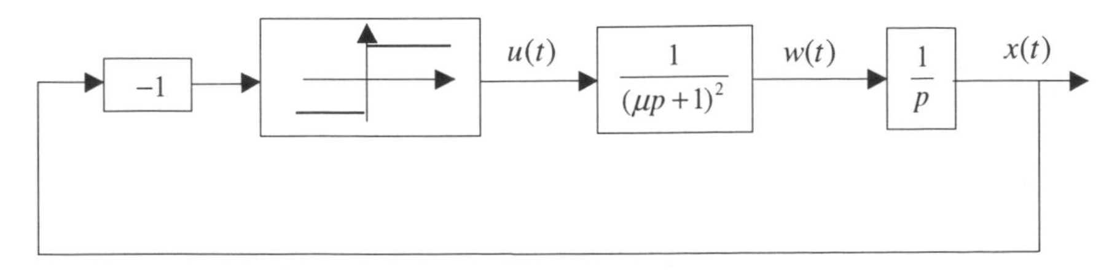

Figure 8.5 Block diagram to illustrate divergence within ε-vicinity of sliding manifold.

The motion equations may be written in the form

$$\begin{aligned} \dot{x} &= w \\ \dot{w} &= v \\ \dot{v} &= -\frac{2}{\mu}v - \frac{1}{\mu^2}w + \frac{1}{\mu^2}u \end{aligned} \tag{8.1.14}$$

For the control $u = -M\,\mathrm{sign}(x)$, the sign-varying Lyapunov function

$$V = xv - 0.5w^2 \tag{8.1.15}$$

has a negative time derivative

$$\dot{V} = x\left(-\frac{2}{\mu}v - \frac{1}{\mu^2}w + \frac{1}{\mu^2}u\right) \tag{8.1.16}$$

for small magnitudes of v and w. This means that the motion is unstable in an $\varepsilon(\mu)$-order vicinity of the manifold $s(x) = x = 0$.

As an alternative to the block diagram of Figure 8.5, system (8.1.14) may be represented by Figure 8.6.

The motion equations may now be written as

$$\begin{aligned} \dot{x}^* &= -M\,\mathrm{sign}(x) \\ \mu^2\ddot{x} + 2\mu\dot{x} + x &= x^* \end{aligned} \tag{8.1.17}$$

Sliding mode cannot occur in the systems since the time derivative $\dot{x}$ is a continuous time function and cannot have its sign opposite to x in the vicinity of the point $x = 0$, where the control undergoes discontinuities.

The value of $\dot{x}^*$ is bounded and, as follows from singular perturbation theory (Kokotovic *et al.*, 1984; Kokotovic, 1996), the difference between x and x^* is of μ-order. The signs of x and x^* coincide beyond the $\varepsilon(\mu)$-vicinity of $s(x) = x = 0$, hence the magnitudes of x^* and x decrease, i.e. the state trajectories converge to this vicinity and after a finite time interval t_1 the state remains in the vicinity. According to the analysis of (8.1.14), the motion in the vicinity of $x = 0$ is unstable.

Local instability explains why chattering may appear in systems with discontinuous controls where there are unmodelled dynamics. The high-frequency oscillations in the discontinuous control system may be analyzed in the time domain as well. These brief periods of divergence occur after switches of the control input variable $u(t)$ when the output $w(t)$ of the actuator is unable to follow the abrupt change of the control command. The proposed solutions to the chattering problem thus focus

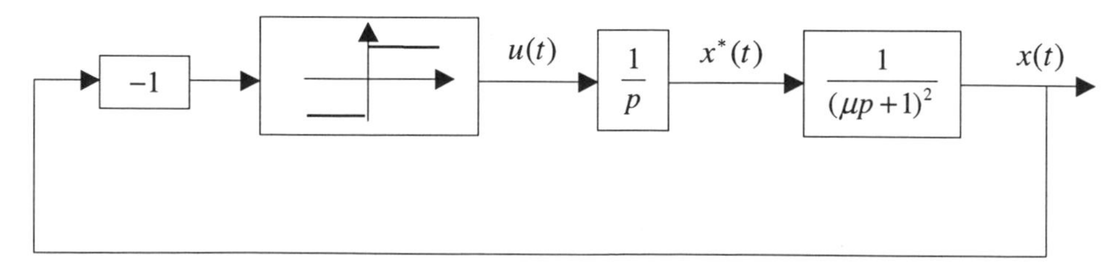

Figure 8.6 Alternative representation for the example of Figure 8.5.

on either avoiding control discontinuities in general or they move the switching action to a controller loop without any unmodelled dynamics. The rest of this chapter discusses various types of chattering prevention schemes and examines their respective benefits.

8.2 Boundary layer solution

The boundary layer solution (Slotine and Sastry, 1983; Slotine 1984), seeks to avoid control discontinuities and switching action in the control loop. Control law (8.1.4) is replaced by a saturation function which approximates the sign(s) term in a boundary layer of the manifold $s(t) = 0$. Numerous types of saturation function sat(s) have been proposed in the literature.

Considering the problem 'in the large', i.e. for $|s(t)| > \varepsilon$, we have $\text{sat}(s) = \text{sign}(s)$. However, in a small ε-vicinity of the origin, the so-called boundary layer, $\text{sat}(s) \neq \text{sign}(s)$ is continuous. As an illustrative example, consider a simple linear saturation function

$$u(t) = \begin{cases} M\ \text{sign}(s(t)) & \text{for } |s(t)| > \varepsilon \\ \dfrac{M}{\varepsilon} s(t) & \text{for } |s(t)| \leq \varepsilon \end{cases} \tag{8.2.1}$$

with linear proportional feedback gain M/ε within the boundary layer in the vicinity of the origin, $|s(t)| \leq \varepsilon$, and symmetrically saturated by M for $|s(t)| > \varepsilon$ outside the boundary layer. A block diagram of this system under control (8.2.1) is shown in Figure 8.7.

For a stability analysis, substitute (8.2.1) into (8.1.6) instead of (8.1.4) to yield

$$\dot{V}(t) \leq \begin{cases} -\xi V^{1/2}(t) & \text{for } |s(t)| > \varepsilon \\ s(t)\left(g^{+} - \dfrac{M}{\varepsilon} s(t)\right) & \text{for } |s(t)| \leq \varepsilon \end{cases} \tag{8.2.2}$$

Direct examination of (8.2.2) shows similar stability properties as (8.1.8) for $|s(t)| > \varepsilon$ and undetermined stability for $|s(t)| \leq \varepsilon$. Hence the system trajectories are guaranteed to converge to the boundary layer. In this simple example the system is continuous and linear within the boundary layer, so linear control theory can be used to further study the stability. Substituting $u(t) = (M/\varepsilon)s(t)$ into (8.1.2) and (8.1.1) with (8.1.3) yields

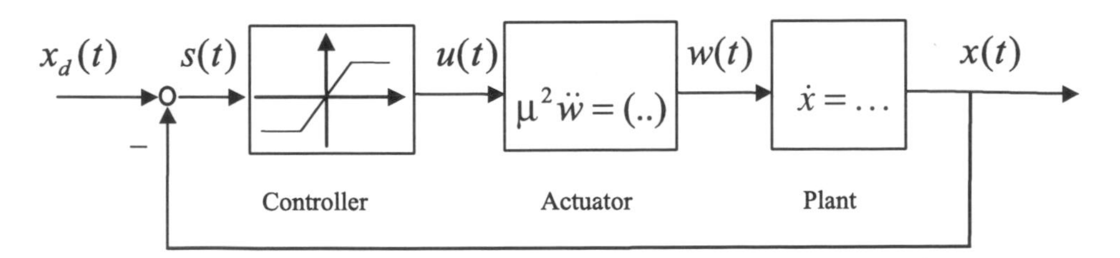

Figure 8.7 Saturation function replaces discontinuous controller. Instead of achieving ideal sliding mode, the system trajectories are confined to a boundary layer of the manifold $s(t) = 0$.

Laplace-domain expression

$$\left(\mu^2 p^3 + \mu(2 - a\mu)p^2 + (1 - 2a\mu)p + \left(b\frac{M}{\varepsilon} - a\right)\right)x(t) = h(x, x_d, t)$$
$$h(x, x_d, t) = b\frac{M}{\varepsilon}x_d(t) + (\mu p + 1)^2 d(x, t) \tag{8.2.3}$$

where $h(x, x_d, t)$ can be interpreted as a disturbance to the left-hand side of the first row in (8.2.3). The Hurwitz stability bounds for the left-hand side of the first equation in (8.2.3) are given as

$$a\mu < 1/2 \tag{8.2.4}$$

and

$$\frac{M}{\varepsilon} < \frac{2}{b\mu}(1 - a\mu)^2 \tag{8.2.5}$$

The first stability bound (8.2.4) states that the unmodelled dynamics have to be stable and faster than the system dynamics (8.1.1) themselves. The stability boundary (8.2.5) defines the highest feedback gain the system (8.1.1) with actuator dynamics (8.1.2) can sustain in the linear sense. Higher gains, in particular theoretically infinite gains of discontinuous sliding mode controllers, result in instability in the vicinity of $s(t) = 0$, causing chattering as shown in Section 8.1.

Furthermore, for oscillation-free trajectories with critically damped eigenvalues in (8.2.3), we require

$$\frac{M}{\varepsilon} < \frac{3\sqrt{2} - 4}{b\mu}(1 - 2a\mu)^2$$

It is interesting to note that in this simple example the boundary layer width ε depends almost linearly on the actuator time constant μ and inverse linearly on the available control resources M. A value $\varepsilon = 0.1$ was chosen for the simulation in Figure 8.8, which leads to stable, but less than critically damped eigenvalues. Consequently, a small overshoot results when $x(t)$ converges to $x_d(t)$.

One of the benefits of the boundary layer approach is that sliding mode control design methodologies can be exploited to derive a *continuous* controller. The invariance property of sliding mode control is partially preserved in the sense that the system trajectories are confined to a $\delta(\varepsilon)$-vicinity of the sliding manifold $s(t) = 0$, instead of exactly to $s(t) = 0$ as in ideal sliding mode (Figures 8.1 and 8.2). Within the $\delta(\varepsilon)$-vicinity, however, the system behaviour is not determined, i.e. further convergence to zero is not guaranteed. This type of control design is part of a class of robust controllers which satisfy the 'globally uniform ultimate boundedness' condition proposed by Leitmann (1981). Note that no *real* sliding mode takes place, since the switching action is replaced by a continuous approximation.

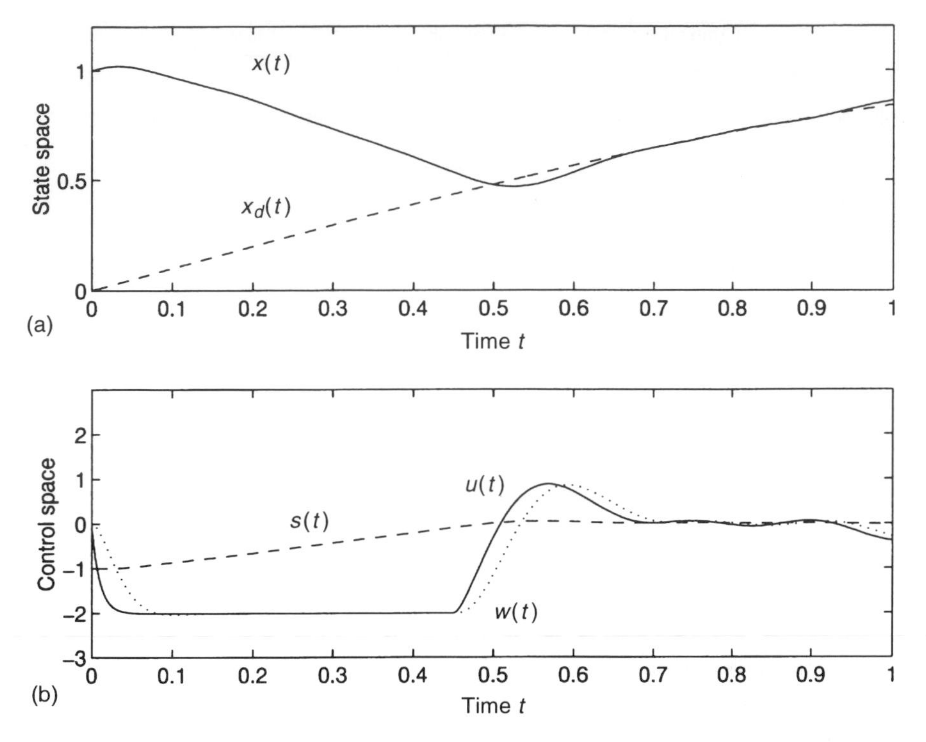

Figure 8.8 Saturation function approximating control discontinuity in the boundary layer yields chattering-free system trajectories. State $x(t)$ converges to $x_d(t)$, but does not track exactly as in ideal sliding mode: (a) output and desired output, (b) control inputs and sliding variable.

8.3 Observer-based solution

The boundary layer approach discussed above avoids generating sliding mode by replacing the discontinuous switching action with a continuous saturation function. But many systems have inherent control discontinuities, e.g. various voltage inputs of power converters or electric drives. When implementing a continuous controller, a technique like pulse-width modulation (PWM) has to be exploited to adapt the control law to the discontinuous system inputs. Recent advances in high-speed circuitry coupled with insufficient linear control methodologies for internally nonlinear high-order plants, such as AC motors, have meant that sliding mode control has become increasingly popular. Commercially available electronic converters enable switching frequencies to be handled in the range of hundreds of kilohertz. Hence it seems unjustified to bypass a system's discontinuous control inputs by converting a continuous controller, e.g. via a PWM scheme. Rather, such system specifications call for alternative methods to prevent chattering while preserving control discontinuities.

An asymptotic observer in the control loop can eliminate chattering despite discontinuous control laws. The key idea, as proposed by Bondarev *et al.* (1985), is to generate ideal sliding mode in an auxiliary observer loop rather than in the main control loop. Ideal sliding mode is possible in the observer loop since it is entirely generated in the control software and thus does not contain any unmodelled dynamics.

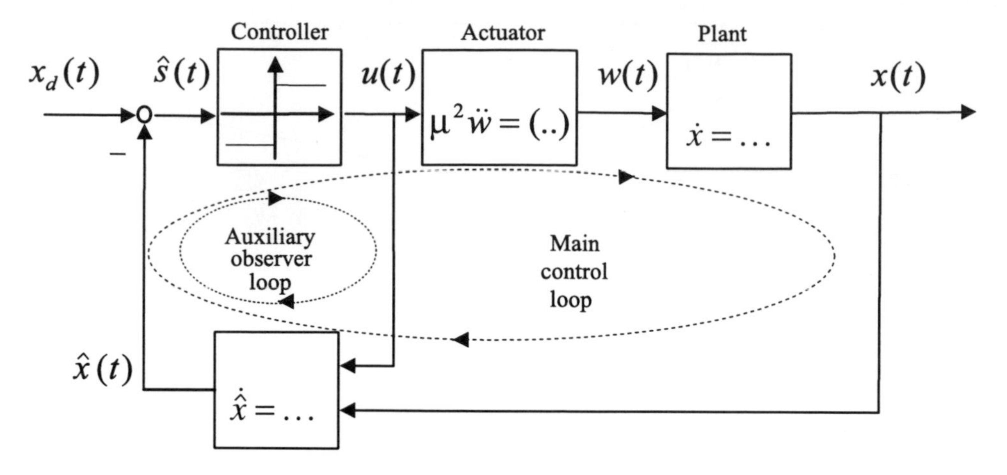

Figure 8.9 Control loop with auxiliary observer loop. Ideal sliding mode occurs in observer manifold $\hat{s}(t) = 0$ since the observer loop is free of unmodelled dynamics. The plant output $x(t)$ follows the observer output $\hat{x}(t)$ without chattering despite discontinuous control $u(t)$ applied to the main loop with actuator dynamics.

The main loop follows the observer loop according to the observer dynamics. Despite applying a discontinuous control signal with switching action to the plant, no chattering occurs and the system behaves as if an equivalent continuous control was applied. Figure 8.9 shows a block diagram for the example used in this chapter plus an auxiliary observer.

Define a first-order observer for system (8.1.1) as

$$\dot{\hat{x}}(t) = ax(t) + bu(t) + L_1\bar{x}(t) \tag{8.3.1}$$

where L_1 is the linear feedback gain for the observation error $\bar{x}(t) = x(t) - \hat{x}(t)$. Exact knowledge of the system parameters a and b is assumed in (8.3.1) for ease of presentation. In the case of parametric uncertainty, parameter estimates $\hat{a}$ and $\hat{b}$ replace a and b in (8.3.1) and a more complex analysis results.

The linear dynamics of the observation error are governed by

$$\dot{\bar{x}}(t) = d(x, t) - L_1\bar{x}(t) \tag{8.3.2}$$

Error $\bar{x}(t)$ in (8.3.2) is stable and bounded by

$$|\bar{x}(t)| \leq \frac{d^+}{L_1} \tag{8.3.3}$$

with the disturbance bounded by $|d(x, t)| \leq d^+$. Introducing an observer sliding manifold

$$\hat{s}(t) = x_d(t) - \hat{x}(t) \tag{8.3.4}$$

allows definition of an ideal sliding mode controller for the observer loop as

$$u(t) = M \text{ sign } \hat{s}(t) \tag{8.3.5}$$

to replace (8.1.3) and (8.1.4). Stability of the auxiliary observer loop is examined via a similar Lyapunov function candidate as in (8.1.5):

$$\hat{V}(t) = \frac{1}{2b}\hat{s}^2(t) \tag{8.3.6}$$

Substitution of (8.3.1) under control (8.3.5) into the time derivative of (8.3.6) reveals

$$\begin{aligned}\dot{\hat{V}}(t) &= \frac{1}{b}\hat{s}(t)\dot{\hat{s}}(t)\\ &= \frac{1}{b}(\dot{x}_d(t) - ax(t) - L_1\bar{x}(t))\hat{s}(t) - M|\hat{s}(t)|\\ &\le \frac{1}{b^-}(v_d^+ + a^+x_d^+ + d^+)|\hat{s}(t)| - M|\hat{s}(t)|\end{aligned} \tag{8.3.7}$$

where observation error bound (8.3.3) was used to reduce the expression. Bound (8.1.7) of the independent observer error system and condition $M \ge g^+ + \xi/\sqrt{2b^-}$ lead to

$$\dot{\hat{V}}(t) \le -\xi\hat{V}^{1/2}(t) \tag{8.3.8}$$

under similar assumptions as for (8.1.8). Sliding mode is established in the observer loop after finite time as in (8.1.9), and $\hat{s}(t) = 0$ holds exactly thereafter.

In order to examine the behaviour of the overall system under sliding mode in the auxiliary observer loop, the equivalent control method is employed. Solving

$$\dot{\hat{s}}(t) = \dot{x}_d(t) - ax(t) - bu(t) - L_1\bar{x}(t) \equiv 0 \tag{8.3.9}$$

for the equivalent control of input $u(t)$ yields

$$bu_{\text{eq}}(t) = \dot{x}_d(t) - ax(t) - L_1\bar{x}(t) \tag{8.3.10}$$

Substitution of (8.3.10) into plant (8.1.1) with actuator dynamics (8.1.2) leads to

$$\begin{aligned}(\mu^2p^3 + \mu(2 - a\mu)p^2 + (1 - 2a\mu)p + L_1)x(t) &= h^*(x, x_d, t)\\ h^*(x, x_d, t) &= (p + L_1)x_d(t) + (\mu p + 1)^2 d(x, t)\end{aligned} \tag{8.3.11}$$

Equation (8.3.11) is similar to (8.2.3) except for two details. First, the desired trajectory $x_d(t)$ enters the disturbance function $h^*(x, x_d, t)$ in a different fashion then $h(x, x_d, t)$. Second, the left-hand side of stability bound (8.2.5) is modified to

$$L_1 + a < \frac{2}{\mu}(1 - a\mu)^2 \tag{8.3.12}$$

The analytical similarities between the boundary layer approach and the asymptotic observer-based solution in this simple first-order example are also apparent in the system trajectories. Compare the simulation of the observer-based approach in Figure 8.10 with the boundary layer approach simulation in Figure 8.8. Both show similar behaviour with a small overshoot when reaching the sliding manifold. Note that for the observer-based solution, the observed state $\hat{x}(t)$ achieves ideal sliding mode with discontinuous switching action (black area in Figure 8.10(b)), whereas the true state $x(t)$ follows the observer dynamics without exhibiting chattering.

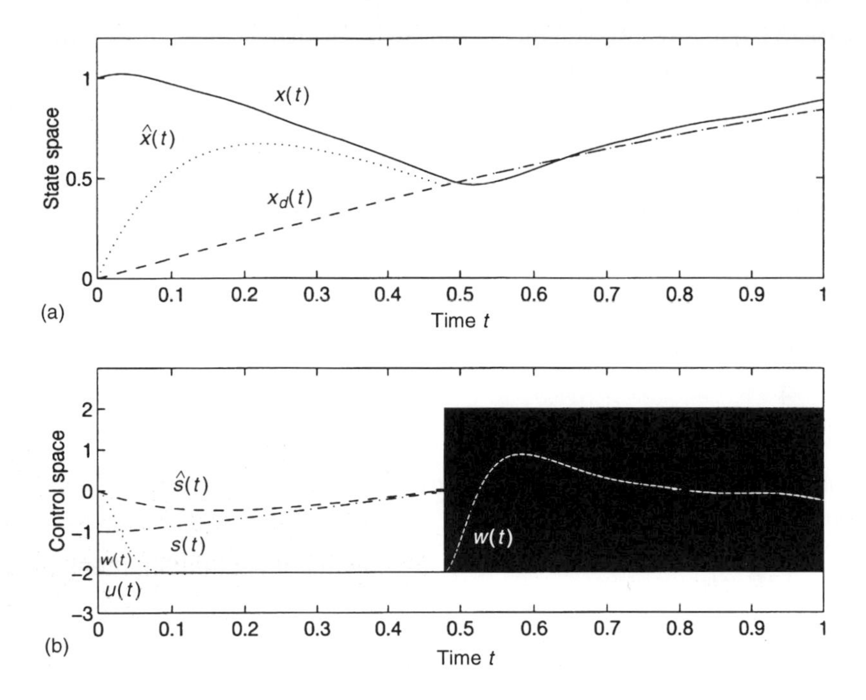

Figure 8.10 Observer in auxiliary control loop enables chattering-free system trajectories despite discontinuous control switching after sliding manifold $\hat{s}(t) = 0$ is reached. Plant output $x(t)$ tracks observer output $\hat{x}(t)$ according to observer error dynamics. (a) output, desired output and observed output; (b) control inputs and sliding variable.

The observer-based solution requires slightly more effort in the control design. However, in many control applications, observers for unmeasurable states are vital parts of the overall system and can be readily included into the control design. Note that the design of the actual observer depends on the system specifications; both full-state and reduced-order observers may be employed. Furthermore, observers provide more flexibility. For instance, in the example studied in this chapter, the observer (8.3.1) may be extended to include an estimate of the disturbance under the assumption that $\dot{d}(t)$ is small:

$$\begin{aligned} \dot{\hat{x}}(t) &= ax(t) + \hat{d}(t) + bu(t) + L_1\bar{x}(t) \\ \dot{\hat{d}}(t) &= \qquad\qquad\qquad\qquad L_2\bar{x}(t) \end{aligned} \tag{8.3.13}$$

where L_1 and L_2 determine the observer dynamics. The simulation in Figure 8.11 shows that the tracking performance of the extended observer (8.3.13) is improved compared to the initial design (8.3.1), i.e. $\hat{x}(t)$ tracks $x(t)$ closer before and after reaching the sliding manifold.

8.4 Regular form solution

Both the boundary layer approach and the observer-based solution to the chattering problem assume that the 'unmodelled' dynamics are completely unknown. In practical applications, however, at least partial information about unmodelled dynamics, in

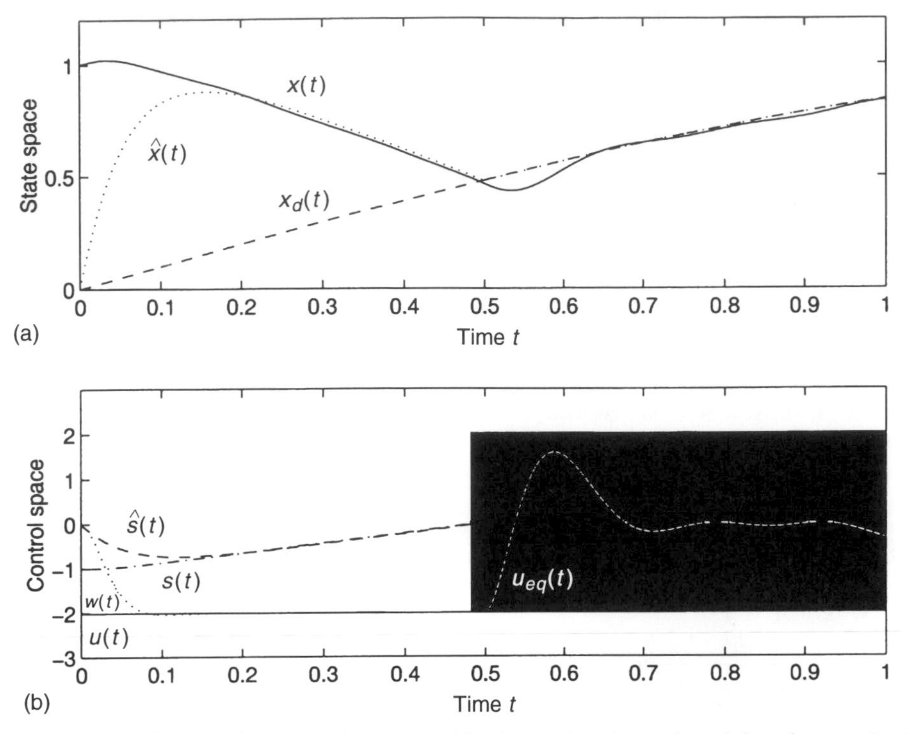

Figure 8.11 Tracking performance is improved by increasing the order of the observer in the auxiliary control loop. After sliding manifold $\hat{s}(t) = 0$ is reached, plant output $x(t)$ closely tracks observer output $\hat{x}(t)$ and desired output $x_d(t)$. (a) output, desired output and observed output; (b) control inputs and sliding variable.

particular of actuators, is often available together with measurements of the actuator outputs. Models are readily available for electric drives, but they may contain uncertain parameters. Thus, in the design of a controller for the overall system, these dynamics can be included into the control design to enhance the performance of the overall system.

Since the actuator dynamics and the plant dynamics are block separated, i.e. the outputs of the actuators are the inputs of the plant, a cascaded control structure can be designed following the regular form approach or the block control principle; see Section 3.3, Section 5.6 or Drakunov *et al.* (1990). The basic idea is to design a cascaded controller in two steps. In the first step, a continuous controller is derived for the plant under the assumption that the plant inputs are the actual control inputs to the overall system, defining 'desired' actuator outputs $w_d(t)$. In the second step, the actuator inputs $u(t)$, i.e. the real control inputs of the system, are used to ensure the actuator outputs track the desired outputs exactly via sliding mode control with $w(t) = w_d(t)$. This approach is a special case of cascaded control structures as applied by the block control principle (Drakunov *et al.*, 1984, 1990) and the integrator backstepping method (Krstic *et al.*, 1995).

The regular form approach to prevent chattering is especially intriguing for systems with electrical actuators where control input discontinuities are often imposed by the system specifications. Particularly for electromechanical systems, the benefits of sliding mode control can be fully exploited based on the wealth of available control designs for electric drives and power converters (Chapters 10 and 11).

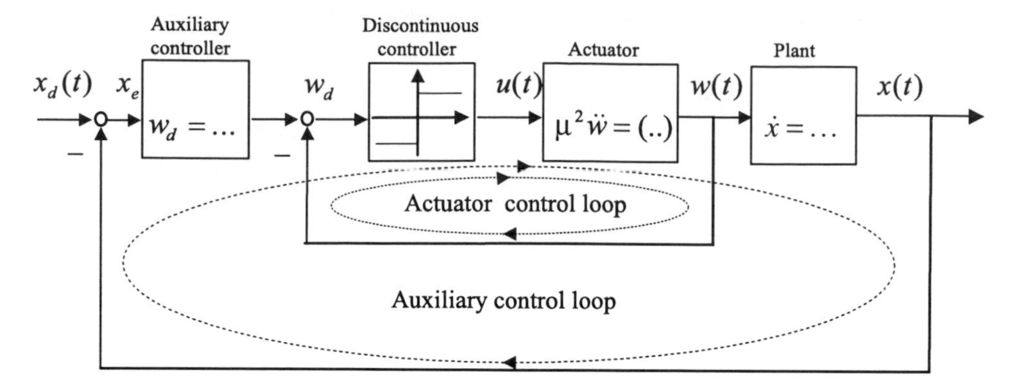

Figure 8.12 Cascaded controller with continuous auxiliary control and discontinuous actuator control loop.

Assume the main source of unmodelled dynamics in a control loop is the actuator, as depicted in Figure 8.3. Further assume availability of a model for actuator (8.1.2), e.g. with uncertain parameters as

$$w(t) = \frac{\hat{\omega}^2}{p^2 + 2\hat{\omega}p + \hat{\omega}^2} u(t) = \frac{1}{(\hat{\mu}p + 1)^2} u(t) \tag{8.4.1}$$

where $\hat{\omega} = 1/\hat{\mu}$ is an estimate for the actuator bandwidth, $u(t)$ is the control input to the overall system and $w(t)$ is the measurable actuator output. For more realistic actuator dynamics, refer to Chapter 10. A block diagram of the complete control system is shown in Figure 8.12. Note that the regular form approach is not applicable to systems with unmodelled dynamics mainly introduced by sensors rather than by actuators because measurement of both inputs and outputs of the unmodelled dynamics is required. Since sensor inputs are usually not available via measurements, other methods such as the observer-based approach should be employed to prevent chattering.

In the first design step, a continuous auxiliary control law $w_d(t)$ is derived for plant (8.1.1) to track desired trajectory $x_d(t)$. Any control design methodology can be employed, linear or nonlinear, but it is important to ensure trackability in the face of limited bandwidth actuator dynamics. In other words, the rate of the auxiliary controller should be bounded as

$$|\dot{w}_d(t)| \leq \dot{w}_d^+ \tag{8.4.2}$$

Depending on the order of the actuator dynamics, additional bounds on higher time derivatives of $w_d(t)$ might be necessary. For system (8.1.1) a first-order linear controller

$$w_d(t) = C(x_d(t) - x(t)) = Cx_e(t) \tag{8.4.3}$$

with proportional gain $C > 0$ may be used to yield error dynamics

$$\dot{x}_e(t) = b(-Cx_e(t) + g(x, x_d, t)) \tag{8.4.4}$$

where an upper bound g^+ for disturbance function $g(x, x_d, t)$ is given in (8.1.7). Error dynamics (8.4.4) are stable, but are disturbed by $g(x, x_d, t)$. If available, the performance of controller (8.4.3) can be improved by feedforward of a disturbance estimate $\hat{g}(x, x_d, t)$ as

$$w_d(t) = Cx_e(t) + \frac{1}{\bar{b}}\hat{g}(x, x_d, t) \tag{8.4.5}$$

The second design step is to drive the error $w_e(t) = w_d(t) - w(t)$ between desired output $w_d(t)$ in (8.4.3) or (8.4.5) and actual output $w(t)$ to zero. Since this inner control loop is free of unmodelled dynamics, a discontinuous sliding mode controller can be designed as

$$u(t) = (\hat{\mu}p + 1)^2 w_d(t) + M \text{ sign } s(t) \tag{8.4.6}$$

with second-order sliding variable

$$s(t) = K\dot{w}_e(t) + w_e(t) \qquad (K > 0) \tag{8.4.7}$$

Control (8.4.6) and (8.4.7) assume that the first and second time derivatives of $w_d(t)$ in (8.4.3) or (8.4.5) and the first time derivative of the actuator output, $\dot{w}(t)$, are available, e.g. from an observer. Controller (8.4.6) leads to the inner loop error dynamics

$$\mu^2\ddot{w}_e(t) + 2\mu\dot{w}_e(t) + w_e(t) + M \text{ sign } s(t) = p(\bar{\mu}^2 p + \bar{\mu})w_d(t) \tag{8.4.8}$$

where $\bar{\mu} = \mu - \hat{\mu}$ is the estimation error of the actuator dynamics and p is the Laplace variable. Following the conventional design methodology of sliding mode control, it can be shown that the values of $s(t)$ and $\dot{s}(t)$ have different signs for bounded $\dot{w}_d(t)$, $\ddot{w}_d(t)$ and a sufficiently high but finite magnitude of control resource M. Hence sliding mode is enforced in the manifold $s(t) = 0$ with $w_e(t)$ decaying to zero as determined by K in (8.4.7). Consequently, desired control (8.4.3) or (8.4.5) is directly implemented to plant (8.1.1).

In the simulation shown in Figure 8.13, the desired actuator output was limited according to the real actuator limits, i.e. $-M \leq w_d(t) \leq M$. The sliding variable in (8.4.7) converges to zero within finite time. Thereafter, $s(t) = 0$ and the control switches at (theoretically) infinite frequency, resulting in the black solid area in Figure 8.13(b). At the same time, $x(t)$ converges to $x_d(t)$, but the tracking is not exact due to linear control (8.4.3).

8.5 Disturbance rejection solution

The regular form solution in the previous section relies on a continuous controller to achieve tracking of the desired trajectory $x_d(t)$ by the output $x(t)$ of plant (8.1.1). The linear controller (8.4.3) was augmented by an estimate of the disturbance $g(x, x_d, t)$ in (8.4.5). Often such an estimate is not readily accessible. The disturbance rejection approach discussed in this section provides a means to obtain an accurate disturbance estimate while avoiding chattering in the main control loop. This approach can be viewed as a special case of so-called integral sliding mode. A more mathematical background of integral sliding mode is described in Sections 6.3 and 7.5; an application example is discussed in Section 12.4.1.

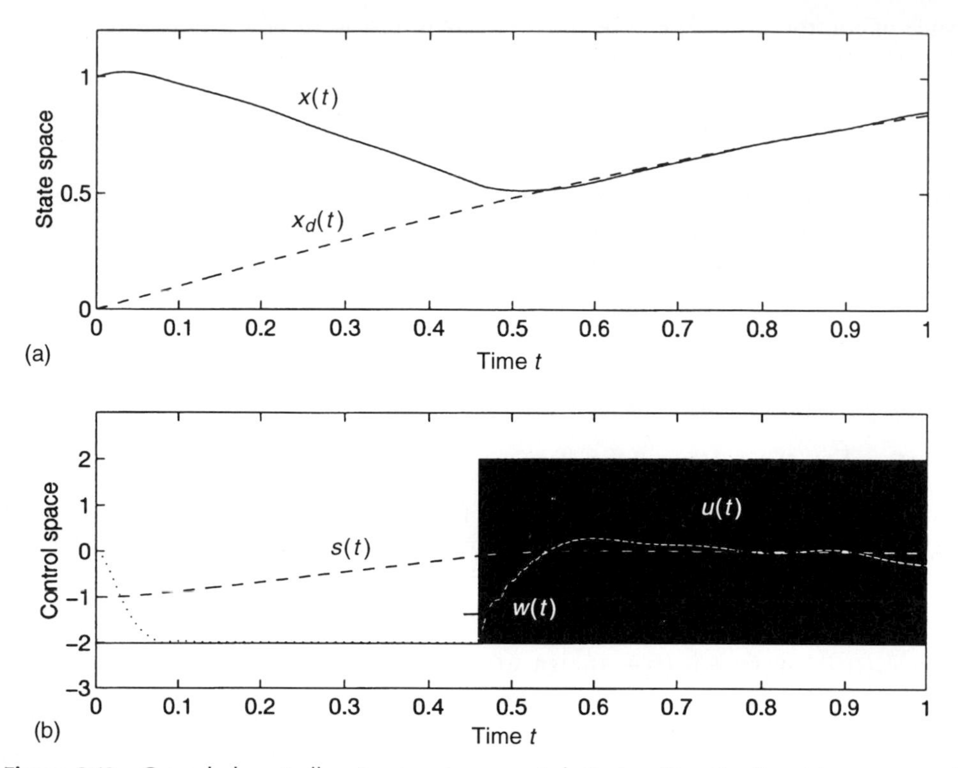

Figure 8.13 Cascaded controller structure to prevent chattering. Despite discontinuous control action in the actuator control loop, plant output $x(t)$ follows desired trajectory $x_d(t)$ without oscillations. (a) output and desired output, (b) control inputs and sliding variable.

The main idea of disturbance rejection via sliding mode is to compose the overall controller of a continuous part and a discontinuous part, i.e.

$$u(t) = u_c(t) + u_d(t) \tag{8.5.1}$$

The idea of combining a continuous part and a discontinuous part for the control input has been employed by many authors; there is a survey in DeCarlo *et al.* (1988). The continuous component $u_c(t)$ is used to control the overall behaviour of the system; the discontinuous component $u_d(t)$ is used to reject disturbances and to suppress parametric uncertainties. A block diagram is shown in Figure 8.14.

For system (8.1.1), assume that the desired trajectory $x_d(t)$ is known but the disturbance $d(x, t)$ is unknown. Also assume that parameter b is known but parameter a is entirely unknown. A continuous controller with linear feedback and feedforward of the desired trajectory $x_d(t)$ can then be designed as

$$u_c(t) = \frac{1}{b}(Cx_e(t) + \dot{x}_d(t)) \tag{8.5.2}$$

where $C > 0$ is a proportional feedback gain for the tracking error $x_e(t) = x_d(t) - x(t)$. Substitute (8.5.2) with the disturbance rejection term $u_d(t)$ set to zero, i.e. $u(t) = u_c(t)$,

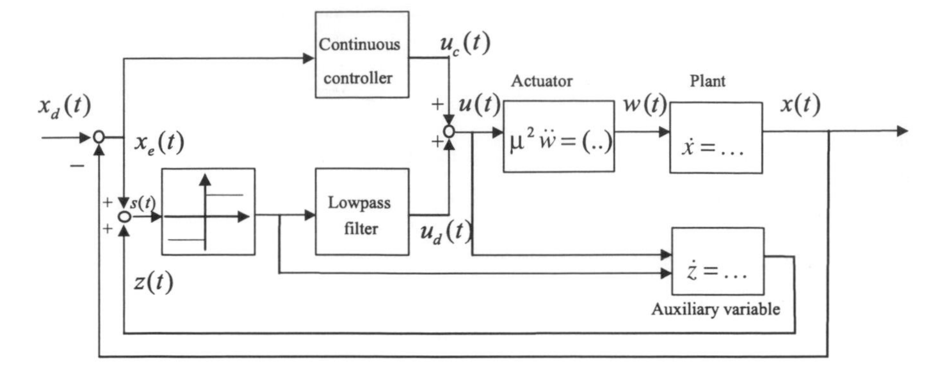

Figure 8.14 Disturbance rejection via sliding mode with auxiliary controller loop to avoid chattering. A continuous controller $u_c(t)$ is augmented by a disturbance rejection controller $u_d(t)$, derived from a lowpass-filtered discontinuous controller for an auxiliary control variable $z(t)$.

into system dynamics (8.1.1) *without* actuator dynamics (8.1.2); this yields stable error dynamics

$$\dot{x}_e(t) + Cx_e(t) = f(x, t) \tag{8.5.3}$$

which are perturbed by the disturbance function $f(x, t) = -ax(t) - d(x, t)$. Since disturbance $f(x, t) \neq 0$, the tracking error $x_e(t)$ does not go to zero. A simulation in Figure 8.15 illustrates this with a 'weak' feedback gain C, leading to imperfect tracking of the desired trajectory $x_d(t)$.

In order to reduce the influence of disturbance $f(x, t)$ on the tracking performance, the second term in controller (8.5.1) is designed as a disturbance estimator using sliding mode control. Define a sliding manifold as

$$s(t) = x_e(t) + z(t) \tag{8.5.4}$$

where $z(t)$ is an auxiliary sliding variable with

$$\dot{z}(t) = -\dot{x}_d(t) = bu(t) - bM \text{ sign } s(t) \tag{8.5.5}$$

Differentiation of the sliding variable $s(t)$ in (8.5.4) and substitution of plant (8.1.1) and auxiliary sliding variable (8.5.5) yields

$$\begin{aligned} \dot{s}(t) &= \dot{x}_e(t) + \dot{z}(t) \\ &= -ax(t) - d(x, t) + b(u(t) - w(t)) - bM \text{ sign } s(t) \\ &< a^+|x(t)| + d^+ + b\left|\left(\frac{\mu^2 p^2 + 2\mu p}{(\mu p + 1)^2}\right)u(t)\right| - bM \text{ sign } s(t) \end{aligned} \tag{8.5.6}$$

The third term in the last row of (8.5.6) is the mismatch between the actuator input and the actuator output; it decays rapidly according to actuator dynamics (8.1.2). This term can be fully eliminated if the actuator output is measurable, allowing (8.5.5) to be rewritten as

$$\dot{z}(t) = -\dot{x}_d(t) + bw(t) - bM \text{ sign } s(t) \tag{8.5.7}$$

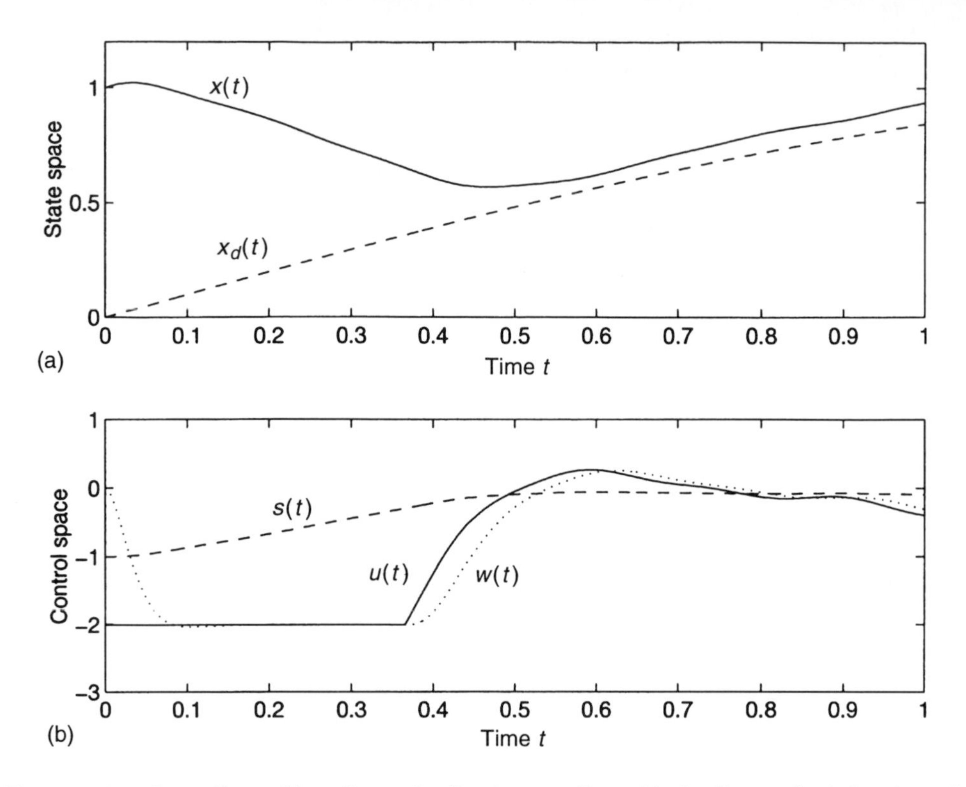

Figure 8.15 Controller $u_c(t)$, a linear feedback controller with feedforward of the desired trajectory, leads to inaccurate tracking of desired trajectory $x_d(t)$, since the closed-loop system (8.5.3) is perturbed by unknown plant dynamics $ax(t)$ and external disturbance $d(x, t)$: (a) output and desired output, (b) control inputs and sliding variable.

Stability of the sliding manifold $s(t)$ can be established using Lyapunov function candidate

$$V(t) = \frac{1}{2}s^2(t) \tag{8.5.8}$$

Differentiation of (8.5.8) and substitution of (8.5.6) yields

$$\begin{aligned} \dot{V}(t) &= s(t)\dot{s}(t) \\ &< |s(t)|\left(a^+|x(t)| + d^+ + b\mu\left|\left(\frac{\mu p^2 + 2p}{(\mu p + 1)^2}\right)u(t)\right|\right) - bM|s(t)| \end{aligned} \tag{8.5.9}$$

Since the actuator time constant μ is assumed small, sliding mode exists for sufficiently large

$$M > \frac{1}{b}(a^+|x(t)| + d^+) \tag{8.5.10}$$

and $s(t) = 0$ after finite time. Choosing the initial conditions of the auxiliary sliding variable $z(t)$ in (8.5.5) as $z(0) = -x_e(0)$ eliminates the reaching phase by setting $s(0) = 0$ in (8.5.4). While the system is in sliding mode, the motion trajectories can be examined using the equivalent control method. Solving

$\dot{s}(t) = 0$ in (8.5.6) under the assumption $w(t) = u(t)$ for the discontinuity term yields the *continuous* equivalent control

$$u_{d_{eq}}(t) = \frac{1}{b}(-ax(t) - d(x,t)) \\ = \frac{f(x,t)}{b} \tag{8.5.11}$$

which delivers an exact estimate of the disturbance perturbing the error dynamics (8.5.3) of the system under continuous control alone. Thus, defining the second term in (8.5.1) as

$$u_d(t) = u_{d_{eq}}(t) \tag{8.5.12}$$

leads to exact tracking with error dynamics

$$\dot{x}_e(t) + Cx_e(t) = 0 \tag{8.5.13}$$

instead of (8.5.3) under $u(t) = u_c(t)$. Equation (8.5.13) assures asymptotic tracking of $x_d(t)$ with disturbances. The equivalent control $u_{d_{eq}}(t)$ can be obtained by averaging the discontinuous switching component on the right-hand side of (8.5.5), e.g. via a lowpass

$$u_{d_{eq}}(t) = u_{d_{ave}}(t) \\ u_{d_{ave}}(t) = \frac{M}{(\varepsilon p + 1)} \text{ sign } s(t) \tag{8.5.14}$$

where p denotes the Laplace variable and $\varepsilon > 0$ is a small time constant. Disturbance rejection controller $u_d(t)$ in (8.5.14) is continuous. It was shown by Utkin (1992) that the lowpass average of the discontinuity term in (8.5.14) equals the equivalent control (8.5.11). In systems with the ability to employ discontinuous control inputs directly, the lowpass filter may be omitted completely.

The simulation in Figure 8.16 shows the improvement of the controller performance achieved by the disturbance estimator $u_d(t)$ as compared to Figure 8.15. Figure 8.16(b) depicts the performance of the disturbance estimator and shows that $s(t) = 0$ at all times, due to setting $z(0) = -x_e(0)$, and thus $s(0) = 0$. Estimation $u_{d_{ave}}(t)$ tracks the disturbance $f(x,t)$ consistently, with a small lag introduced by the averaging lowpass in (8.5.14).

The closed-loop system is free of chattering despite the discontinuity in (8.5.5) since the disturbance rejection term $u_d(t)$ is continuous. However, $x(t)$ tracks $x_d(t)$ exactly, due to the rejection of disturbance $f(x,t)$, which also contains uncertainty in parameter a of system (8.1.1). Note that the rejection of the parametric uncertainty does not assume constant system parameters, but rather is able to also account for parameter variations.

8.6 Comparing the different solutions

In applications of sliding mode control, unmodelled dynamics in the control loop are often excited by the discontinuous switching action of a sliding mode controller, leading to oscillations in the motion trajectory. Due to the acoustic noise these oscillations

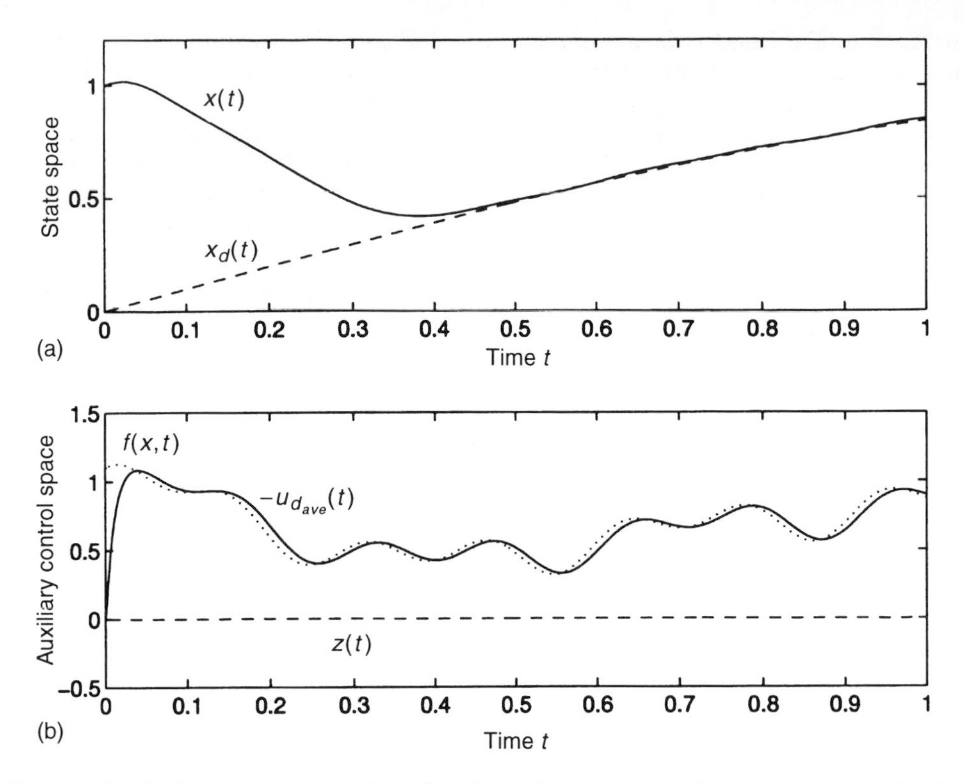

Figure 8.16 Performance of controller $u_c(t)$, a linear feedback controller with feedforward of the desired trajectory, is significantly improved by disturbance rejection controller $u_d(t)$ based on auxiliary variable $z(t)$ for estimating unknown plant dynamics $ax(t)$ and external disturbance $d(x,t)$; this may be summarized as $f(x,t) = -ax(t) - d(x,t)$: (a) output and desired output, (b) auxiliary sliding variable and disturbance/uncertainty estimation.

may cause in mechanical systems, this phenomenon is also known as chattering. This chapter studied the chattering problem and presented four solutions. All four solutions to the chattering problem reliably eliminate chattering in the control loop. In order to successfully prevent chattering, all methods require some estimate of the time constant or the bandwidth of the unmodelled dynamics. Instead of exact tracking performance, as achieved in ideal sliding mode, small tracking errors are tolerated.

In general, the achievable performance of a control system depends on the performance of sensors and actuators, availability of knowledge about the system, i.e. the quality of the system model, and the availability of measurements of system variables. For example, a system with a slow actuator cannot fully reject fast disturbances, regardless of the control design methodology used. A sliding mode controller under ideal conditions is able to fully exploit the system capabilities. Under realistic conditions, a chattering prevention scheme should be selected to meet the system specifications and to ensure good system performance.

The first method substitutes the discontinuity of a sliding mode controller by a saturation function and yields motion in a boundary layer of the sliding manifold instead of true sliding along the manifold. Effectively, sliding mode methodology is used to design a continuous high-gain controller which respects bounds on the control resources.

The second method shifts the switching action of the sliding mode controller into an auxiliary observer loop, thus circumventing unmodelled dynamics in the main loop and achieving ideal sliding mode in the observer loop. The plant follows the ideal trajectory of the observer according to the observer performance. Since the control input to the plant is still discontinuous, this method is ideal for systems which already have an observer in the control structure or for systems with inherently discontinuous control inputs like voltage inputs of electric drives. Implementation of a continuous controller in a system with discontinuous inputs generally requires pulse-width modulation (PWM), whereas direct implementation of sliding mode control with an observer avoids PWM.

The third method is mainly designed for systems where some knowledge of the unmodelled dynamic and intermediate measurements are available, e.g. known actuator dynamics. Such systems consisting of separated blocks may be controlled with a cascaded control structure which avoids chattering by explicitly taking the unmodelled dynamics into account for the control design. In this sense they are no longer 'unmodelled', but rather part of the overall system model.

The last method combines a continuous and a discontinuous controller to achieve good performance without chattering. The continuous part is to control the overall motion and the discontinuous part is to reject the influence of parametric uncertainty and disturbances. This method is a special case of integral sliding mode and is especially useful for systems with large uncertainties and/or disturbances.

All four methods possess advantages and disadvantages which depend on the system specifications. When designing a sliding mode controller for a given system, the chattering prevention method usually requires careful consideration of all details; unfortunately, no textbook solution exists to cope with all systems in a general manner.

References

Bondarev, A. G., Bondarev, S. A., Kostyleva, N. E. and Utkin, V. I., 1985, Sliding modes in systems with asymptotic state observers, *Automation and Remote Control*, **46**(6), 49–64.

DeCarlo, R. A., Zak, S. H. and Matthews, G. P., 1988, Variable structure control of nonlinear multivariable systems: a tutorial, *Proceedings of the IEEE*, **76**, 212–32.

Drakunov, S. V. *et al.*, 1984, A hierarchical principle of the control systems decomposition based on motion separation, *Preprints of the 9th IFAC Congress*, Budapest, Vol. V, pp. 134–39.

Drakunov, S. V. *et al.*, 1990, Block control principle I, II, *Automation and Remote Control*, **51**, 601–9 and **52**, 737–46.

Kokotovic, P. V., O'Malley, R. B. and Sannuti, P., 1976, Singular perturbations and order reduction in control theory, *Automatica*, **12**, 123–32.

Kokotovic, P. V., 1984, Applications of singular perturbation techniques to control problems, *SIAM Review*, **26**, 501–50.

Krstic, M., Kanellakopoulos, I. and Kokotovic, P., 1995, *Nonlinear and Adaptive Control Design*, New York: J. Wiley.

Leitmann, G., 1981, On the efficacy of nonlinear control in uncertain systems, *ASME Journal of Dynamic Systems, Measurement and Control*, **102**, 95–102.

Slotine, J.-J. and Sastry, S. S. 1983, Tracking control of nonlinear systems using sliding surfaces, with application to robot manipulators, *International Journal of Control*, **38**, 465–92.

SLOTINE, J.-J., 1984, Sliding controller design for nonlinear systems, *International Journal of Control*, **40**, 421–34.

UTKIN, V.I., 1992, *Sliding Modes in Control and Optimization,* Berlin: Springer-Verlag.

UTKIN, V.I., 1993, Application-oriented trends in sliding mode theory, *Proceedings of the IEEE Industrial Electronics Conference*, Maui HA, pp. 1937–42.

CHAPTER NINE

Discrete-Time and Delay Systems

Sliding mode is a very powerful tool for control design. So far, this text has concentrated on sliding mode control design for continuous-time systems. However, in many practical problems, controllers are implemented in discrete-time, e.g. employing microprocessors. Similar to the development in linear systems theory in the 1960s and 1970s, the discretization process requires the approach to be rethought. This chapter seeks to develop a general concept for discrete-time sliding mode and presents linear systems as design examples. This new concept is further extended to systems with delays and distributed systems.

9.1 Discrete-time systems

Most sliding mode approaches are based on finite-dimensional continuous-time models and lead to discontinuous control action. Once such a dynamic system is 'in sliding mode', its motion trajectory is confined to a manifold in the state space, i.e. to the sliding manifold. Generally speaking, for continuous-time systems, this reduction of the system order may only be achieved by discontinuous control, switching at theoretically infinite frequency.

When challenged with the task of implementing sliding mode control in a practical system, the control engineer has two options:

- Direct, analogue implementation of a discontinuous control law with a very fast switching device, e.g. with power transistors.
- Discrete implementation of sliding mode control, e.g. with a digital microcontroller.

The first method is only suitable for systems with a voltage input allowing the use of analogue switching devices. Most other systems are usually based on a discrete microcontroller-based implementation. However, a discontinuous control designed for a continuous-time system model would lead to chatter when implemented without modifications in discrete time with a finite sampling rate. This discretization chatter is different from the chattering problem caused by unmodelled dynamics as discussed in Chapter 8. Discretization chatter is due to the fact that the switching frequency is limited to the sampling rate, but correct implementation of sliding mode control requires infinite switching frequency. The following example will illustrate the difference between ideal continuous time sliding mode and direct discrete implementation with discretization chatter. The subsequent sections of this chapter are dedicated to the development of a discrete-time sliding mode concept to eliminate the chatter. This concept is then extended to systems with delays and distributed systems governed by differential-difference equations.

Example 9.1 Ideal sliding mode versus discrete implementation

Examine a first-order example system modelled in continuous time as

$$\dot{x}(t) = g(x(t)) + u(t) \tag{9.1.1}$$

with state $x(t)$, bounded dynamics $|g(x)| \leq \bar{g}$ and control input $u(t)$. To enforce sliding mode on a manifold

$$\sigma = \{x : \ x(t) = 0\} \tag{9.1.2}$$

a discontinuous control law may be designed as

$$u(t) = -u_0 \text{ sign } x(t) \tag{9.1.3}$$

with available control resources $u_0 > \bar{g}$. The usual Lyapunov-based stability analysis examines

$$V = \frac{1}{2}x^2(t) \tag{9.1.4}$$

along the trajectories of system (9.1.1) with control (9.1.3), leading to

$$\begin{aligned} \dot{V}(t) &= x(t)(g(x(t)) - u_0 \text{ sign } x(t)) \\ &\leq -|x(t)|(u_0 - \bar{g}) \end{aligned} \tag{9.1.5}$$

which testifies to convergence to the manifold (9.1.2) within finite time. An example trajectory is shown in Figure 9.1 with $g(t) = \sin t$ and $u_0 = 2$, starting from initial conditions $x(t = 0) = 3$. At $t_{sm} = 2.256$ s, the system reaches the sliding manifold $x = 0$. Thereafter, the motion trajectory is invariantly confined to the manifold (9.1.2) via discontinuously switching control, illustrated by a black rectangle in the lower diagram of Figure 9.1.

A direct discrete implementation with sampling time Δt would result in

$$\begin{aligned} x_{k+1} &= x_k + (g_k + u_k)\Delta t \\ u_k &= -u_0 \text{ sign } x_k \\ k &= 1, 2, \ldots \end{aligned} \tag{9.1.6}$$

where subscript k denotes the sampling points, e.g. the system state x_k at time $t_k = k\Delta t$. The motion trajectory may not reach the manifold $x = 0$ since control u_k is only cal-

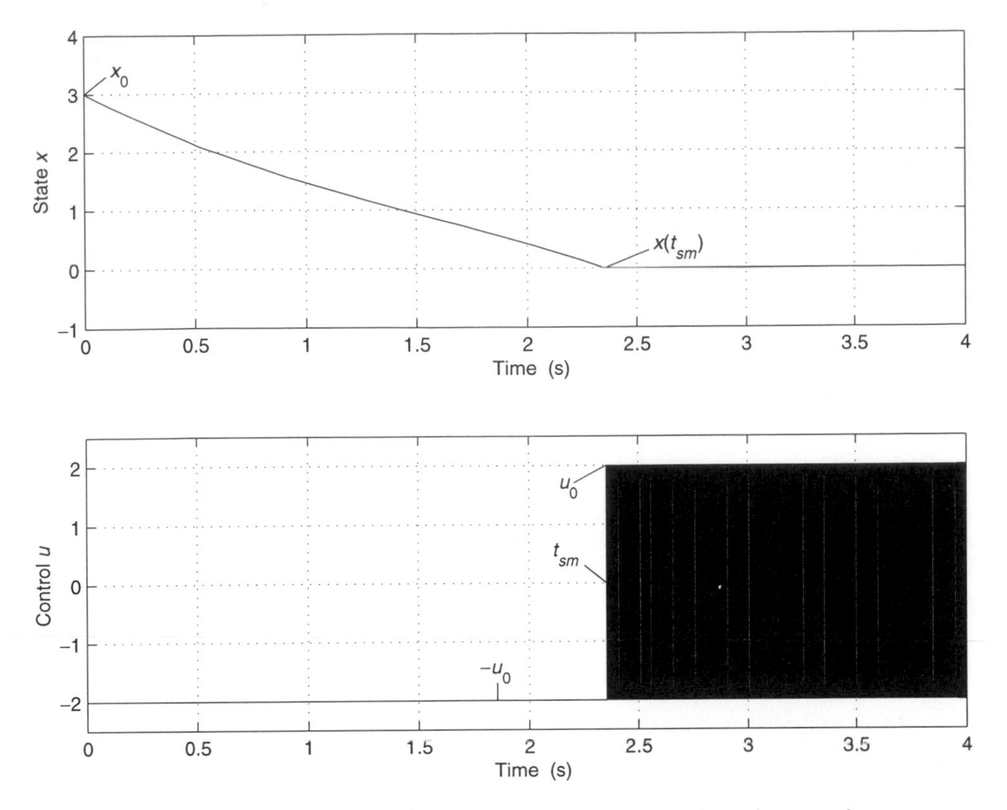

Figure 9.1 Ideal sliding mode in the first-order example achieved via direct analogue implementation of a discontinuous control law with infinitely fast switching.

culated at the sampling points k, i.e. the switching frequency is limited by the sampling rate $1/\Delta t$. During the sampling interval Δt, the control is constant and the system behaves like an open-loop system (Kotta, 1989).

The example with exaggerated sampling timc $\Delta t = 0.1$ s, as depicted in Figure 9.2, illustrates the need to develop a discrete-time sliding mode algorithm rather than implementing (9.1.6).

Note that increasing the sampling rate decreases the amplitude of the discretization chatter and increases its frequency, but may not eliminate this discrete-time phenomenon unless $\Delta t \to 0$. Moreover, the sampling rate of a control system should correspond to the fastest dynamics of the system instead of 'wasting' computational power for the sake of the control algorithm.

9.2 Discrete-time sliding mode concept

Before developing the concept of discrete-time sliding mode, let us revisit the principle of sliding mode in continuous-time systems with ideal discontinuous control from an engineering viewpoint. A more mathematical treatment may be found in Utkin (1993) or in Drakunov and Utkin (1990). Rewrite (9.1.1) as a general continuous-time system

$$\dot{x} = f(x, u, t) \tag{9.2.1}$$

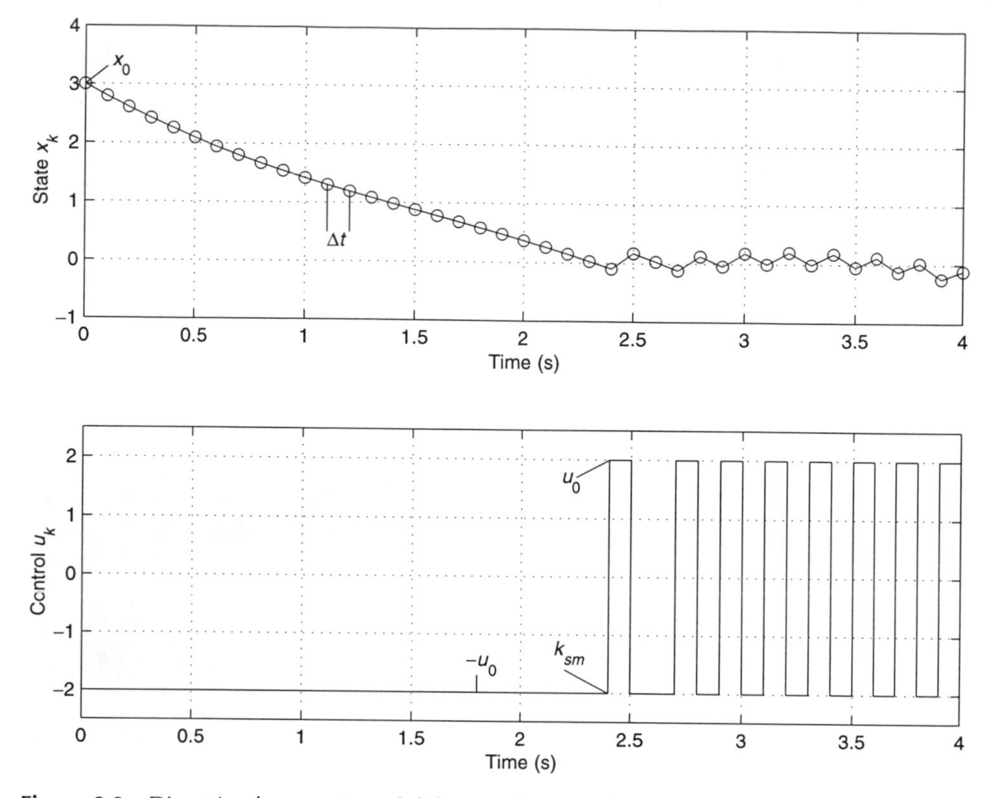

Figure 9.2 Direct implementation of sliding mode control in discrete time. Sampling instances are marked with small circles. Control u_k may only be switched at sampling instances, producing discretization chatter in the motion trajectory after reaching the vicinity of the sliding manifold at $t_{sm} = k_{sm}\Delta t$.

with a discontinuous scalar control law

$$u = \begin{cases} u_0 & \text{if} \quad s(x) \geq 0 \\ -u_0 & \text{if} \quad s(x) < 0 \end{cases} \tag{9.2.2}$$

and sliding mode in some manifold $s(x) = 0$ (Figure 9.3).

Note the following observations characterizing the nature of sliding mode systems:

- The time interval between the initial point $t = 0$ and the reaching of the sliding manifold $\sigma = \{x : s(x) = 0\}$ at t_{sm} is finite, in contrast to systems with a continuous control law, which exhibit asymptotic convergence to any manifold consisting of state trajectories.
- Once the system is 'in sliding mode' for all $t \geq t_{sm}$, its trajectory motion is confined to the manifold $\sigma = \{x : s(x) = 0\}$ and the order of the closed-loop system dynamics is less than the order of the original uncontrolled system.
- After sliding mode has started at t_{sm}, the system trajectory cannot be backtracked beyond the manifold $\sigma = \{x : s(x) = 0\}$ like in systems without discontinuities. In other words, at any point $t_0 \geq t_{sm}$, it is not possible to determine the time t_{sm} or to reverse calculate the trajectory for $t < t_{sm}$ based on information of the system state at t_0.

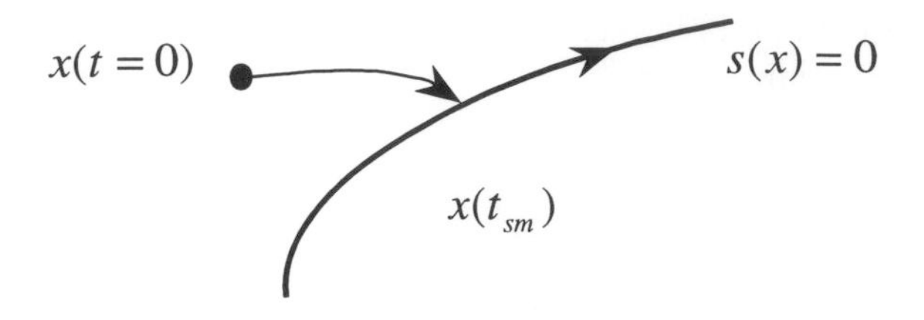

Figure 9.3 Motion trajectory of continuous-time system with scalar sliding mode control. Starting from initial point $x(t=0)$, the trajectory reaches the sliding manifold $s(x)=0$ within finite time t_{sm} and remains on the manifold thereafter.

However, during both time intervals before and after reaching the sliding manifold, the state trajectories are continuous functions of time, and the relation between two values of the state at the ends of a finite time interval $t=[t_0, t_0+\Delta t]$ may be found by solving (9.2.1) as

$$x(t_0+\Delta t) = F(x(t_0)) \tag{9.2.3}$$

where $F(x(t))$ is a continuous function as well. When derived for each sampling point $t_k = k\Delta t$, $k=1,2,\ldots$, equation (9.2.3) is nothing but the discrete-time representation of the continuous-time prototype (9.2.1), i.e.

$$x_{k+1} = F(x_k) \qquad x_k = x(k\Delta t) \tag{9.2.4}$$

Starting from time t_{sm}, the state trajectory belongs to the sliding manifold with $s(x(t))=0$, or for some $k_{sm} \geq t_{sm}/\Delta t$,

$$s(x_k) = 0 \qquad (\forall k \geq k_{sm}) \tag{9.2.5}$$

It seems reasonable to call this motion 'sliding mode in discrete time' or 'discrete-time sliding mode'. Note that the right-hand side of the motion equation of the system with discrete-time sliding mode is a *continuous* state function.

So far, we have generated a discrete-time description of a continuous-time sliding mode system. Next we need to derive a discrete-time control law which may generate sliding mode in a discrete-time system. Let us return to the example (9.2.1) and suppose that for any constant control input u and any initial condition $x(0)$, the solution to (9.2.1) may be found in closed form, i.e.

$$x(t) = F(x(0), u) \tag{9.2.6}$$

Now also assume that control u may be chosen arbitrarily. With the help of (9.2.6), follow the procedure below:

1. At $t=0$ select constant $u(x(t=0), \Delta t)$ for a given time interval Δt such that $s(x(t=\Delta t))=0$.
2. At $t=\Delta t$ find constant $u(x(t=\Delta t), \Delta t)$ such that $s(x(t=2\Delta t))=0$.
3. In general, for each $k=0,1,2,\ldots$, at $t=k\Delta t$ choose constant $u(x_k, \Delta t)$ such that $s(x_{k+1})=0$.

In other words, at each sampling point k, select u_k such that this control input, to be constant during the next sampling interval Δt, will achieve $s(x_{k+1})=0$ at the next sampling point $(k+1)$. During the sampling interval, state $x(k\Delta t < t < (k+1)\Delta t)$

may not belong to the manifold, i.e. $s(x(t)) \neq 0$ is possible for $k\Delta t < t < (k+1)\Delta t$. However, the discrete-time system

$$\begin{aligned} x_{k+1} &= F(x_k, u_k) \\ u_k &= u(x_k) \end{aligned} \tag{9.2.7}$$

hits the sliding manifold at each sampling point, i.e. $s(x_{k+1}) = 0 \; \forall k = 0, 1, 2, \ldots$ is fulfilled.

Since $F(x(0), u)$ tends to $x(0)$ as $\Delta t \to 0$, the function $u(x(0), \Delta t)$ may exceed the available control resources u_0. As a result, the bounded control shown in the lower diagram of Figure 9.4 steers state x_k to zero only after a finite number of steps k_{sm}. Thus the manifold is reached after a finite time interval $t_{sm} = k_{sm}\Delta t$ and thereafter the state x_k remains on the manifold. In analogy to continuous-time systems, this motion may be called 'discrete-time sliding mode'. Note that sliding mode is generated in the discrete-time system with control $-u_0 \leq u \leq u_0$ as a continuous function of the state x_k and is piecewise constant during the sampling interval.

This first-order example clarifies the definition of the term 'discrete-time sliding mode' introduced by Drakunov and Utkin (1990) for an arbitrary finite-dimensional discrete-time system.

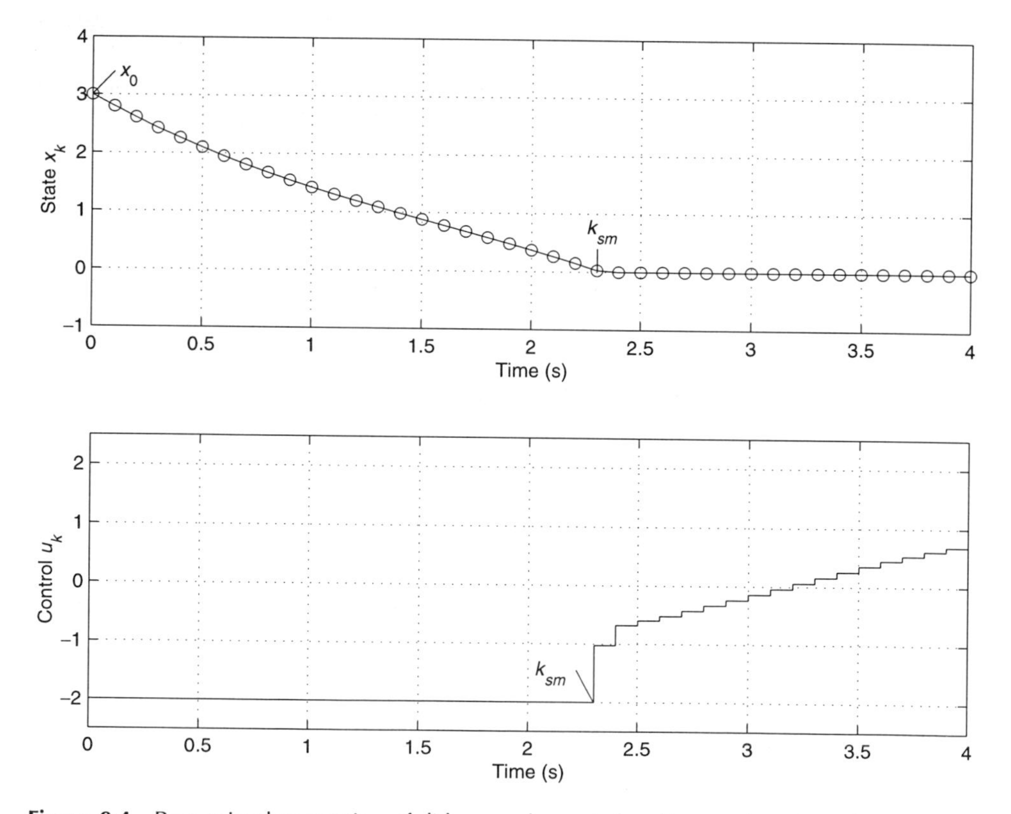

Figure 9.4 Proper implementation of sliding mode control in discrete time. Sampling instances are marked with small circles. Control u_k is selected as $-u_0 \leq u_k \leq u_0$ at each sampling instance to achieve $s(x_{k+1}) = 0$ as quickly as possible in accordance with the available control resources, resulting in chatter-free motion after reaching the sliding manifold at $t_{sm} = k_{sm}\Delta t$.

Definition 9.1: Discrete-time sliding mode

In the discrete-time dynamic system

$$x_{k+1} = F(x_k, u_k) \qquad (x \rightarrow \Re^n, \quad u \in \Re^m, \quad m \leq n) \tag{9.2.8}$$

discrete-time sliding mode takes place on a subset Σ of the manifold $\sigma = \{x: s(x) = 0\}$, $s \in \Re^m$, if there exists an open neighbourhood $\aleph$ of this subset such that for each $x \in \aleph$ it follows that $s(F(x_{k+1})) \in \Sigma$. □

In contrast to continuous-time systems, sliding mode may arise in discrete-time systems with a continuous function on the right-hand side of the closed-loop system equation. Nevertheless, the aforementioned characteristics of continuous-time sliding mode have been transferred to discrete-time sliding mode. The mathematical implications in terms of group theory may be found in Drakunov and Utkin (1990) and in Utkin (1993). Practical issues will be discussed in the subsequent section using linear systems as an example.

9.3 Linear discrete-time systems with known parameters

This section deals with discrete-time sliding mode control for linear time-invariant continuous-time plants. Let us assume that a sliding mode manifold is linear for an nth-order discrete-time system $x_{k+1} = F(x_k)$, i.e. $s_k = Cx_k$, $C \in \Re^{m \times n}$ with m control inputs. According to Definition 9.1, the sliding mode existence condition is of the form

$$s_{k+1} = C(F(x_k)) \tag{9.3.1}$$

for any $x_k \in \aleph$. To design a discrete-time sliding mode control law based on condition (9.3.1), consider the continuous-time representation of the linear time-invariant system

$$\dot{x}(t) = Ax(t) + Bu(t) + Dr(t) \tag{9.3.2}$$

with state vector $x(t) \in \Re^n$, control $u(t) \in \Re^m$, reference input $r(t)$, and constant system matrices A, B and D. Transformation to discrete time with sampling interval Δt yields

$$x_{k+1} = A^* x_k + B^* u_k + D^* r_k \tag{9.3.3}$$

where

$$A^* = e^{A\Delta t} \qquad B^* = \int_0^{\Delta t} e^{A(\Delta t - t)} B \, \mathrm{d}\tau \qquad D^* = \int_0^{\Delta t} e^{A(\Delta t - t)} D \, \mathrm{d}\tau \tag{9.3.4}$$

and the reference input $r(t)$ is assumed to be constant during the sampling interval Δt. In accordance with (9.3.1), discrete-time sliding mode exists if the matrix CB^* has an inverse and the control u_k is designed as the solution of

$$s_{k+1} = CA^* x_k + CD^* r_k + CB^* u_k = 0 \tag{9.3.5}$$

In other words, control u_k should be chosen as

$$u_k = -(CB^*)^{-1}(CA^* x_k + CD^* r_k) \tag{9.3.6}$$

By analogy with continuous-time systems, the control law (9.3.6) yielding motion in the manifold $s = 0$ will be called 'equivalent control'. To reveal the structure of $u_{k_{eq}}$, let us represent it as the sum of two linear functions

$$u_{keq} = -(CB^*)^{-1}s_k - (CB^*)^{-1}((CA^* - C)x_k + CD^*r_k) \tag{9.3.7}$$

and

$$s_{k+1} = s_k + (CA^* - C)x_k + CD^*r_k + CB^*u_k \tag{9.3.8}$$

As in the first-order example of the previous section, $u_{k_{eq}}$ may exceed the available control resources with $\Delta t \to 0$ for initial $s_k \neq 0$, since $(CB^*)^{-1} \to \infty$ means that $(CB^*)^{-1}(CA^* - C)$ and $(CB^*)^{-1}CD^*$ take finite values. Hence the real-life bounds for control u_k should be taken into account.

Suppose that the control can vary within $\|u_k\| \leq u_0$ and the available control resources are such that

$$\|(CB^*)^{-1}\| \cdot \|(CA^* - C)x_k + CD^*r_k\| < u_0 \tag{9.3.9}$$

Note that otherwise, the control resources are insufficient to stabilize the system.

The control

$$u_k = \begin{cases} u_{k_{eq}} & \text{for } \|u_{k_{eq}}\| \leq u_0 \\ u_0 \dfrac{u_{k_{eq}}}{\|u_{k_{eq}}\|} & \text{for } \|u_{k_{eq}}\| > u_0 \end{cases} \tag{9.3.10}$$

complies with the bounds on the control resources. As shown above, $u_k = u_{k_{eq}}$ for $\|u_{k_{eq}}\| \leq u_0$ yields motion in the sliding manifold $s = 0$. To prove convergence to this domain, consider the case $\|u_{k_{eq}}\| > u_0$, but in compliance with condition (9.3.9). From (9.3.7) to (9.3.10) it follows that

$$s_{k+1} = (s_k + (CA^* - C)x_k + CD^*r_k)\left(1 - \frac{u_0}{\|u_{k_{eq}}\|}\right) \quad \text{with} \quad u_0 < \|u_{k_{eq}}\| \tag{9.3.11}$$

Thus

$$\begin{aligned} \|s_{k+1}\| &= \|(s_k + (CA^* - C)x_k + CD^*r_k)\|\left(1 - \frac{u_0}{\|u_{k_{eq}}\|}\right) \\ &\leq \|s_k\| + \|(CA^* - C)x_k + CD^*r_k\| - \frac{u_0}{\|(CB^*)^{-1}\|} \\ &< \|s_k\| \end{aligned} \tag{9.3.12}$$

due to (9.3.9). Hence $\|s_k\|$ decreases monotonically and, after a finite number of steps, $\|u_{k_{eq}}\| < u_0$ is achieved. Discrete-time sliding mode will take place from the next sampling point onwards.

Control (9.3.10) provides chatter-free motion in the manifold $s = 0$ as shown in Figure 9.4, in contrast to the direct implementation of discontinuous control in Figure 9.2, producing discretization chatter in the vicinity of the sliding manifold. Similar to the case of continuous-time systems, the equation $s = Cx = 0$ enables the reduction of system order, and the desired system dynamics 'in sliding mode' can be designed by appropriate choice of matrix C.

9.4 Linear discrete-time systems with unknown parameters

Complete information of the plant parameters is required for implementation of control (9.3.10), which may not be available in practice. To extend the discrete-time sliding mode concept to systems with unknown parameters, suppose that system (9.3.5) operates under uncertainty conditions: the matrices A and D, and the reference input r_k are assumed to be unknown and may vary in some ranges such that condition (9.3.9) holds. Similar to (9.3.10), the control law

$$u = \begin{cases} -(CB^*)^{-1}s_k & \text{for } \|(CB^*)^{-1}s_k\| \le u_0 \\ -u_0\dfrac{(CB^*)^{-1}s_k}{\|(CB^*)^{-1}s_k\|} & \text{for } \|(CB^*)^{-1}s_k\| > u_0 \end{cases} \tag{9.4.1}$$

respects the bounds of the control resources. Furthermore, control (9.4.1) does not depend on the plant parameters A and D and the reference input r_k. Substitution of (9.4.1) into the system equations of the previous section leads to

$$s_{k+1} = s_k\left(1 - \frac{u_0}{\|(CB^*)^{-1}s_k\|}\right) + (CA^* - C)x_k + CD^*r_k \quad \text{for} \quad u_0 < \|(CB^*)^{-1}s_k\| \tag{9.4.2}$$

and, similar to (9.3.12), we obtain

$$\begin{aligned} \|s_{k+1}\| &\le \|s_k\|\left(1 - \frac{u_0}{\|(CB^*)^{-1}s_k\|}\right) + \|(CA^* - C)x_k + CD^*r_k\| \\ &\le \|s_k\| - \frac{u_0\|s_k\|}{\|(CB^*)^{-1}s_k\|} + \|(CA^* - C)x_k + CD^*r_k\| \\ &\le \|s_k\| - \frac{u_0}{\|(CB^*)^{-1}\|} + \|(CA^* - C)x_k + CD^*r_k\| \\ &< \|s_k\| \end{aligned} \tag{9.4.3}$$

Hence, as for the case with complete knowledge of system parameters discussed in Section 9.3, the value of $\|s_k\|$ decreases monotonically and, after a finite number of steps, control $\|u_k\| < u_0$ will be within the available resources. Substituting (9.4.1) into (9.3.8) results in

$$s_{k+1} = (CA^* - C)x_k + CD^*r_k \tag{9.4.4}$$

Since the matrices $(CA^* - C)$ and CD^* are of Δt order, the system motion will be in a Δt-order vicinity of the sliding manifold $s = 0$. Figure 9.5 shows a simulation of the first-order example in Section 9.1.1 for unknown matrices A and D. Convergence to the vicinity of the sliding manifold is achieved in finite time; thereafter, the motion trajectory does not follow the sliding manifold exactly, but rather remains within a Δt-order vicinity. This result is expected from systems operating under uncertainty

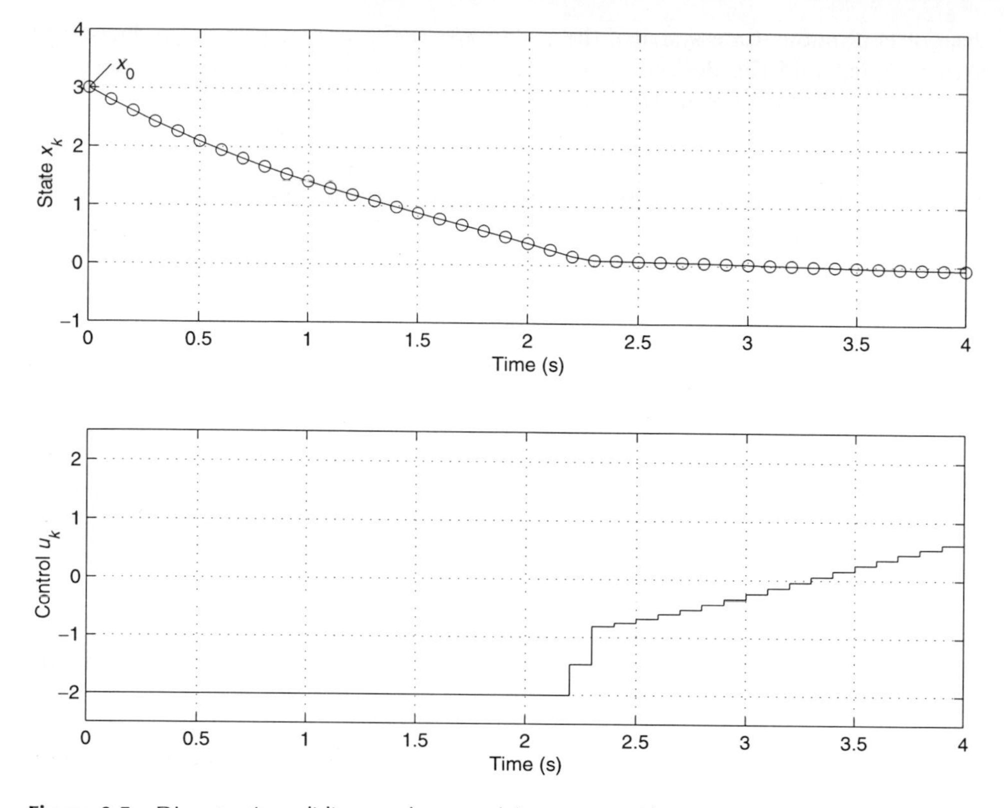

Figure 9.5 Discrete-time sliding mode control for system with uncertain parameters. Sampling instances are marked with small circles. Control u_k is selected as $-u_0 \leq u_k \leq u_0$ at each sampling instance such that $s(x_{k+1})$ approaches the vicinity of the sliding manifold in finite time and remains in the vicinity afterwards with chatter-free motion.

conditions, since we are dealing with an open-loop system during each sampling interval. In contrast to discrete-time systems with direct implementation of discontinuous control as shown in Figure 9.2, this motion is free of discretization chatter.

9.5 Distributed systems and systems with delays

This section discusses design methods for systems described by difference and differential-difference equations. These types of equations may serve as mathematical models for dynamic systems with delays or for distributed systems with finite-dimensional inputs and outputs. Section 9.6 presents the basic sliding mode control design methodology. A flexible shaft as an example for a distributed system is discussed in Section 9.7.

Consider a system modelled by differential-difference equations in regular form as

$$\dot{x}(t) = A_{11}x(t) + A_{12}z(t) \tag{9.5.1}$$

$$z(t) = A_{21}x(t-\tau) + A_{22}z(t-\tau) + B_0u(t-\tau) \tag{9.5.2}$$

where $x \in \Re^n$, $z \in \Re^k$ and $u \in \Re^m$. The pair (A_{11}, A_{12}) is assumed to be controllable and the difference system (9.5.2) is assumed invertible with output $\tilde{A}_{12}z(t)$, where $\tilde{A}_{12}$ consists of the basic rows of A_{12} in (9.5.1).

Recall the sliding mode concept for discrete-time systems presented in Section 9.2 and in Definition 9.1. For a differential-difference system like (9.5.1), (9.5.2) sliding mode can be defined in a similar manner: sliding mode exists in some manifold σ if the state trajectories starting outside this manifold reach σ within finite time and all state trajectories which belong to the manifold σ at some time instant t_{sm} remain in σ for all $t \geq t_{sm}$.

System (9.5.1), (9.5.2) is written in block-control form (Drakunov *et al.*, 1990) for ordinary differential equations; see also Section 3.3. The two-step design procedure first derives a 'desired' control $z_d(x(t))$ for (9.5.1) to yield the desired motion along a manifold $\sigma = \{x: \ S(x) = 0\}$ in this first block, assuming $z = z_d$. The second design step uses 'real' control input $u(x, t)$ in (9.5.2) to ensure this assumption holds by enforcing sliding mode in the second block along manifold $\sigma_0 = \{x: \ z - z_d = 0\}$. In the overall system (9.5.1), (9.5.2) sliding mode exists in the intersection of both manifolds as described by $\sigma \bigcap \sigma_0$.

9.6 Linear systems with delays

This section deals with linear systems with known parameters in analogy to Section 9.3. The extension to systems with unknown parameters follows similar procedures as for discrete-time systems in Section 9.4 and is omitted here.

As an example, consider a time-invariant linear system with a delay in the input variable

$$\dot{x}(t) = Ax(t) + Bu(t - \tau) \tag{9.6.1}$$

where $x \in \Re^n$, $u \in \Re^m$, $t > 0$ and the initial conditions are denoted by $x(0) = x_0$ and $u(\xi) = u_0(\xi)$ for $-\tau < \xi < 0$. System (9.6.1) can be represented in differential-difference block form (9.5.1), (9.5.2) by setting $A_{11} = A$, $A_{12} = B$, $A_{21} = 0_{m\times n}$, $A_{22} = 0_{m\times m}$ and $B_0 = I_{m\times m}$:

$$\dot{x}(t) = A_{11}x(t) + A_{12}z(t) \tag{9.6.2}$$

$$z(t) = u(t - \tau) \tag{9.6.3}$$

First, design a smooth function $S(x) = (s_1(x), \ldots, s_i(x), \ldots, s_m(x)) \in \Re^m$ and a discontinuous control $z_d = (z_{d_1}(x), \ldots, z_{d_i}(x), \ldots, z_{d_m}(x)) \in \Re^m$ with components

$$z_{d_i}(x(t)) = \begin{cases} z_{d_i}^+(x(t)) & \text{for } s_i(x(t)) > 0 \\ z_{d_i}^-(x(t)) & \text{for } s_i(x(t)) < 0 \end{cases} \qquad (i = 1, 2, \ldots, m) \tag{9.6.4}$$

such that, after a finite time interval, every trajectory belongs to the intersection $\sigma = \bigcap_{i=1}^m \sigma_i$ of the surfaces $\sigma_i = \{x: \ s_i(x) = 0\}$ and sliding mode exists thereafter. Second, to implement control (9.6.4) in (9.6.2), assign

$$u(t) = z_d(x(t + \tau)) \tag{9.6.5}$$

for (9.6.3) to enforce sliding mode on the manifold $\sigma_0 = \{x : \; z - z_d = 0\}$. The values of $x(t+\tau)$ have to be extrapolated from the solution of (9.6.2) as

$$x(t+\tau) = e^{At}x(t) + \int_0^{\tau} e^{At}Bu(t-\xi)\,\mathrm{d}\xi \tag{9.6.6}$$

Note that control $u(t)$ needs to be stored in the microcontroller for the time interval $[t, t-\xi]$.

Similar to the discrete-time systems discussed in Sections 9.2 to 9.4, control $u(t)$ is designed with preview such that $z(t+\tau) = z_d(t+\tau)$ at some future point, τ ahead of the current time instance t. In contrast to discrete-time systems, time t is continuous rather than sampled at discrete time instances $k\Delta t$.

The motion of system (9.6.2), (9.6.3) in sliding mode along manifold σ_0 is described by

$$\dot{x}(t) = Ax(t) + Bz_d(x(t)) \tag{9.6.7}$$

Once sliding mode also occurs in the intersection $\sigma = \bigcap_{i=1}^{m} \sigma_i$ by design of (9.6.4), sliding mode exists in the intersection of all $(m+1)$ sliding manifolds $\Sigma = \sigma \bigcap \sigma_0 = \bigcap_{i=0}^{m} \sigma_i$ with the desired dynamics of system (9.6.1).

9.7 Distributed systems

This section discusses a flexible shaft as an example for a distributed system. Since distributed systems can be described by similar differential-difference equations as systems with delays, the design methodology of Section 9.6 will be employed.

Consider a flexible shaft with lengths l and inertial load J acting as a torsion bar (Figure 9.6). Let $e(t)$ be the absolute coordinate of the left end of the bar with input torque M and let $d(x, t)$ be the relative deviation at location $0 \leq x \leq l$ at time t. Hence the absolute deviation of a point $0 < x < l$ at time t is described by $q(x, t) = e(t) + d(x, t)$ and governed by

$$\frac{\partial^2 q(x,t)}{\partial t^2} = a^2 \frac{\partial^2 q(x,t)}{\partial x^2} \tag{9.7.1}$$

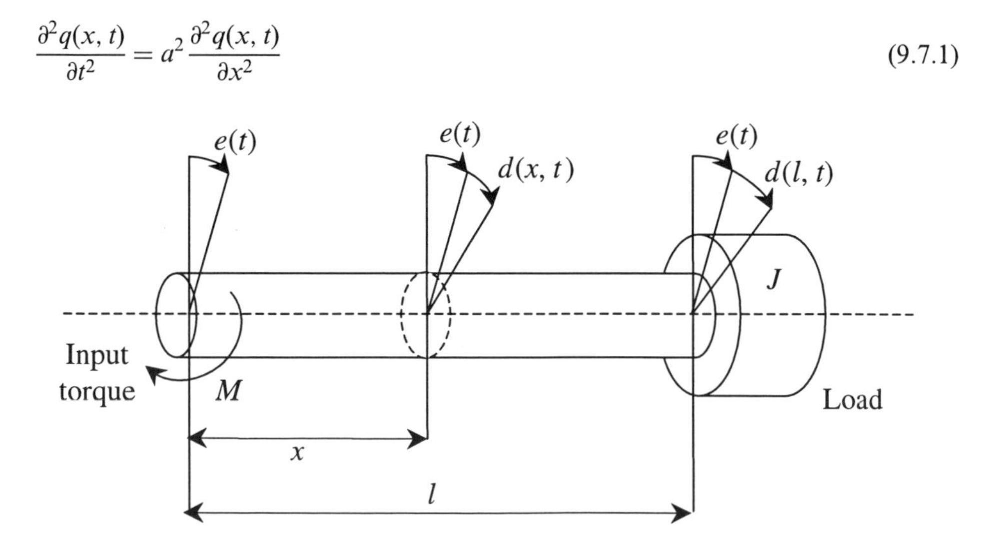

Figure 9.6 Flexible shaft with input torque M and load J acting as a torsion bar.

where a is the flexibility constant depending on the geometry and the material of the bar. The boundary conditions corresponding to the input torque M and the load inertia J are

$$M = -a^2 \frac{\partial q(0, t)}{\partial x} \qquad J\frac{\partial^2 q(l, t)}{\partial t^2} = -a^2 \frac{\partial^2 q(l, t)}{\partial x^2} \tag{9.7.2}$$

Consider the input torque M as the control input $u(t)$ and the load position $q(l, t)$ as the system output. To find the transfer function $W(p)$ via Laplace transform, assume zero initial conditions

$$q(x, 0) = 0 \qquad \frac{\partial q(x, 0)}{\partial t} = 0 \tag{9.7.3}$$

to yield

$$\begin{aligned} &a^2 Q(0, p) = -U(p) \\ &p^2 Q(x, p) = a^2 Q''(x, p) \\ &a^2 Q'(l, p) = -Jp^2 Q(l, p) \end{aligned} \tag{9.7.4}$$

where $Q(x,p)$ denotes the Laplace transform of $q(x, t)$ with spatial derivatives $Q'(x,p) = \partial Q(x,p)/\partial x$ and $Q''(x,p) = \partial^2 Q(x,p)/\partial x^2$, and $U(p)$ represents the Laplace transform of input variable $u(t)$. The solution to the boundary value problem (9.7.4) is given by

$$Q(x, p) = -\frac{\left(1 - \frac{J}{a}p\right)e^{-\frac{l-x}{a}p} + \left(1 + \frac{J}{a}p\right)e^{\frac{l-x}{a}p}}{ap\left(-\left(1 - \frac{J}{a}p\right)e^{-\frac{l}{a}p} + \left(1 + \frac{J}{a}p\right)e^{\frac{l}{a}p}\right)} U(p) \tag{9.7.5}$$

from which $W(p)$ may be found by setting $x = l$ to yield

$$W(p) = \frac{2e^{-\tau p}}{ap\left(1 + \frac{J}{a}p\right) + \left(1 - \frac{J}{a}p\right)e^{-2\tau p}} \tag{9.7.6}$$

where $\tau = l/a$ describes the 'delay' between the left end and the right end of the bar. The corresponding differential-difference equation may be written in the form

$$J\ddot{q}(t) + J\ddot{q}(t - \tau) + a\dot{q}(t) - a\dot{q}(t - 2\tau) = 2u(t - \tau) \tag{9.7.7}$$

Denoting $x_1(t) = q(t)$, $x_2(t) = \dot{q}(t)$ and $z(t) = J\ddot{q}(t) + a\dot{q}(t)$, we obtain the motion equations as

$$\begin{aligned} &\dot{x}_1(t) = x_2(t) \\ &\dot{x}_2(t) = -(ax_2(t) + z(t))/J \end{aligned} \tag{9.7.8}$$

$$z(t) = 2ax_2(t - 2\tau) - z(t - 2\tau) + 2u(t - \tau) \tag{9.7.9}$$

CHAPTER TEN

Electric Drives

In recent years, a lot of research effort has been devoted to the application of sliding mode control techniques to power electronic equipment and electrical drives. Interest in this control approach has emerged due to its potential for circumventing parameter variation effects under dynamic conditions with a minimum of implementation complexity. In electric drive systems, the existence of parameter changes caused by, for instance, winding temperature variation, converter switching effect and saturation, is well recognized, though infrequently accounted for. In servo applications, significant parameter variations arise from often unknown loads; for example, in machine tool drives and robotics, the moment of inertia represents a variable parameter depending on the load of the tool or the payload. Among the distinctive features claimed for sliding mode control are order reduction, disturbance rejection, strong robustness and simple implementation by means of power converters. Hence sliding mode is attributed high potentials as a prospective control methodology for electric drive systems. The experience gained so far testifies to its efficiency and versatility. In fact, control of electric drives is one of the most challenging applications due to increasing interest in using electric servomechanisms in control systems, the advances of high-speed switching circuitry, as well as insufficient linear control methodology for inherently nonlinear high-order multivariable systems such as AC motors.

Implementation of sliding mode control by means of the most common electric components has turned out to be simple enough. Commercially available power converters enable several dozen kilowatts to be handled at frequencies of several hundred kilohertz. When using converters of this type, confining their function to pulse-width modulation seems unjustified, and it is reasonable to turn to algorithms with discontinuous control actions. Introduction of discontinuities is dictated by the very nature of power converters.

This chapter consists of three main parts: sliding mode control of DC motors, permanent-magnet synchronous motors and induction motors. These three types of electric motor are the most commonly used drive systems in industrial applications. All these drive systems have much in common: current control, speed control, observer design and issues of sensorless control. As will be shown in each section, sliding mode control techniques are used flexibly to achieve the desired control performance, not only in controller design, but also in observer design and estimation processes.

The basic framework of this chapter has been given in Utkin (1993) and Sabanovic *et al.* (1993). However, the content and the theoretical aspect have been extended considerably.

10.1 DC motors

10.1.1 Introduction

To show the effectiveness of sliding mode techniques in the control of electric drives, we start with the simplest drive systems, i.e. DC motor systems. This section is an extension of the sliding mode approaches applied to DC motors given in Utkin (1993). Moreover, some implementation aspects will be enhanced, aiming at real-life applications of theoretical derivations given in the earlier chapters.

10.1.2 Model of the DC motor

From the viewpoint of controllability, a DC motor with a constant excitation is the simplest electric drive. Figure 10.1 shows the structure of the electric circuit of a permanently excited DC motor. The motor dynamics are governed by two coupled first-order equations with respect to armature current and shaft speed:

$$\begin{aligned} L\frac{di}{dt} &= u - Ri - \lambda_0\omega \\ J\frac{d\omega}{dt} &= k_t i - \tau_l \end{aligned} \tag{10.1.1}$$

where

i = armature current	u = terminal voltage
ω = shaft speed	J = inertia of the motor rotor and load
R = armature resistance	L = armature inductance
λ_0 = back-EMF constant	k_t = torque constant
τ_l = load torque	

Throughout this section the motor parameters used to verify the design principles are $L = 1.0\,\text{mH}$, $R = 0.5\,\Omega$, $J = 0.001\,\text{kg}\,\text{m}^2$, $k_t = 0.008\,\text{Nm}\,\text{A}^{-1}$, $\lambda_0 = 0.001\,\text{V}\,\text{s}\,\text{rad}^{-1}$

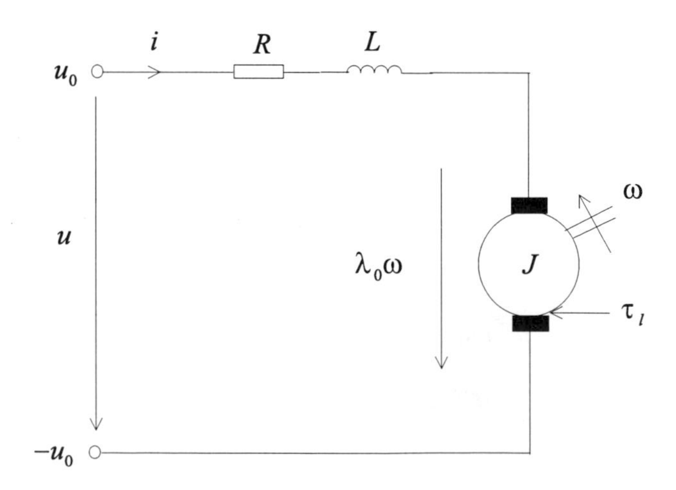

Figure 10.1 Electric model of a DC motor with permanent excitation.

and $\tau_l = B\omega$ where B is the coefficient of viscous friction equal to 0.01 Nm s rad^{-1}. The supplied link voltage is $u_0 = 20$ V.

For speed control of a DC motor, a cascaded control structure is usually preferred, with an inner current control loop and an outer speed control loop. Control input u may be continuous or discontinuous, depending on the output power of the DC motor. A low-power system can use continuous control. A high-power system requires discontinuous control, perhaps in the form of *pulse-width modulation* (PWM), since a continuously controlled voltage is difficult to generate while providing a large current.

We concentrate on discontinuous control in the rest of this chapter, since control discontinuities are the very nature of sliding mode control. Furthermore, discontinuous control of DC motors is universal in the sense that it can be employed for low-power systems and for high-power systems. Figure 10.2 shows the typical control structure of a drive system based on a DC motor.

10.1.3 Current control

At first, assume there exists an outer control loop providing a desired current i^*. Let us consider the current control problem by defining switching function

$$s = i^* - i \tag{10.1.2}$$

as the error between the real, measured current i and the reference current i^* determined by the outer-loop controller. Design the discontinuous control as

$$u = u_0 \operatorname{sign}(s) \tag{10.1.3}$$

where u_0 denotes the supplied link voltage. As discussed in the previous chapters, to enforce the sliding mode, control gain u_0 should be selected such that $s\dot{s} < 0$. Now check this condition by evaluating $s\dot{s}$ and select u_0 as

$$\begin{aligned} s\dot{s} &= s\left(\frac{di^*}{dt} + \frac{R}{L}i + \frac{\lambda_0}{L}\omega\right) - \frac{1}{L}u_0|s| \\ u_0 &> \left|L\frac{di^*}{dt} + Ri + \lambda_0\omega\right| \end{aligned} \tag{10.1.4}$$

then sliding mode can be enforced.

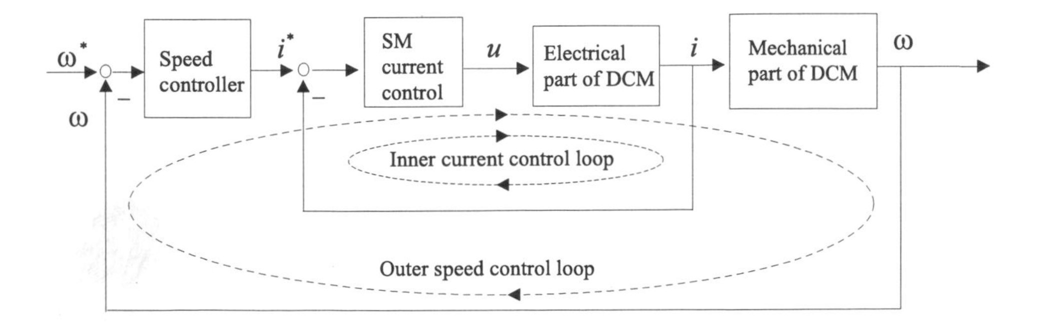

Figure 10.2 Cascaded control structure of DC motors.

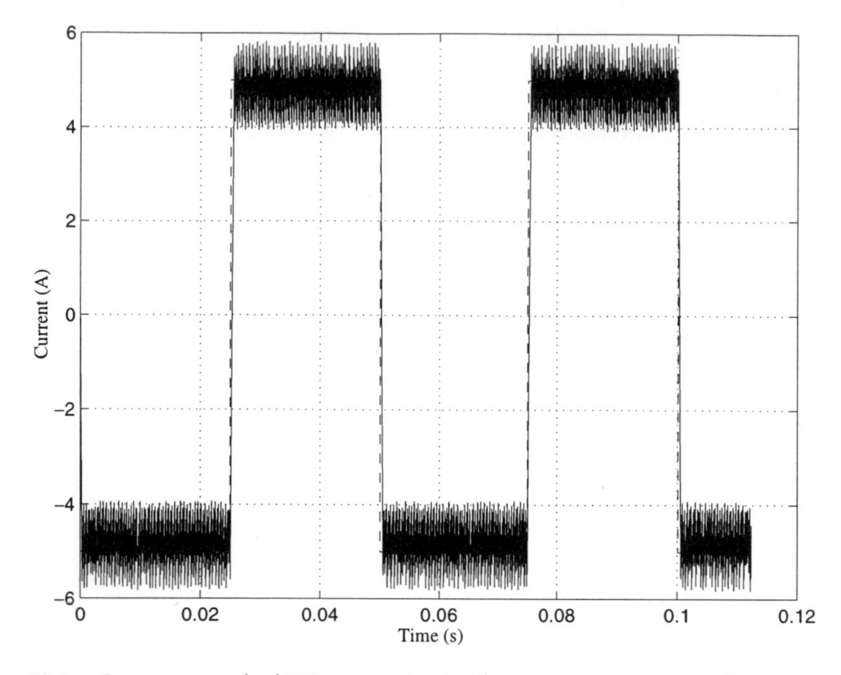

Figure 10.3 Current control of DC motor: (- - -) reference current, (—) real current.

Examine inequality (10.1.4). If the reference current is constant, the link voltage u_0 needed to enforce sliding mode should be higher than the voltage drop at the armature resistance plus the induced back-EMF, otherwise the reference current i^* cannot be followed. Furthermore, reference current i^* may not vary arbitrarily; its time derivative di^*/dt should be bounded to ensure existence of sliding mode for a given link voltage u_0. Figure 10.3 depicts a simulation result of the proposed current controller.

10.1.4 Speed control

For the speed controller in an outer loop, the current control loop may be treated as an ideal current source, i.e. given a reference current i^*, it will be tracked immediately. This assumption may become true only for systems in which the electrical time constant is much smaller than the mechanical time constant, or for systems where the dynamic response of the speed control is not a critical problem. Any control design technique, linear or nonlinear, may be used for speed control: PID (proportional integral derivative) control or a more sophisticated methodology, but without discontinuities like a sliding mode controller. The reason is that a sliding mode controller has already been employed in the inner current control loop; thus, if we were to use another sliding mode controller for speed control, the output of the speed controller i^* would be discontinuous, implying an infinite di^*/dt and therefore destroying inequality (10.1.4) for any implementable u_0.

In many industrial systems, PI (proportional integral) controllers are used with or without feedforward compensation, depending on the nature of the controlled system and the performance requirements. This type of controller is simple, but it may be sensitive to disturbances in the mechanical subsystem.

Suppose an exponential stability of the speed tracking error is desired, and design

$$c(\omega^* - \omega) + (\dot{\omega}^* - \dot{\omega}) = 0 \tag{10.1.5}$$

where c is a positive constant determining the convergence rate. As follows from the motor mechanical equation in (10.1.1), the reference current i^* feeding to the inner current controller should be selected as

$$i^* = \frac{J}{k_t} c(\omega^* - \omega) + \frac{J}{k_t} \dot{\omega}^* - \frac{1}{k_t} \tau_l \tag{10.1.6}$$

However, implementation of speed controller (10.1.6) requires knowledge of the motor parameters J, k_t and the load torque τ_l, which are normally unknown.

10.1.5 Integrated structure for speed control

To overcome the problems of the cascade control structure, we propose another control approach based on the sliding mode control principle to track a given speed trajectory. In this new control structure, the inner current controller is removed. Current control is achieved in an implicit manner. The advantages of this control structure lie in the fast dynamic response and high robustness with respect to disturbances in both the electrical and mechanical subsystems.

Let $\omega^*(t)$ be the reference shaft speed and $e = \omega^* - \omega$ be the speed tracking error. Define state variables $x_1 = e$ and $x_2 = \dot{e}$. The motion equation of the DC motor with respect to the states x_1, x_2 is given by

$$\begin{aligned} \dot{x}_1 &= x_2, \\ \dot{x}_2 &= -a_1 x_1 - a_2 x_2 + f(t) - bu \end{aligned} \tag{10.1.7}$$

where $a_1 = k_t \lambda_0/(JL)$, $a_2 = R/L$ and $b = k_t/(JL)$ are constant values. The linear part of (10.1.7) is perturbed by $f(t) = \ddot{\omega}^* + a_2\dot{\omega}^* + a_1\omega^* + R\tau_1/JL + \dot{\tau}_l/J$, depending on the desired speed and load torque disturbances. Since (10.1.7) is a second-order system, the switching function is designed as

$$s = cx_1 + \dot{x}_1 \tag{10.1.8}$$

with c being a positive constant. The associated controller is defined as

$$u = u_0 \operatorname{sign}(s) \tag{10.1.9}$$

where u_0 is the link voltage. According to these equations, the speed tracking error x_1 decays exponentially after sliding mode occurs in the manifold $s = 0$, i.e.

$$s = cx_1 + \dot{x}_1 = 0 \tag{10.1.10}$$

where constant c determines the rate of the convergence. The system motion in sliding mode is independent of parameters a_1, a_2, b and disturbances in $f(t)$. Similar to the case of current control, the link voltage u_0 should satisfy the condition

$$u_0 > \frac{1}{b} |cx_2 - a_1 x_1 - a_2 x_2 + f(t)| \tag{10.1.11}$$

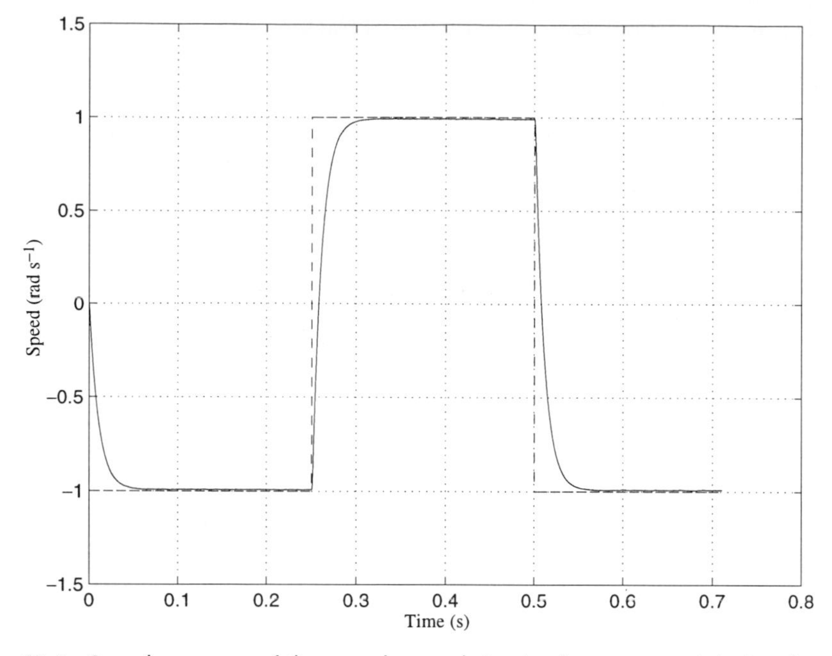

Figure 10.4 Speed response of the speed control: (- - -) reference speed, (—) real speed.

for sliding mode to exist. Then, after a finite time interval, the system state will reach the sliding manifold (10.1.10). Thereafter, the system response depends only on the design parameter c.

Figures 10.4 to 10.6 show the simulation results of the proposed speed controller. The control gain c is selected as $c = 100$. Figure 10.4 depicts the response of the speed control. As illustrated, the speed transition time is about 0.05 s. Figure 10.5 shows the waveform of the motor current. Bear in mind that the motor current is not controlled explicitly, the ordinary behaviour of the current is the result of the acceleration control. Figure 10.6 gives the response of the sliding variable s.

10.1.6 Observer design

For the implementation of the control algorithm given in (10.1.8) and (10.1.9), angular acceleration of the shaft is needed for calculating the sliding variable s in (10.1.8) due to $x_2 = \dot{\omega}^* - \dot{\omega}$. However, in practice, the angular acceleration is not measured. Instead, the motor current i and the shaft velocity ω are normally available. Numerical differentiation of the speed signal may result in a high level of noise in the calculated acceleration signal. In this case, the acceleration signal may be calculated as

$$\dot{\omega} = \frac{1}{J}k_t i - \frac{1}{J}\tau_l \tag{10.1.12}$$

where the load torque τ_l is assumed to be known; parameters J and k_t may be found by an identification process.

However, if the load torque is unknown or is varying under different working conditions, we may estimate it by making some assumptions which correspond to real-life conditions. For example, we may assume that the load torque changes very slowly, i.e.

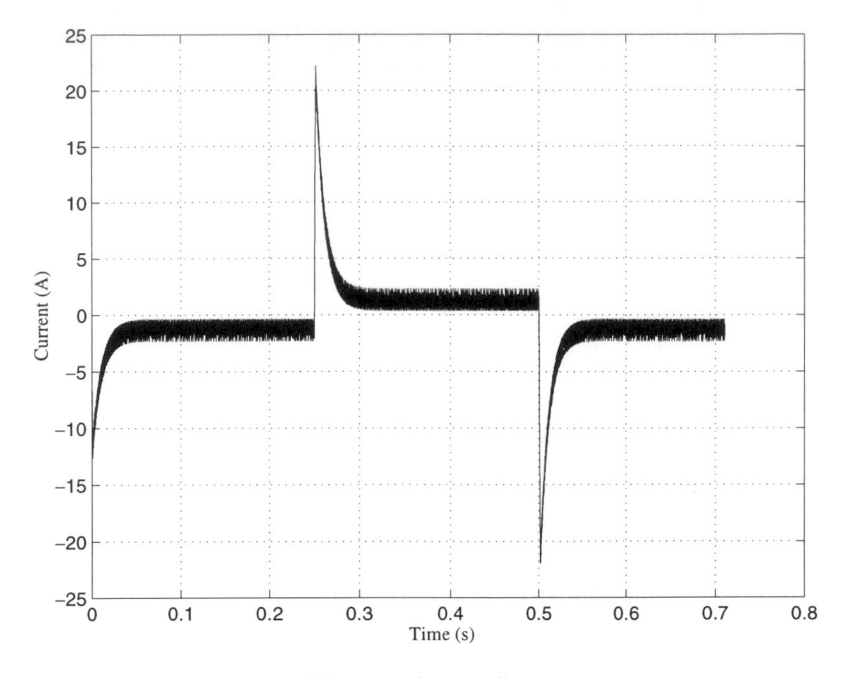

Figure 10.5 Current response of the speed controller.

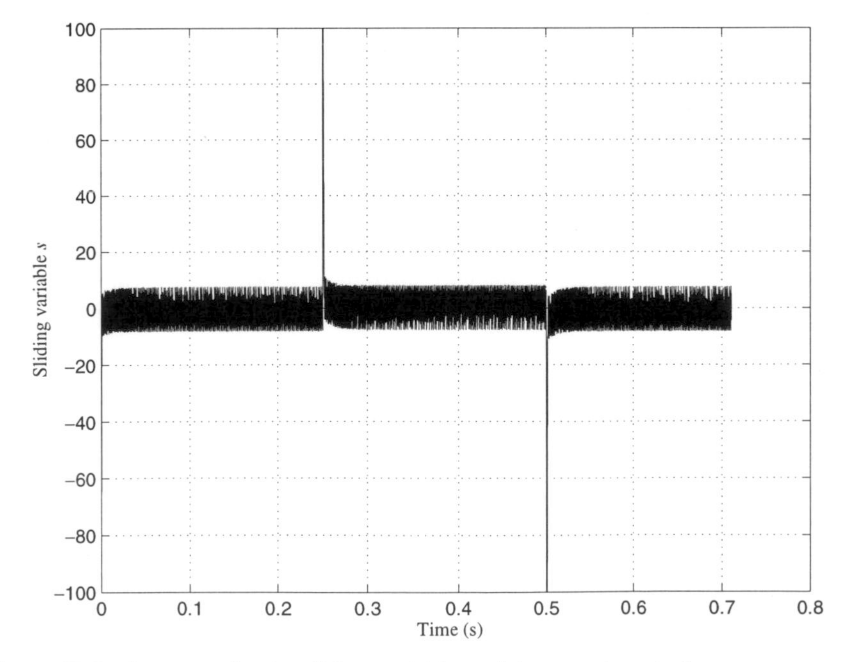

Figure 10.6 Response for the sliding variable s of the speed controller.

$\dot{\tau}_l = 0$. This condition means the load torque is assumed to be constant. Practically, this assumption has great significance, since it provides an equation for the observer design and enables one to estimate the load torque which varies 'slowly' with the time. The word 'slowly' is relative to the mechanical and the electrical time constant of the DC motor. An observer is proposed here for estimation of the load torque.

To involve the measured current and the velocity signals into the observer design, we employ the Luenberger reduced-order asymptotic observer (Section 6.1). An intermediate variable is introduced:

$$z = \tau_1 + l\omega \tag{10.1.13}$$

where l is a constant observer gain. The motion equation for z, under the assumption $\dot{\tau}_l = 0$, is of form

$$\frac{dz}{dt} = l\dot{\omega} = \frac{1}{J}(-lz + l^2\omega + lk_t i) \tag{10.1.14}$$

Design an observer for the intermediate variable z as

$$\frac{d\hat{z}}{dt} = \frac{1}{J}(-l\hat{z} + l^2\omega + lk_t i) \tag{10.1.15}$$

The solution to the mismatch equation

$$\frac{d\bar{z}}{dt} = -\frac{l}{J}\bar{z} \quad \text{with} \quad \bar{z} = \hat{z} - z \tag{10.1.16}$$

tends to zero exponentially and the rate of convergence can be selected by a proper choice of the observer gain l. As a result, $\hat{z}$ will converge to z asymptotically and the load torque can be estimated as

$$\tau_l = \hat{z} - l\omega \tag{10.1.17}$$

Now we are able to calculate the acceleration signal using equation (10.1.12).

Estimation of the shaft acceleration can also be achieved using the following Luenberger observer (Section 6.1) without explicitly involving knowledge of load torque:

$$\begin{aligned}\frac{d\hat{z}_1}{dt} &= \hat{z}_2 - l_1(\hat{\omega} - \omega)\\ \frac{d\hat{z}_2}{dt} &= \frac{k_t}{JL}(u - Ri - \lambda_0\hat{z}_1) - l_2(\hat{\omega} - \omega)\end{aligned} \tag{10.1.18}$$

where $\hat{\omega}$ is an estimate of the shaft speed, $\hat{z}_1 = \hat{\omega}$ and $\hat{z}_2 = \dot{\hat{\omega}}$, and l_1, l_2 are observer gains. Assumption $\dot{\tau}_1 = 0$ was also used when deriving (10.1.18). Denoting $\bar{z}_1 = \hat{z}_1 - z_1$ and $\bar{z}_2 = \hat{z}_2 - z_2$ as the mismatches between observed and real quantities, the mismatch dynamics can be obtained as

$$\begin{aligned}\frac{d\bar{z}_1}{dt} &= \bar{z}_2 - l_1\bar{z}_1\\ \frac{d\bar{z}_2}{dt} &= -\left(\frac{k_t\lambda_0}{JL} + l_2\right)\bar{z}_1\end{aligned} \tag{10.1.19}$$

The characteristic polynomial of the above system is given by

$$p^2 + l_1 p + \left(\frac{k_t \lambda_0}{JL} + l_2\right) = 0 \tag{10.1.20}$$

Obviously, the poles of the observer system can be placed arbitrarily by adjusting the observer gains l_1, l_2.

Figures 10.7 and 10.8 show the simulation results of the proposed estimation algorithms (10.1.12) to (10.1.17) for estimating the load torque as well as the shaft acceleration. The observer gain is designed to be $l = 40$. The moment of inertia J in the observer model is selected to have 10% difference from the real inertia in the motor model, in order to generate a difference in the figures. Otherwise, the observer outputs and the model outputs would be too close to be distinguished.

10.1.7 Speed control with reduced-order model

Speed control in a DC motor requires either control of the motor current or control of the motor acceleration. However, some industrial systems have employed simple relay-controlled DC motors based only on speed measurement. In this section we will design simplified sliding mode control based only on speed measurement Furthermore, the control methods proposed below will solve the chattering problem often encountered in those industrial systems.

The mechanical motion of a DC motor is normally much slower than the electromagnetic dynamics, implying that the relation $L \ll J$ holds. Suppose the speed tracking error is $\omega_e = \omega^* - \omega$, we may rewrite the DC motor model in terms of ω_e:

$$\begin{aligned} L\frac{di}{dt} &= u - Ri - \lambda_0(\omega^* - \omega_e) \\ J\frac{d\omega_e}{dt} &= -k_t i + \tau_l + J\dot{\omega}^* \end{aligned} \tag{10.1.21}$$

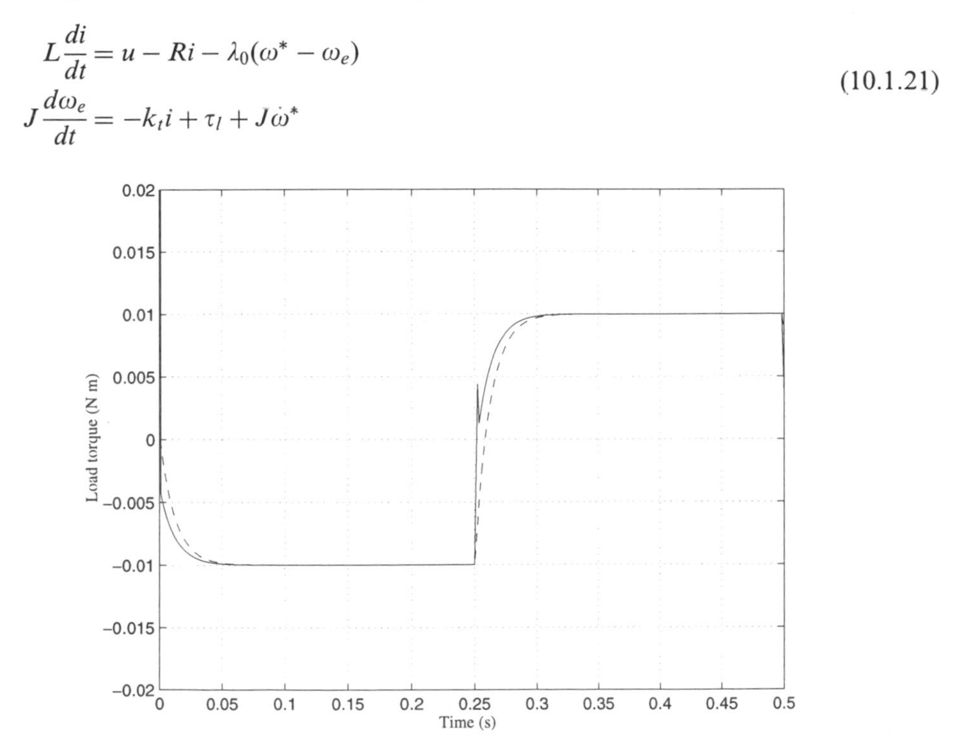

Figure 10.7 Load torque: (- - -) real, (—) observed.

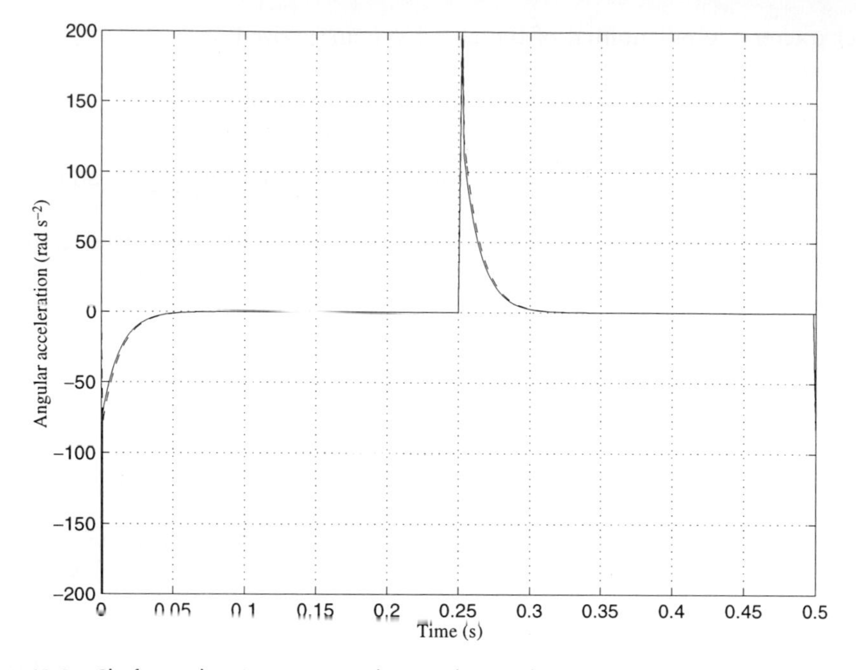

Figure 10.8 Shaft acceleration: (- - -) real, (—) observed.

Calling upon the theory of *singularly perturbed systems* (Kokotovic *et al.*, 1976), we may formally let L be equal to zero due to $L \ll J$. Solving for i from the resulting algebraic equation of (10.1.21) yields

$$i = -\frac{\lambda_0}{R}(\omega^* - \omega_e) + \frac{1}{R}u \tag{10.1.22}$$

Substituting (10.1.22) into (10.1.21) results in

$$J\frac{d\omega_e}{dt} = \frac{k_t\lambda_0}{R}(\omega^* - \omega_e) - \frac{k_t}{R}u + \tau_l + J\dot{\omega}^* \tag{10.1.23}$$

This equation is a reduced-order (first-order) model of the DC motor system and can be used for the purpose of speed control without involving the knowledge of current and acceleration. The controller is designed as

$$u = u_0 \ \text{sign}(\omega_e) \tag{10.1.24}$$

and the existence condition for the sliding mode $\omega_e = 0$ will be

$$u_0 > \left|\lambda_0\omega + \frac{\tau_l R}{k_t} + \frac{JR}{k_t}\dot{\omega}^*\right| \tag{10.1.25}$$

However, in real-life systems under sliding mode control, neglected dynamics in the closed loop may result in the chattering phenomenon. In the case of the speed controller (10.1.24), current i is assumed to be a linear function of the input voltage u in equation (10.1.22) due to the assumption $L \approx 0$. From a macroscopic perspective, this assumption holds true. For high-frequency switching of the discontinuous input voltage u according to (10.1.24), however, the electrical dynamics prevent ideal tracking of voltage u by armature current i; this leads to chattering (Chapter 8). Moreover, the order reduction technique of *singular perturbation theory* is formally not applicable to differential equations with discontinuous right-hand sides.

In a sliding mode control system with unmodelled dynamics, chattering can be bypassed by constructing the sliding manifold using observed states rather than direct measurements (Section 8.3). Bearing in mind the assumption $\dot{\tau}_l = 0$, design an asymptotic observer for estimating ω_e and τ_l as

$$\begin{aligned} J\frac{d\hat{\omega}_e}{dt} &= \frac{k_t\lambda_0}{R}(\omega^* - \hat{\omega}_e) - \frac{k_t}{R}u + \hat{\tau}_l + J\dot{\omega}^* - l_1(\hat{\omega}_e - \omega_e) \\ \frac{d\hat{\tau}}{dt} &= -l_2(\hat{\omega}_e - \omega_e) \end{aligned} \tag{10.1.26}$$

where l_1 and l_2 are positive observer gains. The mismatch dynamics of the observer can be obtained as

$$J\ddot{\bar{\omega}}_e + \left(\frac{k_t\lambda_0}{R} + l_1\right)\dot{\bar{\omega}}_e + \frac{l_2}{J}\bar{\omega}_e = 0 \tag{10.1.27}$$

with $\bar{\omega}_e = (\hat{\omega}_e - \omega_e)$. Since all the coefficients in this equation are positive, $\bar{\omega}_e$ will tend to zero asymptotically and the desired convergence rate can be provided by selection of the constants l_1 and l_2. The discontinuous control using estimated state $\hat{\omega}_e$ will be

$$u = u_0 \operatorname{sign}(\hat{\omega}_e) \tag{10.1.28}$$

The ideal sliding mode can be enforced if

$$u_0 > \left|\lambda_0(\omega^* - \hat{\omega}_e) + \frac{\hat{\tau}_l R}{k_t} + \frac{JR}{k_t}\dot{\omega}^* + \frac{l_1 R}{k_t}\bar{\omega}_e\right| \tag{10.1.29}$$

Under this control scheme, chattering is eliminated, but the robustness provided by the sliding mode control is preserved within an accuracy of L/J. (Remember that the reduced-order equation was obtained under the assumption $L/J \ll 1$.) The sliding mode occurs in the observer loop, which does not contain unmodelled dynamics. The observer gains l_1 and l_2 should be chosen to yield mismatch dynamics (10.1.27) slower than the electrical dynamics of the DC motors to prevent chattering. Since the estimated $\hat{\omega}$ is close to ω, the real speed ω tracks the desired value ω^*. Figure 10.9 shows the control structure based on the reduced-order model and observed state.

Figure 10.10 shows a simulation result of the reduced-order speed control with measured speed (10.1.24). The high-frequency chatter is due to neglecting the fast dynamics, i.e. the dynamics of the electric part. Figure 10.11 depicts the response of the speed control with observed speed using (10.1.26). The observer gains are selected as $l_1 = l_2 = 20$. Notice that the high-frequency chattering has now disappeared, confirming the theory.

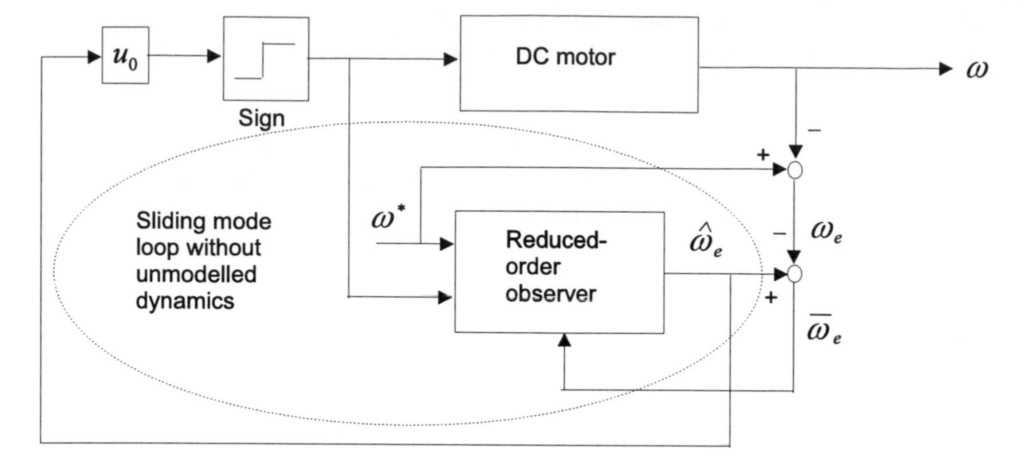

Figure 10.9 Speed control based on reduced-order model and observed state.

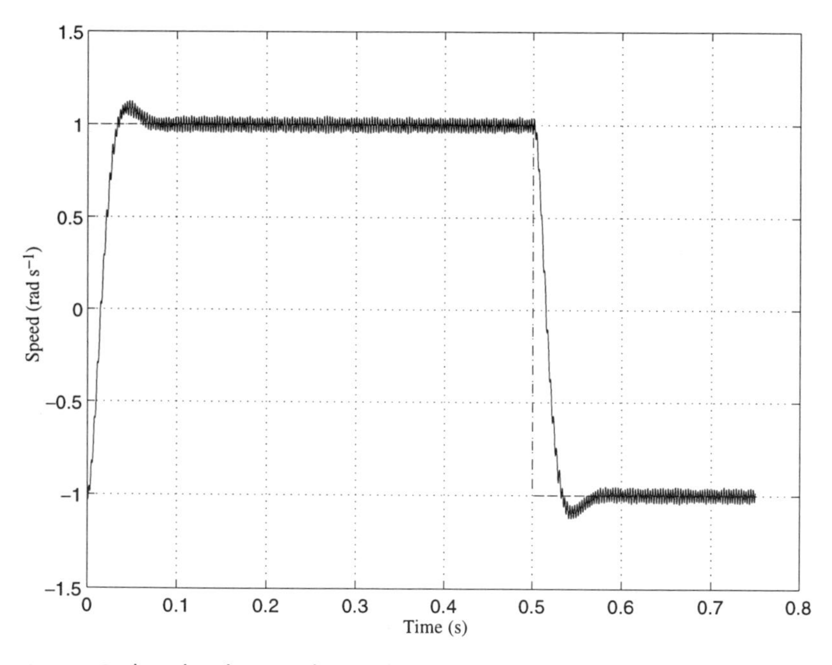

Figure 10.10 Reduced-order speed control with measured speed: (- - -) reference speed, (—) real speed.

10.1.8 Observer design for sensorless speed control

Strictly speaking, the word 'sensorless' is not correct, since one must sense or measure some variable to obtain some information as the basis of estimating the unknown variables. Normally, sensorless control of an electric drive implies that no sensor for any mechanical variable is necessary, but electrical variables like motor current and voltage should be available. In the following, we treat the problem of estimating the motor speed and/or the load torque based on the motor current and voltage. The methodology used here is again based on the sliding mode design principle.

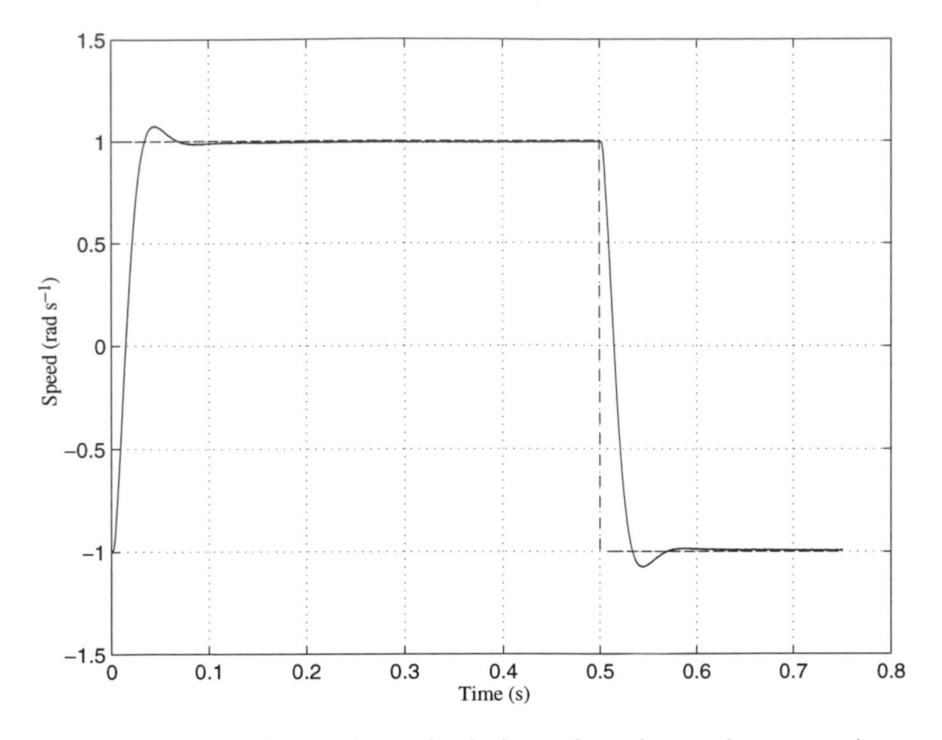

Figure 10.11 Reduced-order speed control with observed speed: (- - -) reference speed, (—) real speed.

Estimation of the shaft speed

Design a current observer as follows:

$$L\frac{d\hat{i}}{dt} = u - R\hat{i} - l_1 \ \text{sign}(\bar{i}) \tag{10.1.30}$$

where $\bar{i} = \hat{i} - i$ and $l_1 > 0$ is a constant observer gain. The dynamics of the mismatch $\bar{i}$ can be obtained by subtracting (10.1.30) from the model equation given in (10.1.1):

$$L\frac{d\bar{i}}{dt} = -R\bar{i} + \lambda_0\omega - l_1 \ \text{sign}(\bar{i}) \tag{10.1.31}$$

Now select the constant l_1 to enforce sliding mode ($s = \bar{i} = 0$) by applying the existence condition $s\dot{s} < 0$ to yield

$$l_1 > |\lambda_0\omega - R\bar{i}| \tag{10.1.32}$$

under which $\bar{i}$ will decay to zero in finite time. Using the concept of equivalent control (Section 2.3), we obtain

$$L\frac{d\bar{i}}{dt} = -R\bar{i} + \lambda_0\omega - (l_1 \ \text{sign}(\bar{i}))_{eq} = 0 \tag{10.1.33}$$

and

$$(l_1 \text{ sign}(\bar{i}))_{eq} = -R\bar{i} + \lambda_0 \omega \tag{10.1.34}$$

After the reaching phase of sliding mode, $\bar{i}$ is equal to zero and for known back-EMF constant λ_0 (which can be determined during the parameter identification), the motor speed can be obtained as

$$\omega = (l_1 \text{ sign}(\bar{i}))_{eq}/\lambda_0 \tag{10.1.35}$$

Now the problem is how to determine the value of $(l_1 \text{ sign}(\bar{i}))_{eq}$. To determine an equivalent control, a lowpass filter may be employed (Section 2.3). Only first-order filters were analyzed in Section 2.3. In fact, the same methodology is applicable for higher-order lowpass filters, e.g. an l-order lowpass filter like

$$\begin{aligned} \mu\dot{x} &= Ax + B(l_1 \text{ sign}(\bar{i})) \\ y &= C^T x \end{aligned} \tag{10.1.36}$$

where $A \in \Re^{l \times l}$, $B \in \Re^l$ and $C \in \Re^l$ are the filter parameters which satisfy $C^T A^{-1} B = -1$; $x \in \Re^l$ represents the state vector of the filter and y is the filter output; small positive parameter $\mu \ll 1$ represents the filter time constant. Calling upon the theory of singularly perturbed systems, if $\mu \to 0$, then $y \to (l_1 \text{ sign}(\bar{i}))_{eq}$.

So far, we have discussed the problem of estimating the speed signal. Once the speed signal is available, a speed controller may be constructed. Bearing in mind that the speed signal is obtained by passing the discontinuous control through a lowpass filter, high-frequency components of the discontinuous control may not be filtered out completely due to the limitation on the filter time constant (i.e. μ should be small enough). A sliding mode controller will no longer be suitable for speed control due to the high-frequency disturbances. It would be helpful if the speed control included an integral term, e.g. a PI type controller is appropriate. In this case the controlled real speed will follow the mean value of the estimated speed.

Estimation of load torque

As the speed signal is available, estimated or measured, the load torque can also be obtained by applying the sliding mode technique to an observer. First, design a speed observer as follows:

$$J\frac{d\hat{\omega}}{dt} = k_t i - l_2 \text{ sign}(\bar{\omega}), \quad \text{with} \quad \bar{\omega} = \hat{\omega} - \omega \tag{10.1.37}$$

Suppose that the parameters J and k_t are known, then the mismatch $\bar{\omega}$ is governed by

$$J\frac{d\bar{\omega}}{dt} = \tau_l - l_2 \text{ sign}(\bar{\omega}) \tag{10.1.38}$$

where l_2 is a constant observer gain. Applying the existence condition for sliding mode to occur in $\bar{\omega} = 0$ yields

$$\bar{\omega}\dot{\bar{\omega}} = \frac{\tau_l}{J}\bar{\omega} - \frac{l_2}{J}|\bar{\omega}| < 0 \tag{10.1.39}$$

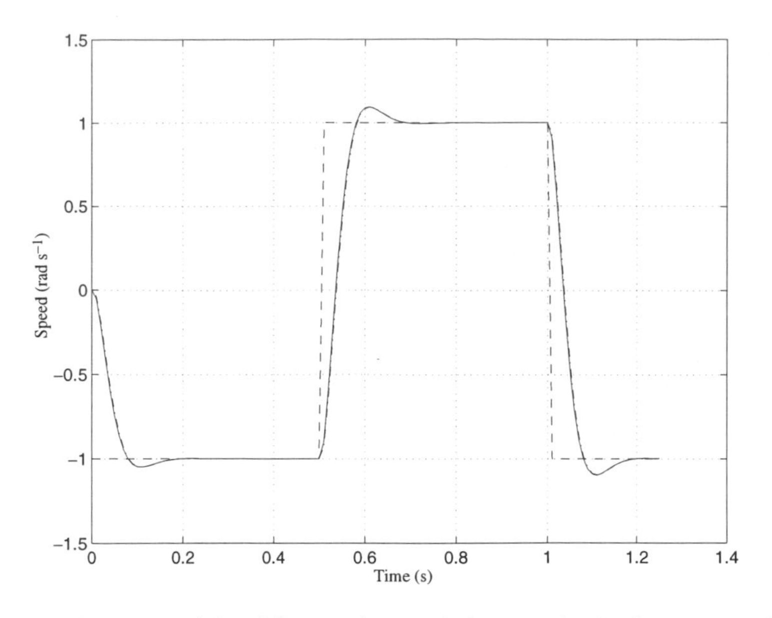

Figure 10.12 Response of the sliding mode speed observer: (- - -) reference speed, (—) real speed; (- · -) observed speed.

implying that $l_2 > |\tau_l|$ should hold. Since l_2 appears only in the observer algorithm, it can be selected high enough. Once sliding mode occurs, the load torque is equal to the equivalent control of the discontinuous term $l_2 \operatorname{sign}(\bar{\omega})$, i.e.

$$\tau_l = (l_2 \operatorname{sign}(\bar{\omega}))_{eq} \tag{10.1.40}$$

To extract the equivalent control given above, we need a lowpass filter again.

Note that, during the derivations of the load torque observer, no assumption like $\dot{\tau}_l = 0$ was made. Actually, this observer design scheme works as long as the spectrum of the load torque does not intersect with the spectrum of the high-frequency components of the switching action.

Figures 10.12 to 10.14 show the simulation results of the proposed observer algorithms for sensorless control. The observer gains are chosen as $l_1 = l_2 = 20\,000$. The lowpass filters for extracting the motor speed and the load torque are both of Butterworth type with second order. The cutoff frequencies of the lowpass filters are, 1000 rad s^{-1} and 550 rad s^{-1}, respectively. The motor is commanded to follow a block-form reference speed. As shown in Figure 10.12, the estimated speed tracks the real speed very closely, as do the estimated acceleration and the estimated load torque (Figure 10.13 and Figure 10.14).

10.1.9 Discussion

Generally speaking, uncertainties in the plant model of any observer design will cause the observed state to differ from the real state. If this observed state is used within the control loop, it may result in a control error. However, if the observer error is within a small range, the control error is also bounded by a similar small range, implying that the observed state may still be useful for increasing the control performance.

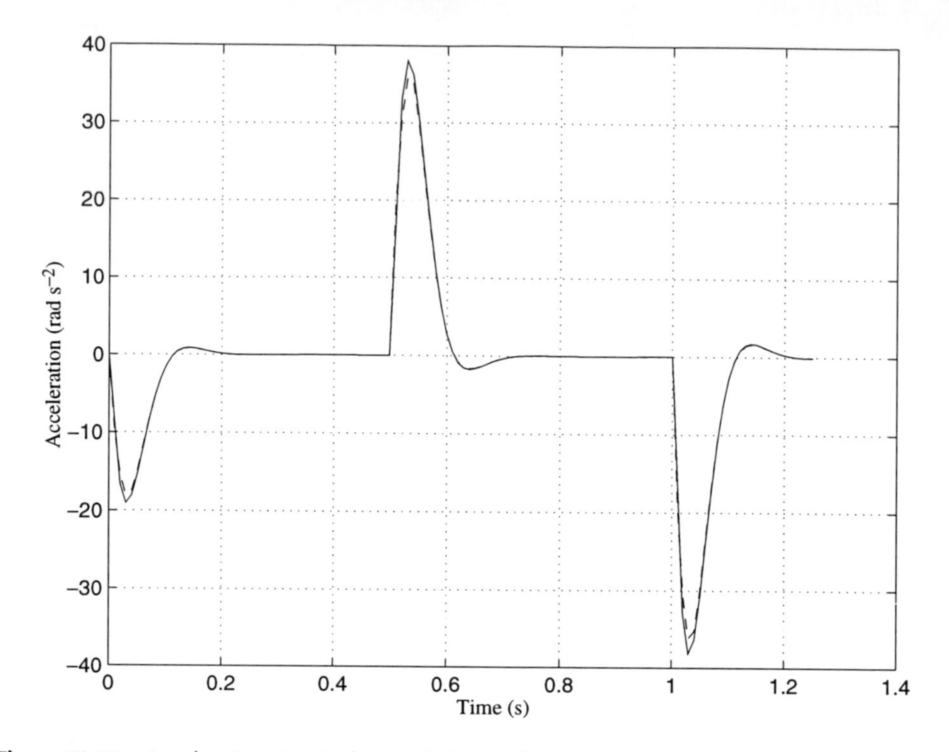

Figure 10.13 Acceleration: (- - -) observed, (—) real.

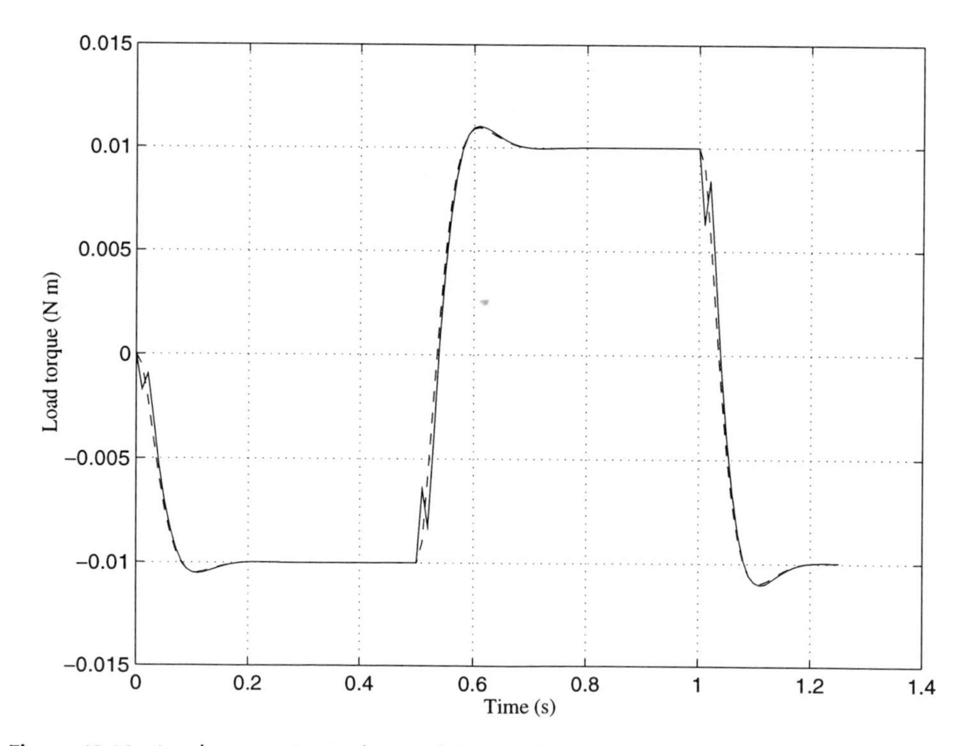

Figure 10.14 Load torque. (- - -) observed, (—) real.

For implementation of an observer, the observer equation should be integrated in real time using a microprocessor or signal processor with a fixed sampling period. As a result, the system to be controlled is continuous in time, but the observer in the control computer is a discrete-time system. Four situations will lead to a large error of integration:

- The sampling period of the microprocessor is too large.
- The integration algorithm is primitive.
- The observer gain is too high.
- The time constant of the observer system is too small.

The system accuracy may be improved by decreasing the sampling interval. However, for a larger sampling interval, more complex and efficient control algorithms may be implemented. The control engineer has to find a suitable trade-off. Readers particularly interested in microprocessor implementation of sliding mode controllers and observers are encouraged to study Chapter 9 on discrete-time systems.

10.2 Permanent-magnet synchronous motors

10.2.1 Introduction

Permanent-magnet synchronous motors (PMSMs) belong to the category of alternating current (AC) drives. The term 'brushless DC motor', often used in the fields of machine tools and robotics, actually refers to a current-controlled PMSM.

Note that the discussion of the control problem of PMSMs in the field-oriented reference frame, called the (d, q) frame, is independent of the number of phases; two-phase motors and three-phase motors have the same structure in this coordinate frame. The differences lie only in the forward and back transformations of the states and controls (here currents and voltages). In fact, a model in the (d, q) frame is a uniform description of an AC device based on the electromagnetic principle, including all types of AC drives and power converters. Control design methods performed in the (d, q) coordinates are called *field-oriented control* (FOC). FOC is sometimes called *vector control*, implying that a current vector should be controlled rather than only one current component, as in the case of a pure DC motor. The principle of FOC was developed more than 20 years ago (Blaschke, 1974). From a control viewpoint, this approach uses a state transformation after which the decoupling and linearization tasks can be performed easily. Recently, owing to rapid progress in semiconductor technology, high-speed microprocessors and signal processors have become much more common in electric drive systems. As a result, the concept of FOC has been implemented in many drive systems of AC machines. The remaining problem is robustness associated with parameter variations and load variations.

Theoretically, sliding mode control provides advantages over conventional control designs such as decoupling, linearization, linear control and PWM, due to the simple control structure and the robustness property of the sliding mode control principle. However, it is recognized that in microprocessor-based implementations, the sampling interval, and thus the minimum switching interval of the sliding mode controller, is much larger than the resolution of a hardware-based PWM. As a result, some deterioration in control performance was observed. This problem may be solved by *discrete-time sliding mode control* (Chapter 9) in conjunction with hardware PWM techniques. Moreover, since the structure of a sliding mode controller is very simple, a nonmicroprocessor, purely hardware-based implementation is possible and often leads to favourable results. Hence the effectiveness of the sliding mode control can be fully demonstrated without using PWM units.

10.2.2 Modelling of the PMSH motor

The structure of a PMSM-based drive system is shown in Figure 10.15. For sliding mode control design, it is convenient if the control inputs take values from the discrete set $\{-u_0, u_0\}$ instead of *on-off* signals from the discrete set $\{0, 1\}$. Let the six on-off signals be $\boldsymbol{s}_w = [s_{w1}\ s_{w2}\ s_{w3}\ s_{w4}\ s_{w5}\ s_{w6}]^T$ with $s_{w4} = 1 - s_{w1}$, $s_{w5} = 1 - s_{w2}$, $s_{w6} = 1 - s_{w3}$, and the control inputs for sliding mode control design be $\boldsymbol{U}_{gate} = [u_1\ u_2\ u_3]^T$, then the following relation holds:

$$\boldsymbol{U}_{gate} = u_0 G_w \boldsymbol{s}_w, \quad \text{with} \quad G_w = \begin{bmatrix} 1 & 0 & 0 & -1 & 0 & 0 \\ 0 & 1 & 0 & 0 & -1 & 0 \\ 0 & 0 & 1 & 0 & 0 & -1 \end{bmatrix} \tag{10.2.1}$$

The backward transformation can be obtained as

$$\begin{aligned} s_{w1} &= 0.5(1 + u_1/u_0) \qquad & s_{w4} &= 1 - s_{w1} \\ s_{w2} &= 0.5(1 + u_2/u_0) \qquad & s_{w5} &= 1 - s_{w2} \\ s_{w3} &= 0.5(1 + u_3/u_0) \qquad & s_{w6} &= 1 - s_{w3} \end{aligned} \tag{10.2.2}$$

In general, the dynamic model of an AC motor can be established using physical laws:

$$\begin{aligned} \boldsymbol{U} &= \boldsymbol{RI} + \frac{d\boldsymbol{\Psi}}{dt} \\ \boldsymbol{\Psi} &= \boldsymbol{LI} + \boldsymbol{\Psi}_M \end{aligned} \tag{10.2.3}$$

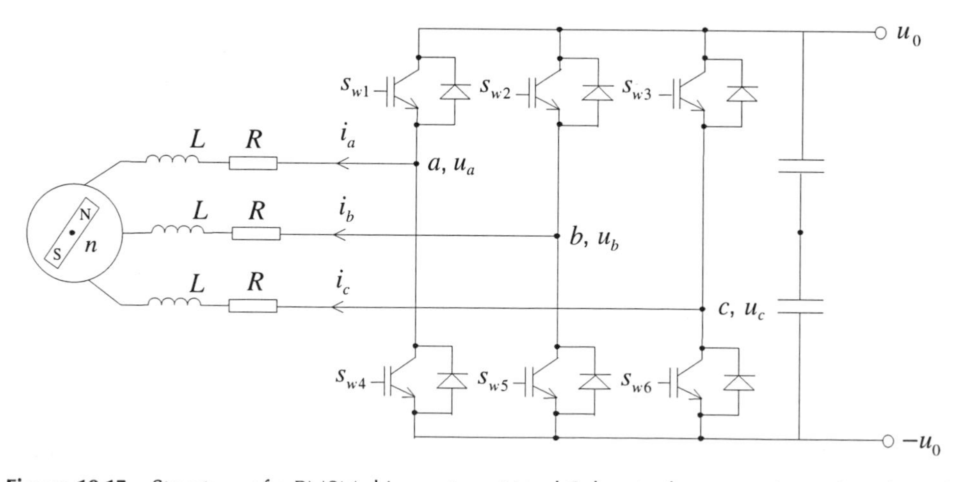

Figure 10.15 Structure of a PMSM drive system. N and S denote the magnetic north and south, respectively; n is the neutral point of the stator windings; u_a, u_b and u_c are the potential differences between points a, b and c and the neutral point n, respectively.

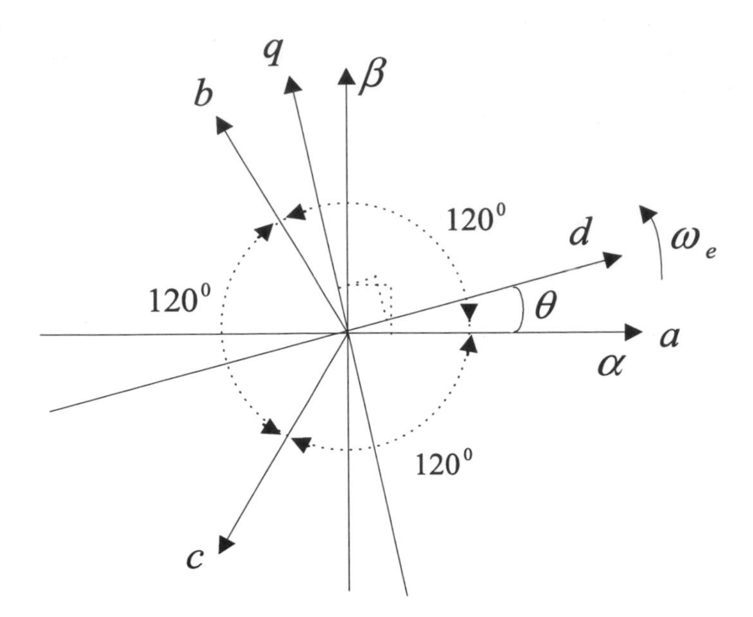

Figure 10.16 Coordinate systems of PMSM.

where $\boldsymbol{U}$, $\boldsymbol{I}$ and $\boldsymbol{\Psi}$ are the voltage vector, the current vector and the flux vector, respectively; $\boldsymbol{R}$, $\boldsymbol{L}$ are the resistance matrix and the inductance matrix, respectively; $\boldsymbol{\Psi}_M$ is the flux vector caused by the permanent magnet, if applicable.

The equation system (10.2.3) is a general description of the electromagnetic effects and is independent of the employed coordinate system. For PMSMs, three reference frames are normally used to describe the dynamic behaviour of a motor (Figure 10.16): the phase frame, i.e. the (a, b, c) coordinate frame; the stator frame, i.e. the (α, β) coordinate frame; and the field-oriented frame, i.e. the (d, q) coordinate frame (also called the rotor coordinate frame for PMSMs).

For a symmetrical PMSM in the (a, b, c) coordinate system, the flux components generated by the permanent magnet are given as

$$\begin{aligned}\Psi_{ma} &= \lambda_0 \cos\theta_e \\ \Psi_{mb} &= \lambda_0 \cos(\theta_e + 2\pi/3) \\ \Psi_{mc} &= \lambda_0 \cos(\theta_e - 2\pi/3)\end{aligned} \tag{10.2.4}$$

where λ_0 is the flux linkage of the permanent magnet and θ_e is the electrical angular position of the motor rotor. The electrical motion equations of a PMSM, neglecting the reluctance effects, can be written as

$$\begin{aligned}\frac{di_a}{dt} &= -\frac{R}{L}i_a - \frac{1}{L}e_a + \frac{1}{L}u_a \\ \frac{di_b}{dt} &= -\frac{R}{L}i_b - \frac{1}{L}e_b + \frac{1}{L}u_b \\ \frac{di_c}{dt} &= -\frac{R}{L}i_c - \frac{1}{L}e_c + \frac{1}{L}u_c\end{aligned} \tag{10.2.5}$$

where R is the winding resistance and L is the winding inductance; i_a, i_b, i_c are the phase currents and u_a, u_b, u_c the phase voltages. Furthermore, e_a, e_b, e_c are the induced EMF components of the following form:

$$\begin{aligned} e_a &= \frac{d\Psi_{ma}}{dt} = -\lambda_0\omega_e \sin\theta_e \\ e_b &= \frac{d\Psi_{mb}}{dt} = -\lambda_0\omega_e \sin(\theta_e + 2\pi/3) \\ e_c &= \frac{d\Psi_{mc}}{dt} = -\lambda_0\omega_e \sin(\theta_e - 2\pi/3) \end{aligned} \tag{10.2.6}$$

in which $\omega_e = d\theta_e/dt$ is the electrical angular speed of the motor rotor.

The relationship between the phase voltages u_a, u_b, u_c and the discontinuous controls u_1, u_2, u_3 is given by

$$[u_a \quad u_b \quad u_c]^T = A_{abc}^{123}[u_1 \quad u_2 \quad u_3]^T \tag{10.2.7}$$

where matrix A_{abc}^{123} is defined as

$$A_{abc}^{123} = \frac{1}{3}\begin{bmatrix} 2 & -1 & -1 \\ -1 & 2 & -1 \\ -1 & -1 & 2 \end{bmatrix} \tag{10.2.8}$$

Note that matrix A_{abc}^{123} is a singular matrix implying that the phase voltages u_a, u_b, u_c are not independent. As shown later, the sum of u_a, u_b, u_c is equal to zero due to the physical configuration given in Figure 10.15.

The motor model in (α, β) coordinates can be obtained either by applying the definition given in (10.2.3) or by transforming the motor model from the (a, b, c) coordinate frame into the (α, β) frame. The electrical part of the motor model in this coordinate frame is

$$\begin{aligned} \frac{di_\alpha}{dt} &= -\frac{R}{L}i_\alpha - \frac{1}{L}e_\alpha + \frac{1}{L}u_\alpha \\ \frac{di_\beta}{dt} &= -\frac{R}{L}i_\beta - \frac{1}{L}e_\beta + \frac{1}{L}u_\beta \end{aligned} \tag{10.2.9}$$

where $e_\alpha = -\lambda_0\omega_e \sin\theta_e$ and $e_\beta = -\lambda_0\omega_e \cos\theta_e$ are the induced EMF components in the (α, β) coordinate frame; the stator currents $[i_\alpha \quad i_\beta]^T$ and the stator voltages $[u_\alpha \quad u_\beta]^T$ are defined as follows:

$$\begin{aligned} [i_\alpha \quad i_\beta]^T &= A_{\alpha,\beta}^{a,b,c}[i_a \quad i_b \quad i_c]^T \\ [u_\alpha \quad u_\beta]^T &= A_{\alpha,\beta}^{a,b,c}[u_a \quad u_b \quad u_c]^T \end{aligned} \tag{10.2.10}$$

where $A_{\alpha,\beta}^{a,b,c}$ denotes the transformation matrix

$$A_{\alpha,\beta}^{a,b,c} = \frac{2}{3}\begin{bmatrix} 1 & -1/2 & -1/2 \\ 0 & \sqrt{3}/2 & -\sqrt{3}/2 \end{bmatrix} \tag{10.2.11}$$

Since the rank of this matrix is 2, the backward transformation has no unique solution. For the backward transformation of (10.2.10), the concept of a *pseudoinverse* matrix (also called the Moore-Penrose inverse) can be used. The pseudoinverse of matrix $A_{\alpha,\beta}^{a,b,c}$, denoted as $(A_{\alpha,\beta}^{a,b,c})^{+}$, is calculated as

$$(A_{\alpha,\beta}^{a,b,c})^{+} = (A_{\alpha,\beta}^{a,b,c})^{T}(A_{\alpha,\beta}^{a,b,c}(A_{\alpha,\beta}^{a,b,c})^{T})^{-1} = \frac{3}{2}(A_{\alpha,\beta}^{a,b,c})^{T} = \begin{bmatrix} 1 & -1/2 & -1/2 \\ 0 & \sqrt{3}/2 & -\sqrt{3}/2 \end{bmatrix}^{T} \tag{10.2.12}$$

It is easy to prove that $A_{\alpha,\beta}^{a,b,c}(A_{\alpha,\beta}^{a,b,c})^{+}$ is a 2×2 unit matrix. As a result, the backward transformation of the stator currents and voltages can be given as

$$\begin{aligned} [i_a \quad i_b \quad i_c]^T &= (A_{\alpha,\beta}^{a,b,c})^{+}[i_\alpha \quad i_\beta] \\ [u_a \quad u_b \quad u_c]^T &= (A_{\alpha,\beta}^{a,b,c})^{+}[u_\alpha \quad u_\beta]^T \end{aligned} \tag{10.2.13}$$

The motor model in the (d, q) coordinate frame, which rotates synchronously with the motor rotor, can also be obtained by transforming the motor model from the (α, β) coordinate frame to (d, q) coordinates:

$$\begin{aligned} \frac{di_d}{dt} &= -\frac{R}{L}i_d + \omega_e i_q + \frac{1}{L}u_d \\ \frac{di_q}{dt} &= -\frac{R}{L}i_q - \omega_e i_d - \frac{1}{L}\lambda_0\omega_e + \frac{1}{L}u_q \end{aligned} \tag{10.2.14}$$

where i_d and i_q are the stator currents in the (d, q) coordinate frame; u_d and u_q are the stator voltages in the same coordinate frame. Term $\lambda_0\omega_e = e_q$ is the q-component of the induced EMF generated by the permanent magnet; the d-component of the EMF e_d is equal to zero. Note the second equation of (10.2.14). If the current component i_d could be made equal to zero, we would get exactly the behaviour of a constant-excited DC motor. This is the main idea of field-oriented control: to decouple the motor dynamics such that the resulting system behaves like a DC motor. The current vector $[i_d \quad i_q]^T$ and the voltage vector $[u_d \quad u_q]^T$ can be transformed from the (α, β) coordinate frame

$$\begin{aligned} [i_d \quad i_q]^T &= A_{d,q}^{\alpha,\beta}[i_\alpha \quad i_\beta]^T \\ [u_d \quad u_q]^T &= A_{d,q}^{\alpha,\beta}[u_\alpha \quad u_\beta]^T \end{aligned} \tag{10.2.15}$$

in which matrix $A_{d,q}^{\alpha,\beta}$ is defined as

$$A_{d,q}^{\alpha,\beta} = \begin{bmatrix} \cos\theta_e & \sin\theta_e \\ -\sin\theta_e & \cos\theta_e \end{bmatrix} \tag{10.2.16}$$

This matrix is an orthogonal matrix whose inverse is equal to its transpose. As a result, the backward transformation can be written as

$$\begin{aligned} [i_\alpha \quad i_\beta]^T &= (A_{d,q}^{\alpha,\beta})^T[i_d \quad i_q]^T \\ [u_\alpha \quad u_\beta]^T &= (A_{d,q}^{\alpha,\beta})^T[u_d \quad u_q]^T \end{aligned} \tag{10.2.17}$$

The relationship between the (d, q) coordinate frame and the (a, b, c) coordinate frame can also be established. The phase currents i_a, i_b, i_c and the phase voltages u_a, u_b, u_c are transformed into the (d, q) coordinates as follows:

$$\begin{aligned} [i_d \quad i_q]^T &= A_{d,q}^{a,b,c}[i_a \quad i_b \quad i_c]^T \\ [u_d \quad u_q] &= A_{d,q}^{a,b,c}[u_a \quad u_b \quad u_c]^T \end{aligned} \tag{10.2.18}$$

where matrix $A_{d,q}^{a,b,c}$ depends on the electrical angular position of the rotor θ_e and is defined as

$$A_{d,q}^{a,b,c} = A_{d,q}^{\alpha,\beta}A_{\alpha,\beta}^{a,b,c} \tag{10.2.19}$$

Matrix $A_{d,q}^{a,b,c}$ is a 2×3 matrix, hence for the backward transformation we again need its pseudoinverse given as

$$(A_{d,q}^{a,b,c})^+ = (A_{d,q}^{a,b,c})^T(A_{d,q}^{a,b,c}(A_{d,q}^{a,b,c})^T)^{-1} = \frac{3}{2}(A_{d,q}^{a,b,c})^T \tag{10.2.20}$$

resulting in

$$\begin{aligned} [i_a \quad i_b \quad i_c]^T &= (A_{d,q}^{a,b,c})^+[i_d \quad i_q]^T \\ [u_a \quad u_b \quad u_c]^T &= (A_{d,q}^{a,b,c})^+[u_d \quad u_q]^T \end{aligned} \tag{10.2.21}$$

The relationship between the control voltages u_d, u_q and the discontinuous controls u_1, u_2, u_3 can be established as

$$[u_d \quad u_q]^T = A_{d,q}^{1,2,3}[u_1 \quad u_2 \quad u_3]^T \tag{10.2.22}$$

where matrix $A_{d,q}^{1,2,3}$ is defined as

$$A_{d,q}^{1,2,3} = A_{d,q}^{\alpha,\beta}A_{\alpha,\beta}^{a,b,c}A_{a,b,c}^{1,2,3} \tag{10.2.23}$$

Matrices $A_{\alpha,\beta}^{a,b,c}$ and $A_{a,b,c}^{1,2,3}$ satisfy the condition $A_{\alpha,\beta}^{a,b,c}A_{a,b,c}^{1,2,3} = A_{\alpha,\beta}^{a,b,c}$, therefore

$$A_{d,q}^{1,2,3} = A_{d,q}^{\alpha,\beta}A_{\alpha,\beta}^{a,b,c} = A_{d,q}^{a,b,c} \tag{10.2.24}$$

However, to maintain clarity, we prefer to use matrix $A_{d,q}^{1,2,3}$ to denote the transformation between the discontinuous controls u_1, u_2, u_3 and the control voltages u_d, u_q. Matrix $A_{d,q}^{1,2,3}$ is a 2×3 matrix, hence for the backward transformation we need its pseudoinverse as well:

$$(A_{d,q}^{1,2,3})^+ = (A_{d,q}^{1,2,3})^T(A_{d,q}^{1,2,3}(A_{d,q}^{1,2,3})^T)^{-1} = \frac{3}{2}(A_{d,q}^{1,2,3})^T \tag{10.2.25}$$

resulting in

$$[u_1 \quad u_2 \quad u_3]^T = (A_{d,q}^{1,2,3})^+[u_d \quad u_q]^T \tag{10.2.26}$$

Finally, the electrical torque τ_e and the mechanical power P of the motor are given by

$$\begin{aligned} \tau_e &= K_t i_q \\ P &= \tau_e \omega_r \end{aligned} \tag{10.2.27}$$

in which K_t is the torque constant, assumed to be equal to $(3/2)\lambda_0 N_r$ with N_r being the number of pole pairs of the motor, and ω_r is the mechanical angular speed of the motor rotor. In developing the motor models, we assume there is no reluctance torque in the PMSM motor. Under this assumption, the output torque of the motor is proportional to the q-axis stator current i_q. The mechanical motion equation of the motor can be written as

$$\begin{aligned} J\frac{d\omega_r}{dt} &= \tau_e - \tau_l \\ \frac{d\theta_r}{dt} &= \omega_r \end{aligned} \tag{10.2.28}$$

where τ_l and θ_r denote the load torque and the mechanical angular position of the motor rotor. For the electrical angular position/speed and the mechanical angular position/speed, the following relations hold:

$$\begin{aligned} \omega_e &= N_r \omega_r \\ \theta_e &= N_r \theta_r \end{aligned} \tag{10.2.29}$$

Rotor position θ_r is usually measured; ω_r, θ_e, ω_e are calculated according to (10.2.28) and (10.2.29).

For stator windings connected at the neutral point n, the following *balance conditions* hold:

$$\begin{aligned} I_a + I_b + I_c &= 0 \\ e_a + e_b + e_c &= 0 \\ u_a + u_b + u_c &= 0 \end{aligned} \tag{10.2.30}$$

Throughout this section the motor parameters used to verify the design principles are $L = 1.0\,\text{mH}$, $R = 0.5\,\Omega$, $J = 0.001\,\text{kg}\,\text{m}^2$, $B = 0.01\,\text{N}\,\text{m}\,\text{s}\,\text{rad}^{-1}$, $N_r = 4$, $\lambda_0 = 0.001\,\text{V}\,\text{s}\,\text{rad}^{-1}$, $K_t = (3/2)\lambda_0 N_r\,\text{N}\,\text{m}\,\text{A}^{-1}$. The supplied link voltage is $u_0 = 20\,\text{V}$.

So far we have discussed the model descriptions in the different coordinate systems and the transformations of the state variables and/or the control signals between these reference frames. In the following sections we deal with the sliding mode control issues of PMSMs.

10.2.3 Current Control

The goal of the current control is to design a current controller to track the desired currents which are normally provided by an outer-loop speed/position controller. A current controller can be implemented either with pure hardware or with a microprocessor. The so-called *chopper control* and *hysteresis control* are the hardware versions of a current controller due to the simplicity in the implementation. For field-oriented current control, however, a microprocessor-based implementation is recommended. Current control based on the sliding mode approach can be implemented either with pure hardware or within a microprocessor, and for both implementations, the control performance provided by the field orientation concept can be achieved.

In the context of sliding mode control of AC drives, there are two methods to determine the discontinuous controls u_1, u_2, u_3 as well as the on-off signals s_{w1}, $s_{w2}, s_{w3}, s_{w4}, s_{w5}, s_{w6}$. The on-off signals may also be called switching patterns. The switching patterns are the control signals feeding to the gates of the power converters, e.g. a voltage source inverter as shown in Figure 10.15.

The first method implies that the control voltages u_d and u_q are designed using the existence condition of sliding mode and mapping the resulting controls to the switching patterns of the inverter. The second method determines the switching patterns directly using the method of switching surface transformation (Section 3.2). Both methods are able to generate switching commands for the voltage source inverter without involving the traditional PWM technique.

First method for current control

Since the model in (d, q) coordinates gives clear physical interpretation in terms of a DC motor, we prefer to start with this model. Model (10.2.14) is still a coupled nonlinear dynamic system with two control inputs u_d and u_q. However, if we are able to reduce the current component i_d to zero, we can get exactly the same behaviour as for a DC motor with constant excitation flux. Another difference from a DC motor is that all variables with subscripts d and q cannot be measured directly, but rather are transformed from the variables measured in the (a, b, c) coordinate frame. For performing this transformation, the electrical angular position of the rotor θ_e is required.

Like a DC motor, for current control of a PMSM we may design the switching functions as the difference between the desired and real currents. Select the switching functions for both current components i_d and i_q as

$$\begin{aligned} s_d &= i_d^* - i_d \\ s_q &= i_q^* - i_q \end{aligned} \tag{10.2.31}$$

where i_d^* and i_q^* denote the desired value for the currents i_d and i_q, respectively.

Conventional field-oriented control of PMSMs employs linear control design techniques. The nonlinear model (10.2.14) has to be decoupled and linearized before linear control techniques can be applied. However, since the sliding mode approach belongs to the category of nonlinear control techniques, no such decoupling and linearization procedures are necessary. Nevertheless, it can be shown that the controlled errors, i.e. s_d and s_q, vanish after finite time. As to the switching functions given by (10.2.31), the existence condition of sliding mode (2.4.1) may be applied to find the controls.

Select the control voltages as

$$\begin{aligned} u_d &= u_{d0}\ \text{sign}(s_d) \\ u_q &= u_{q0}\ \text{sign}(s_q) \end{aligned} \tag{10.2.32}$$

where u_{d0}, u_{q0} are the amplitudes of u_d, u_q, respectively. It follows from

$$s_d\dot{s}_d = s_d\left(\frac{di_d^*}{dt} + \frac{R}{L}i_d - \omega_e i_q\right) - \frac{1}{L}u_{d0}|s_d|$$
$$s_q\dot{s}_q = s_q\left(\frac{di_q^*}{dt} + \frac{R}{L}i_q + \omega_e i_d + \frac{1}{L}\lambda_0\omega_e\right) - \frac{1}{L}u_{q0}|s_q| \qquad (10.2.33)$$

that the deviation from each of the switching surfaces and its time derivative have opposite signs if

$$u_{d0} > \left|L\frac{di_d^*}{dt} + Ri_d - L\omega_e i_q\right|$$
$$u_{q0} > \left|L\frac{di_q^*}{dt} + Ri_q + L\omega_e i_d + \lambda_0\omega_e\right| \qquad (10.2.34)$$

Unlike DC motors, in addition to back-EMF the control voltages should suppress the coupling terms which are proportional to the angular speed of the motor rotor.

Remark 10.1

1. In the case of discontinuous reference currents i_d^* and i_q^* with associated 'large' time derivatives, the real currents will not be able to follow their reference values. This can happen if the outer-loop controllers use high feedback gains. Due to the inductive nature of electric drives, discontinuous reference currents cannot be followed with any controller. However, a sliding mode controller is able to immediately utilize the full available control resources u_{d0} and u_{q0} such that the fastest possible response is guaranteed.
2. It is also possible to perform the decoupling and linearization procedure first and then apply the sliding mode technique to the resulting linear system. In this case the resulting control voltages u_d and u_q will no longer be in the form of sign functions, but will contain an additive continuous part due to the decoupling procedure. Since the decoupling procedure involves the motor parameters, the resulting control is sensitive to these parameters. □

Once the control voltages u_d and u_q are obtained, the next step is to map them into the switching patterns of the inverter. Lookup table techniques are often used for this purpose with the electrical angular position θ_e as the input of the table (Sabanovic and Utkin, 1994). This solution is simple, but cannot implement arbitrary voltage vectors $[u_d \quad u_q]^T$ due to the finite resolution of a lookup table. In the following, we propose a novel PWM technique using the sliding mode principle, hence it may be called *sliding mode PWM*.

The basic problem is that the controls u_1, u_2 and u_3, required by the inverter, take only discrete values $-u_0$ or u_0. As a result, a direct implementation of the transformation $(A_{d,q}^{1,2,3})^+$ is not possible. Instead, a second set of switching functions is defined for u_1, u_2 and u_3. Define a set of desired controls according to transformation $(A_{d,q}^{1,2,3})^+$ in (10.2.25) as

$$\begin{bmatrix} u_1^* \\ u_2^* \\ u_3^* \end{bmatrix} = (A_{d,q}^{1,2,3})^+ \begin{bmatrix} u_{d0} \operatorname{sign}(s_d) \\ u_{q0} \operatorname{sign}(s_q) \end{bmatrix} \tag{10.2.35}$$

These controls cannot yet be applied to the inverter directly since matrix $(A_{d,q}^{1,2,3})^+$ is a function of the electrical angular position θ_e, so they do not take the values from the discrete set $\{-u_0, +u_0\}$. The second set of switching functions for controls u_1, u_2 and u_3 is chosen as

$$s_i^* = \int_0^t (u_i^*(\zeta) - u_i(\zeta))\, d\zeta \qquad (i = 1, 2, 3) \tag{10.2.36}$$

Note that the time derivatives $\dot{s}_i^*$ are linear in the inputs u_i $(i = 1, 2, 3)$. Consequently, control laws

$$u_i = u_0 \operatorname{sign}(s_i^*) \qquad (i = 1, 2, 3) \tag{10.2.37}$$

enforce sliding mode in manifold $s_i^* = 0$ $(i = 1, 2, 3)$ if

$$u_0 > \max(|u_1^*|, |u_2^*|, |u_3^*|) \tag{10.2.38}$$

The sliding condition (10.2.38) is derived from the existence condition $s_i^* \dot{s}_i^* < 0$. Employing the equivalent control method (Section 2.3) by setting $\dot{s}_i^* = 0$ and solving for inputs u_i (for each $i = 1, 2, 3$) yields

$$u_{ieq} = u_i^* \qquad (i = 1, 2, 3) \tag{10.2.39}$$

which implies that the coordinate transformation (10.2.35) was implemented exactly by the proposed method.

As the last step of the control design, the resulting discontinuous controls u_1, u_2, u_3 will be transformed to the switching patterns $s_{w1}, s_{w2}, s_{w3}, s_{w4}, s_{w5}, s_{w6}$ using (10.2.2). This completes the design procedure for the first version of the current control. For the implementation, phase currents i_a, i_b, i_c should be available. The design sequence assumes that an outer control loop provides desired currents i_d^* and i_q^*, and can be summarized in the following steps:

1. Transform the measured phase currents i_a, i_b, i_c into the (d, q) coordinate frame to obtain the torque current i_q and the field current i_d using (10.2.18). In practice, only two components of the phase currents are measured; the third one is calculated according to the *balance condition*, e.g. $i_c = -(i_a + i_b)$.
2. Design switching functions $s_d = i_d^* - i_d$, $s_q = i_q^* - i_q$, and associated control voltages u_d, u_q, according to (10.2.32) and (10.2.34). Since the maximum value of $\sqrt{u_{d0}^2 + u_{q0}^2}$ is limited by the DC link voltage, the selection of u_{d0} and u_{q0} is not arbitrary, i.e. u_{d0} and u_{q0} should be selected such that inequalities (10.2.34) and $u_0 > |u_i^*|$ $(i = a, b, c)$ are satisfied.
3. Transform the resulting discontinuous controls u_d and u_q to u_1^*, u_2^*, u_3^* according to (10.2.35).

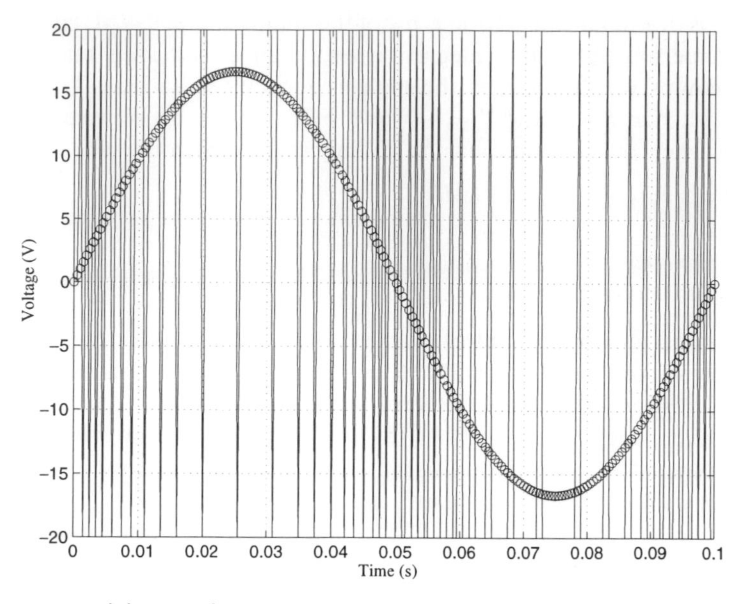

Figure 10.17 A sliding mode PWM.

4. Calculate the integral of (10.2.36) to obtain the new switching functions s_1^*, s_2^*, s_3^*; compute the discontinuous controls u_1, u_2, u_3 using (10.2.37).
5. Apply (10.2.2) to the resulting controls u_1, u_2, u_3 to derive the switching patterns of the inverter $s_{w1}, s_{w2}, s_{w3}, s_{w4}, s_{w5}, s_{w6}$.

Figure 10.17 shows an example of the proposed sliding mode PWM technique which modulates a sinusoidal voltage waveform to the width of a switching signal. Figure 10.18 gives the simulation result of the proposed current controller. The relatively high

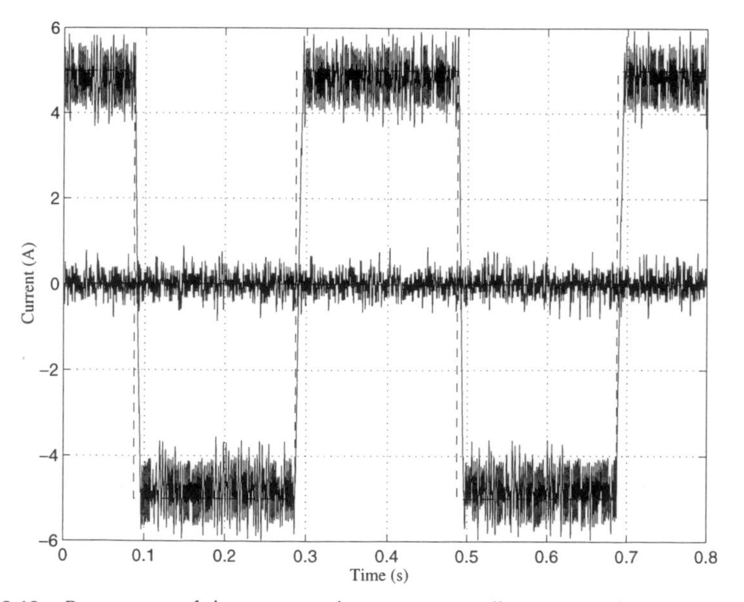

Figure 10.18 Responses of the proposed current controller: (- - -) reference currents, (—) real currents.

switching noise is due to the integration error of the sliding mode PWM, caused by the finite sampling period. Also, the integration action in closed loop may introduce additional dynamics which enlarges the high-frequency oscillations. The reference currents are selected as $i_d^* = 0$ and $i_q^* = 5.0\,\text{sqw}(15.7t)$, where sqw represents a square wave function. The control gains for the sliding mode controller are designed as $u_{d0} = 5\,\text{V}$ and $u_{q0} = \sqrt{u_0^2 + u_{d0}^2} = 19.36\text{V}$.

Second method for current control

The second method for current control is able to generate the discontinuous controls u_1, u_2, u_3 directly without involving a PWM technique. Moreover, this design method is theoretically compact and straightforward (although no PWM technique is used, the implementation does not imply that the role of a power converter is reduced to amplifying some continuous control signal). This method may need a high enough DC link voltage u_0 for every time instant.

The second method also uses the (d, q) coordinate frame for ease of presentation. Given the desired currents i_d^* and i_q^*, provided by an outer control loop, design switching functions

$$\begin{aligned} s_d &= L(i_d^* - i_d) \\ s_q &= L(i_q^* - i_q) \end{aligned} \tag{10.2.40}$$

where L is the inductance of the stator windings. Note that L does not change the sign of s_d and s_q; it is involved here only for simplifying the derivations.

The time derivatives of s_d and s_q in matrix–vector form are given by

$$\dot{\boldsymbol{S}}_{dq} = \boldsymbol{F}_{dq} - \boldsymbol{U}_{dq} \tag{10.2.41}$$

where $\boldsymbol{S}_{dq} = [s_d \quad s_q]^T$, $\boldsymbol{U}_{dq} = [u_d \quad u_q]^T$ and

$$\boldsymbol{F}_{dq} = \begin{bmatrix} F_d \\ F_q \end{bmatrix} = \begin{bmatrix} L\dfrac{di_d^*}{dt} \\ L\dfrac{di_d^*}{dt} + \lambda\omega_e \end{bmatrix} + \begin{bmatrix} R & -L\omega_e \\ L\omega_e & R \end{bmatrix}\begin{bmatrix} i_d \\ i_q \end{bmatrix} \tag{10.2.42}$$

Substitution of (10.2.22) into (10.2.41) yields

$$\dot{\boldsymbol{S}}_{dq} = \boldsymbol{F}_{dq} - A_{d,q}^{1,2,3}\boldsymbol{U}_{gate} \tag{10.2.43}$$

where $\boldsymbol{U}_{gate} = [u_1 \quad u_2 \quad u_3]^T$. Equation (10.2.43) establishes the direct relation between the controlled errors s_d, s_q and the discontinuous controls u_1, u_2, u_3. Design controls u_1, u_2, u_3 as follows:

$$\boldsymbol{U}_{gate} = u_0\,\text{sign}(\boldsymbol{S}^*) \tag{10.2.44}$$

where $\boldsymbol{S}^* = [s_1^* \quad s_2^* \quad s_3^*]^T$ is a vector of transformed switching functions to be determined later, and

$$\text{sign}(\boldsymbol{S}^*) = [\text{sign}(s_1^*) \quad \text{sign}(s_2^*) \quad \text{sign}(s_3^*)]^T \tag{10.2.45}$$

Controls u_1, u_2, u_3 take values from the discrete set $\{-u_0, +u_0\}$. The transformed vector

S^* should be designed such that under controls (10.2.44) s_d and s_q vanish in finite time. A proper candidate for S^* is

$$S^* = (A_{d,q}^{1,2,3})^+ S_{dq} = \frac{3}{2}(A_{d,q}^{1,2,3})^T S_{dq} \tag{10.2.46}$$

where matrix $(A_{d,q}^{1,2,3})^+$ can be found as

$$(A_{d,q}^{1,2,3})^+ = \begin{bmatrix} \cos\theta_a & -\sin\theta_a \\ \cos\theta_b & -\sin\theta_b \\ \cos\theta_c & -\sin\theta_c \end{bmatrix} \tag{10.2.47}$$

with $\theta_a = \theta_e$, $\theta_b = \theta_e - 2\pi/3$ and $\theta_c = \theta_e + 2\pi/3$. Matrix $A_{d,q}^{1,2,3}(A_{d,q}^{1,2,3})^+$ is a 2×2 identity matrix.

Theorem 10.1

For high enough link voltage u_0, system (10.2.43) under control (10.2.44) via transformation (10.2.46) converges to its origin $s_d = 0$, $s_q = 0$ in finite time. □

Proof

Design a Lyapunov function candidate

$$V = \frac{1}{2} S_{dq}^T S_{dq} \tag{10.2.48}$$

and take the time derivative of V along the solutions of (10.2.43) to yield

$$\dot{V} = (S^*)^T F^* - (S^*)^T (A_{d,q}^{1,2,3})^T A_{d,q}^{1,2,3} U_{gate} \tag{10.2.49}$$

where F^* is defined as $F^* = [F_1^* \quad F_2^* \quad F_3^*]^T = (A_{d,q}^{1,2,3})^T F_{dq}$. Substitution of controls (10.2.44) into (10.2.49) results in

$$\dot{V} = (S^*)^T F^* - u_0 (S^*)^T (A_{d,q}^{1,2,3})^T A_{d,q}^{1,2,3} \operatorname{sign}(S^*) \tag{10.2.50}$$

where matrix $(A_{d,q}^{1,2,3})^T A_{d,q}^{1,2,3}$ is a *singular* matrix and can be calculated as

$$(A_{d,q}^{1,2,3})^T A_{d,q}^{1,2,3} = \frac{4}{9} \begin{bmatrix} 1 & -\frac{1}{2} & -\frac{1}{2} \\ -\frac{1}{2} & 1 & -\frac{1}{2} \\ -\frac{1}{2} & -\frac{1}{2} & 1 \end{bmatrix} \tag{10.2.51}$$

Depending on the signs of s_1^*, s_2^* and s_3^*, there are eight possible combinations of values of $\operatorname{sign}(s_1^*)$, $\operatorname{sign}(s_2^*)$ and $\operatorname{sign}(s_3^*)$. Evaluation of (10.2.46) shows that $\operatorname{sign}(s_1^*)$, $\operatorname{sign}(s_2^*)$ and $\operatorname{sign}(s_3^*)$ can never be all +1 or all −1 simultaneously. The remaining six combinations can be summarized as

$$\operatorname{sign}(s_l^*) \neq \operatorname{sign}(s_m^*) = \operatorname{sign}(s_n^*) \quad \text{with} \quad l \neq m \neq n \text{ and } l, m, n \in \{1, 2, 3\} \tag{10.2.52}$$

Starting from this notion, equation (10.2.50) can be expanded based on (10.2.51) to yield

$$\dot{V} = (s_1^* F_1^* + s_2^* F_2^* + s_3^* F_3^*) - (2/3)^2 u_0 (2|s_l^*| + |s_m^*| + |s_n^*|) \tag{10.2.53}$$

with $l \neq m \neq n$ and $l, m, n \in \{1, 2, 3\}$.

As long as $\|\boldsymbol{S}^*\| \neq 0$ and the DC link voltage u_0 satisfies

$$u_0 > (3/2)^2 \max(|F_1^*|, |F_2^*|, |F_3^*|) \tag{10.2.54}$$

$\dot{V} < 0$ will be guaranteed for all possible l, m, n. This proof implies that $s_d = 0, s_q = 0$ can be reached in finite time by directly switching the sign of discontinuous controls u_1, u_2, u_3 according to (10.2.44). ■

Similar to the first method, the next step is to map the resulting controls u_1, u_2, u_3 into the switching patterns applied to the inverter. Equation (10.2.2) can be used for this purpose. For the implementation, matrix $(A_{d,q}^{1,2,3})^+$ is needed to carry out transformation (10.2.46), whereas the exact values of $\boldsymbol{F}_{dq}$ are not required.

What is the physical meaning of the transformed switching functions s_1^*, s_2^*, s_3^*? From (10.2.46) and (10.2.21) we have

$$[s_1^* \quad s_2^* \quad s_3^*]^T = L[i_a^* - i_a \quad i_b^* - i_b \quad i_c^* - i_c]^T \tag{10.2.55}$$

where i_a^*, i_b^*, i_c^* are the reference currents in the (a, b, c) coordinate frame transformed from the reference currents i_d^*, i_q^*, usually determined by an outer-loop speed controller. From (10.2.55) and (10.2.44), controls u_1, u_2, u_3 can be found as

$$\begin{aligned} u_1 &= u_0 \operatorname{sign}(i_a^* - i_a) \\ u_2 &= u_0 \operatorname{sign}(i_b^* - i_b) \\ u_3 &= u_0 \operatorname{sign}(i_c^* - i_c) \end{aligned} \tag{10.2.56}$$

The reader familiar with conventional current control of AC motors will immediately recognize that (10.2.55) and (10.2.56) are similar to *hysteresis control* and *chopper control* for reducing heat loss and electromagnetic disturbances caused by the switching actions. Hysteresis control adds a hysteresis circuit after each sign function; chopper control uses a fixed clock source to synchronize the switching actions.

The sliding mode method presented here gives the condition, inequality (10.2.54), for the current controller to be effective. Furthermore, it provides a way of transforming the reference currents i_d^*, i_q^* to the (a, b, c) coordinate frame to track the reference currents i_a^*, i_b^*, i_c^*; see equation (10.2.46). In the (d, q) coordinate frame, reference current i_d^* is set to zero if no field-weakening operation (Section 10.2.4) is required. For field-weakening operations of a PMSM, reference current i_d^* should be varied depending on the current motor speed. Reference current i_q^*, provided by an outer-loop controller, corresponds to the required motor torque.

Figure 10.19 show the simulation results of the second version of the current control. The reference currents are the same as for the first version of the current control. But the high-frequency switching noise is much lower in the second version; this is because it has no integration action.

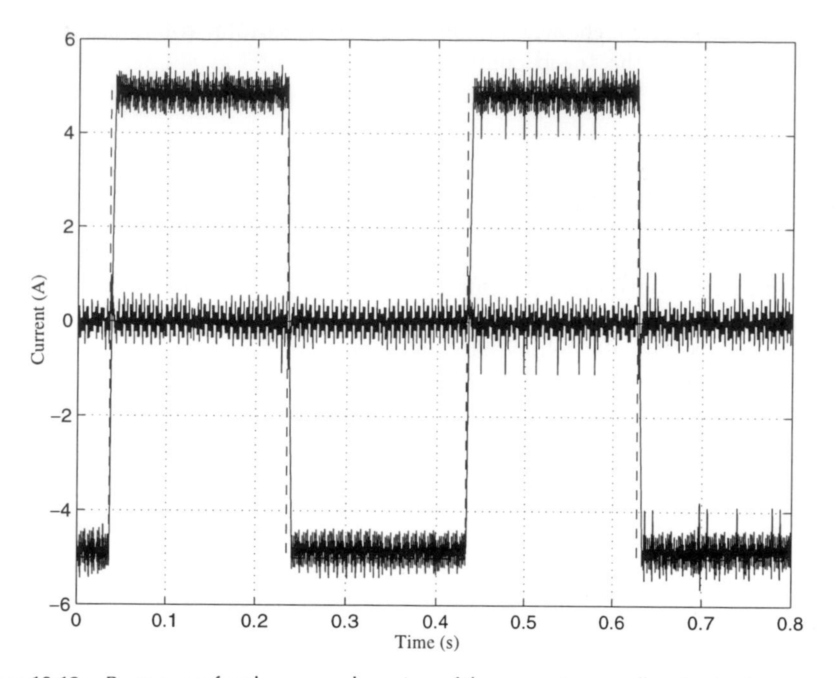

Figure 10.19 Responses for the second version of the current controller: (- - -) reference currents, (—) real currents.

10.2.4 Speed control

Similar to DC motors, speed control of permanent-magnet synchronous motors can be realized in a cascaded control structure, with a current controller in an inner loop and a speed controller in an outer loop, providing the reference currents i_d^* and i_q^* for the inner loop. Reference current i_d^* is normally set to zero so the motor is able to provide a constant torque. However, in some applications, high torque or high speed is desired. For high-speed operation, the reference current i_d^* may be varied from zero to some negative value depending on the current motor speed. This technique is called *field weakening* and is able to extend the operating range of the motor speed so it becomes several times larger than for zero i_d^*. Shi and Lu (1996) give details of the field-weakening operation of PMSMs.

For the outer speed control loop, most control techniques may be applied, with the exception of high-gain and sliding mode techniques, since the reference input of the inner control loop should have bounded time derivative. This may explain why some researchers complain that sliding mode control techniques do not work properly for some mechanical systems such as robotic manipulators. In general, a high-gain controller or a sliding mode controller may only be used once in a cascaded control structure.

A suitable solution for designing the speed controller is to apply sliding mode control directly to the motor model to achieve the discontinuous controls u_1, u_2, u_3 without explicitly involving a current control loop. This control scheme may be called *direct speed control*. The advantages of this control scheme have already been given in Section 10.1.5. Since we deal with sliding mode control applications, we are interested here in the details of *direct speed control* for permanent-magnet synchronous motors.

Let ω_r^* be the desired mechanical angular speed of the motor rotor, and design two switching functions for the speed control as

$$
\begin{aligned}
s_d &= \frac{K_t}{J}(i_d^* - i_d) \\
s_\omega &= c(\omega_r^* - \omega_r) + \frac{d}{dt}(\omega_r^* - \omega_r)
\end{aligned}
\tag{10.2.57}
$$

where c is a positive constant determining the motion performance in sliding mode; positive coefficient K_t/J is introduced to simplify the derivations; reference current i_d^* is normally set to zero; for field-weakening operation, i_d^* is a function of the motor speed and motor parameters (Shi and Lu, 1996). From these equations, if $s_\omega \equiv 0$ can be achieved after a finite time interval then ω_r will converge to ω_r^* exponentially.

The motion projection of the system on the subspaces s_ω, s_d is

$$
\dot{\boldsymbol{S}}_{d\omega} = \boldsymbol{F}_{d\omega} - D_\omega \boldsymbol{U}_{gate} \tag{10.2.58}
$$

where $\boldsymbol{U}_{gate}$ is the same as for the current control, $\boldsymbol{S}_{d\omega} = [s_d \quad s_\omega]^T$ and

$$
\begin{aligned}
\boldsymbol{F}_{d\omega} &= \begin{bmatrix} F_d \\ F_\omega \end{bmatrix} = \begin{bmatrix} \dfrac{K_t}{J}\dfrac{di_d^*}{dt} \\ c(\dot{\omega}_r^* - \ddot{\omega}_r^*) + \ddot{\omega}_r^* + \dfrac{\dot{\tau}_l}{J} + \dfrac{K_t}{JL}\lambda N_r \omega_r \end{bmatrix} + \frac{K_t}{J}\begin{bmatrix} \dfrac{R}{L} & -N_r\omega_r \\ N_r\omega_r & \dfrac{R}{L} \end{bmatrix}\begin{bmatrix} i_d \\ i_q \end{bmatrix} \\
D_\omega &= \frac{K_t}{JL} A_{d,q}^{1,2,3}
\end{aligned}
\tag{10.2.59}
$$

where matrix $A_{d,q}^{1,2,3}$ is defined by (10.2.23).

From this stage we may follow the same procedure as in the second version of the current control. At first, the switching function transformation should be performed by

$$
\boldsymbol{S}^* = D_\omega^+ \boldsymbol{S}_{d\omega} \tag{10.2.60}
$$

where $\boldsymbol{S}^* = [s_1^* \quad s_2^* \quad s_3^*]^T$ is again the vector representing the transformed switching functions; D_ω^+ is the pseudoinverse of matrix D_ω and is given as

$$
D_\omega^+ = D_\omega^T (D_\omega D_\omega^T)^{-1} = \frac{3}{2}\frac{JL}{K_t}(A_{d,q}^{1,2,3})^T \tag{10.2.61}
$$

Matrix D_ω^+ can be expanded as

$$
D_\omega^+ = \frac{3}{2}\frac{JL}{K_t}(A_{d,q}^{1,2,3}) = \frac{JL}{K_t}\begin{bmatrix} \cos\theta_a & -\sin\theta_a \\ \cos\theta_b & -\sin\theta_b \\ \cos\theta_c & -\sin\theta_c \end{bmatrix} \tag{10.2.62}
$$

where $\theta_a = \theta_e, \theta_b = \theta_e - 2\pi/3$ and $\theta_c = \theta_e + 2\pi/3$. Matrix $D_\omega D_\omega^+$ is a 2×2 identity matrix.

The controls u_1, u_2, u_3 have the same form as the second version of the current control:

$$U_{gate} = u_0 \ \text{sign}(\boldsymbol{S}^*) \tag{10.2.63}$$

with $\text{sign}(\boldsymbol{S}^*) = [\text{sign}(s_1^*) \quad \text{sign}(s_2^*) \quad \text{sign}(s_3^*)]^T$.

Theorem 10.2

For high enough link voltage u_0, system (10.2.58) under control (10.2.63) via transformation (10.2.60) converges to its origin $s_d = 0, s_\omega = 0$ in finite time. □

Proof

Following the same procedure as for the current control, select a Lyapunov function candidate as

$$V = \frac{1}{2} \boldsymbol{S}_{d\omega}^T \boldsymbol{S}_{d\omega} \tag{10.2.64}$$

Differentiation of V along the system trajectories yields

$$\dot{V} = (s_1^* F_1^* + s_2^* F_2^* + s_3^* F_3^*) - (2a/3)^2 u_0 (2|s_l^*| + |s_m^*| + |s_n^*|) \tag{10.2.65}$$

with $l \neq m \neq n$ and $l, m, n \in \{1, 2, 3\}$. F_1^*, F_2^*, F_3^* are defined by $D_\omega^T \boldsymbol{F}_{d\omega} = [F_1^* \quad F_2^* \quad F_3^*]^T$. Reference current i_d^* is normally set to zero, so that $di_d^*/dt = 0$. Assume the time derivative of load torque $\dot{\tau}_l$ and the time derivative of the desired acceleration $\ddot{\omega}_r^*$ are bounded, $\|D_\omega^T \boldsymbol{F}_{d\omega}\|$ is bounded as well (10.2.59).

Obviously, as long as $\|\boldsymbol{S}^*\| \neq 0$ and the DC link voltage u_0 satisfies

$$u_0 > \left(\frac{3}{2}\frac{JL}{K_t}\right)^2 \max(|F_1^*|, |F_2^*|, |F_3^*|) \tag{10.2.66}$$

the condition $\dot{V} < 0$ holds for all possible l, m, n. It means that the surfaces $s_d = 0, s_\omega = 0$, are reached in finite time in the system with the discontinuous controls u_1, u_2, u_3 given in (10.2.63). ■

The phrase 'high enough u_0' means the proposed speed controller is robust with respect to the terms included in $D_\omega^T \boldsymbol{F}_{d\omega}$. For example, under condition (10.2.66), the speed controller is robust with respect to the load torque variation, since $\dot{\tau}_l$ is contained in $D_\omega^T \boldsymbol{F}_{d\omega}$. Again, it is not necessary to know exact values of $\boldsymbol{F}_{d\omega}$ to produce the implementation. Only matrix D_ω^+ is relevant, since we need this matrix for calculating the switching function transformation.

In order to construct the sliding surface $s_\omega = 0$, the acceleration signal of the motor rotor is needed. Normally, this variable cannot be obtained directly, but has to be obtained by a state observer based on the measured rotor speed and stator currents. The acceleration signal can be retrieved using techniques similar to those used for DC motors (Section 10.1.6). Figures 10.20 and 10.21 show the simulation results of the proposed sliding mode speed controller. The transition time of the speed control is about 0.1 s. Figure 10.21 gives the smoothed waveforms of the motor currents. Bear

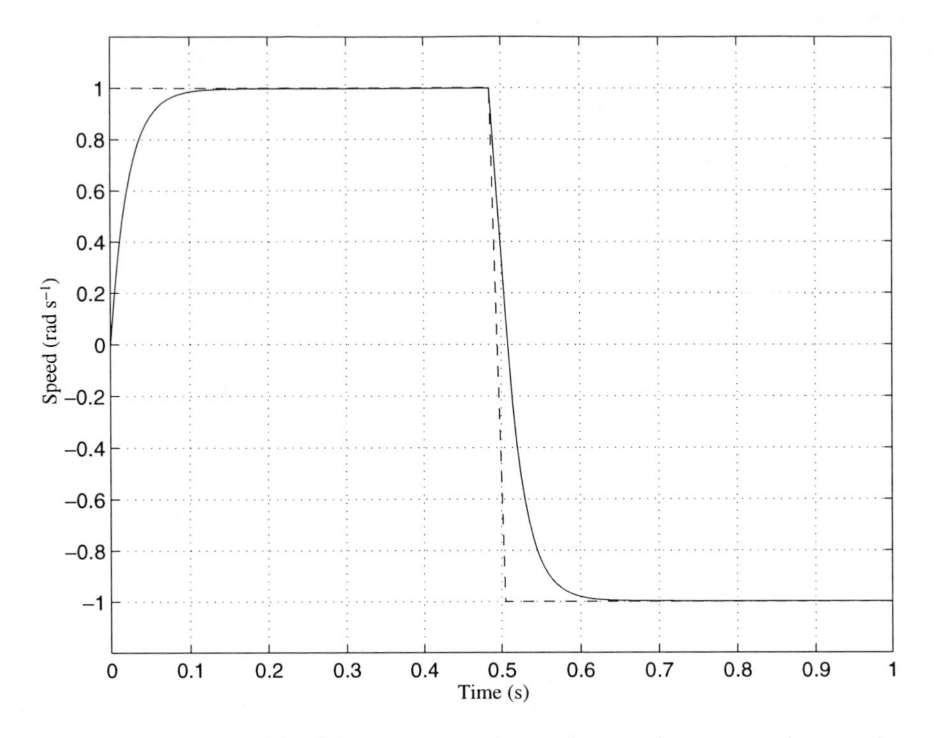

Figure 10.20 Response of the sliding mode speed control: (- - -) reference speed, (—) real speed.

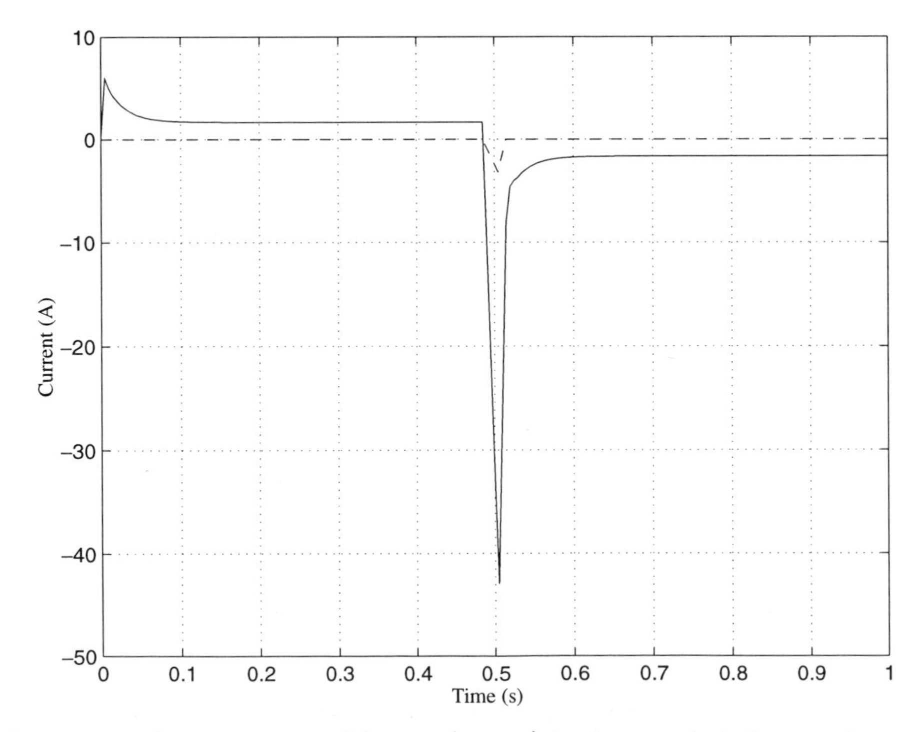

Figure 10.21 Current response of the speed control: (- - -) current i_d, (—) current i_q.

in mind that the current i_q is not controlled explicitly; the ordinary behaviour of the current components is the result of the acceleration control. The control gain c in equation (10.2.57) is selected as $c = 50$.

10.2.5 Current observer

From the viewpoint of practical applications, it is interesting to design a control system with no current transducers. In practice there are two difficulties to obtaining the stator currents i_d and i_q in the field-oriented coordinate frame. The first difficulty is due to the noise in the measured phase currents i_a and i_b; the second difficulty is due to the coordinate transformation. The noise problem is evident, whereas the second difficulty lies in the speed limitation of the calculation unit used for the transformation. Usually i_a and i_b are high-frequency signals compared to the current components i_d and i_q, especially for PMSMs with a large number of pole pairs. These high-frequency signals must be sampled with a high enough sampling rate, otherwise information will be lost. Furthermore, to transform the measured phase currents into the field-oriented coordinate frame, a rotor position signal with high enough resolution must be available. This is normally achieved by involving a high-resolution optical encoder.

As a result, it is meaningful to design a current observer using the rotor speed measurement. Examination of the motor model leads to the conclusion that on-line simulation of the motor model suffices to achieve a stable observation of the stator currents. Let $\hat{i}_d$ and $\hat{i}_q$ be the estimates of the stator currents, then the observer equations are just a copy of the motor model (10.2.14):

$$\begin{aligned}\frac{d\hat{i}_d}{dt} &= -\frac{R}{L}\hat{i}_d + \omega_e \hat{i}_q + \frac{1}{L}u_d \\ \frac{d\hat{i}_q}{dt} &= -\frac{R}{L}\hat{i}_q - \omega_e \hat{i}_d - \frac{1}{L}\lambda_0\omega_e + \frac{1}{L}u_q\end{aligned} \tag{10.2.67}$$

It is proven below that this current observer is stable, and the observer errors will tend to zero asymptotically. The mismatch dynamics can be obtained as

$$\begin{aligned}\frac{d\bar{i}_d}{dt} &= -\frac{R}{L}\bar{i}_d + \omega_e \bar{i}_q \\ \frac{d\bar{i}_q}{dt} &= -\frac{R}{L}\bar{i}_q - \omega_e \bar{i}_d\end{aligned} \tag{10.2.68}$$

where $\bar{i}_d = \hat{i}_d - i_d$, $\bar{i}_q = \hat{i}_q - i_q$. Under the assumption that the electrical rotor speed $\omega_e = N_r\omega_r$ varies much more slowly than the current observation errors $\bar{i}_d$ and $\bar{i}_q$, equation (10.2.68) can be treated as a linear system with characteristic polynomial

$$p^2 + 2\frac{R}{L}p + \frac{R^2}{L^2} + \omega_e^2 = 0 \tag{10.2.69}$$

which is stable for any physically plausible $R, L > 0$. Thus $\hat{i}_d$ and $\hat{i}_q$ will tend to i_d and i_q exponentially.

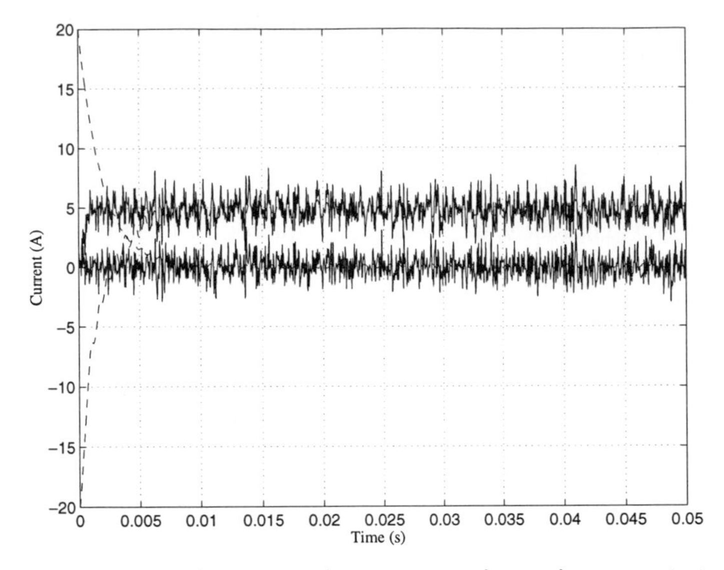

Figure 10.22 Responses of the current observer: ( ) observed currents, (—) real currents.

Figure 10.22 shows the responses of the proposed current observer. The real currents i_d and i_q should follow their given references: $i_d^* = 0$ A and $i_q^* = 5.0$ A. The initial conditions of the observed currents are selected as $\hat{i}_d(0) = 20$ A and $\hat{i}_q(0) = -20$ A to see the convergence process. Since the current observer is just the on-line simulation of the motor model, no observer is to be adjusted.

10.2.6 Observer design for sensorless speed control

Section 10.1.8 discussed the issue of sensorless control. Coupling effects in the motor dynamics mean that sensorless control of AC machines is more difficult than sensorless control of DC motors.

The term 'sensorless control' indicates that some internal states of a dynamic system are available, measured or estimated, but the output measurements are unknown; however, these unmeasured output variables serve as principal control variables. In the case of PMSMs, the phase voltages (or the line voltages) and the phase currents are normally measured, whereas the motor speed ω_r and the motor position θ_r are unknown, despite being the principal control variables of a PMSM drive system.

When estimating these variables, it is rare to use the motor model in the field-oriented coordinate frame, i.e. the (d, q) coordinate frame; this is because the state variables in this coordinate frame are transformed from the fixed coordinate frames, i.e. from the (a, b, c) coordinate frame through the (α, β) coordinate frame, using transformation matrix $A_{dq}^{a,b,c}$ defined by (10.2.19). It is therefore essential to know the electrical angular position of the motor rotor θ_e. As a result, the model in the (α, β) coordinate frame is usually used for sensorless control design. Rewriting the motor model in this coordinate system yields

$$\begin{aligned} \frac{di_\alpha}{dt} &= -\frac{R}{L}i_\alpha - \frac{1}{L}e_\alpha + \frac{1}{L}Bu_\alpha \\ \frac{di_\beta}{dt} &= -\frac{R}{L}i_\beta - \frac{1}{L}e_\beta + \frac{1}{L}Bu_\beta \end{aligned} \tag{10.2.70}$$

where $e_\alpha = -\lambda_0\omega_e \sin\theta$ and $e_\beta = -\lambda_0\omega_e \cos\theta$ are the induced EMF components. Assuming the motor speed changes slowly, implying that $\dot{\omega}_e \approx 0$, the model of these induced EMF components is

$$\begin{aligned} \dot{e}_\alpha &= \omega_e e_\beta \\ \dot{e}_\beta &= -\omega_e e_\alpha \end{aligned} \tag{10.2.71}$$

Conventional approaches to sensorless control of PMSMs usually follow a two-step procedure. First, the induced EMF components are observed. The electrical angular speed and position are then derived in a second step. Furthermore, in conventional sensorless control designs, the mechanical motion equation is normally involved in the observation algorithms. This model contains inaccurate mechanical parameters as well as the unknown load torque, both of them recognized as major obstacles. Here we propose a design technique based on a sliding mode observer and using only the electrical motion equations.

Current observer for EMF components

Design a set of observer equations for model (10.2.70) as

$$\begin{aligned} \frac{d\hat{i}_\alpha}{dt} &= -\frac{R}{L}\hat{i}_\alpha + \frac{1}{L}u_\alpha - \frac{l_1}{L}\ \mathrm{sign}(\hat{i}_\alpha - i_\alpha) \\ \frac{d\hat{i}_\beta}{dt} &= -\frac{R}{L}\hat{i}_\beta + \frac{1}{L}u_\beta - \frac{l_1}{L}\ \mathrm{sign}(\hat{i}_\beta - i_\beta) \end{aligned} \tag{10.2.72}$$

where $l_1 > 0$ is a constant observer gain. Assume the parameters R and L are known exactly and thus are identical with those in the model. Subtracting the above equations from the model equations (10.2.70) yields the mismatch dynamics

$$\begin{aligned} \frac{d\bar{i}_\alpha}{dt} &= -\frac{R}{L}\bar{i}_\alpha + \frac{1}{L}e_\alpha - \frac{l_1}{L}\ \mathrm{sign}(\bar{i}_\alpha) \\ \frac{d\bar{i}_\beta}{dt} &= -\frac{R}{L}\bar{i}_\beta + \frac{1}{L}e_\beta - \frac{l_1}{L}\ \mathrm{sign}(\bar{i}_\beta) \end{aligned} \tag{10.2.73}$$

where $\bar{i}_\alpha = \hat{i}_\alpha - i_\alpha$ and $\bar{i}_\beta = \hat{i}_\beta - i_\beta$ denote the observation errors. The mismatch dynamics are disturbed by the unknown induced EMF components. However, since the EMF components are bounded, they may be suppressed by discontinuous inputs with $l_1 > \max(|e_\alpha|, |e_\beta|)$. Then sliding mode with $\bar{i}_\alpha = 0$, $\bar{i}_\beta = 0$ will occur after a finite time interval. In order to examine the system behaviour during sliding mode, the equivalent control method (Section 2.3) is exploited. We formally set $d\bar{i}_\alpha/dt = 0$ and $d\bar{i}_\beta/dt = 0$ in (10.2.73), giving

$$\begin{aligned} (l_1\ \mathrm{sign}(\bar{i}_\alpha))_{eq} &= e_\alpha \\ (l_1\ \mathrm{sign}(\bar{i}_\beta))_{eq} &= e_\beta \end{aligned} \tag{10.2.74}$$

Again, to extract e_α, e_β from the corresponding equivalent control values in (10.2.74), we use lowpass filters with z_α, z_β as the filter outputs:

$$\begin{aligned} z_\alpha(t) &= e_\alpha(t) + \Delta_\mu(t) \\ z_\beta(t) &= e_\beta(t) + \Delta_\mu(t) \end{aligned} \tag{10.2.75}$$

where $\Delta_\mu(t)$ is the error determined by the distortions of both slow and fast components of the discontinuous filter input. Filter time constant μ directly determines the amount of error $\Delta_\mu(t)$. It should be chosen small enough to have filter dynamics faster than those of the system (10.2.70) and at the same time not too small, in order not to filter out the high frequency components.

Observer for EMF components

For high-performance applications, z_α, z_β cannot be used directly as the estimation of the induced EMF components, since they contain disturbance $\Delta_\mu(t)$. Model equations (10.2.71) are thus needed to design the observer for better filtering and for simultaneously estimating the rotation speed. The following observer is designed to undertake this filtering task:

$$\begin{aligned}\dot{\hat{e}}_\alpha &= \hat{\omega}_e \hat{e}_\beta - l_2(\hat{e}_\alpha - z_\alpha)\\ \dot{\hat{e}}_\beta &= -\hat{\omega}_e \hat{e}_\alpha - l_2(\hat{e}_\beta - z_\beta)\\ \dot{\hat{\omega}}_e &= (\hat{e}_\beta - z_\beta)\hat{e}_\alpha - (\hat{e}_\alpha - z_\alpha)\hat{e}_\beta\end{aligned} \tag{10.2.76}$$

where $l_2 > 0$ is a constant observer gain. The observer has the structure of an extended *Kalman filter* and is expected to have high filtering properties. Therefore, further analysis will be performed under the assumption $\Delta_\mu(t) = 0$. Bearing in mind that $\omega_e = \text{const}$, write down the mismatch equations

$$\begin{aligned}\dot{\bar{e}}_\alpha &= \hat{\omega}_e \hat{e}_\beta + \omega_e e_\beta - l_2(\hat{e}_\alpha - e_\alpha)\\ \dot{\bar{e}}_\beta &= -\hat{\omega}_e \hat{e}_\alpha - \omega_e e_\alpha - l_2(\hat{e}_\beta - e_\beta)\\ \dot{\bar{\omega}}_e &= (\hat{e}_\beta - e_\beta)\hat{e}_\alpha - (\hat{e}_\alpha - e_\alpha)\hat{e}_\beta\end{aligned} \tag{10.2.77}$$

where $\bar{e}_\alpha = \hat{e}_\alpha - e_\alpha$, $\bar{e}_\beta = \hat{e}_\beta - e_\beta$ and $\bar{\omega}_e = \hat{\omega}_e - \omega_e$ are the observation errors. Substituting $\hat{\omega}_e = \omega_e + \bar{\omega}_e$ into (10.2.77) yields a simplified equation system:

$$\begin{aligned}\dot{\bar{e}}_\alpha &= \bar{\omega}_e \hat{e}_\beta + \omega_e \bar{e}_\beta - l_2(\hat{e}_\alpha - e_\alpha)\\ \dot{\bar{e}}_\beta &= -\bar{\omega}_e \hat{e}_\alpha - \omega_e \bar{e}_\alpha - l_2(\hat{e}_\beta - e_\beta)\\ \dot{\bar{\omega}}_e &= (\hat{e}_\beta - e_\beta)\hat{e}_\alpha - (\hat{e}_\alpha - e_\alpha)\hat{e}_\beta\end{aligned} \tag{10.2.78}$$

To prove the convergence of observer (10.2.76), we design a Lyapunov function candidate

$$V = \frac{1}{2}(\bar{e}_\alpha^2 + \bar{e}_\beta^2 + \bar{\omega}_e^2) \tag{10.2.79}$$

Its time derivative along the solutions of (10.2.77) can be calculated as

$$\begin{aligned}\dot{V} &= -l_2\{\bar{e}_\alpha(\hat{e}_\alpha - e_\alpha) + \bar{e}_\beta(\hat{e}_\beta - e_\beta)\} - \bar{\omega}_e\{\hat{e}_\beta[\bar{e}_\alpha - (\hat{e}_\alpha - e_\alpha)] - \hat{e}_\alpha[\bar{e}_\beta - (\hat{e}_\beta - e_\beta)]\}\\ \text{or} \quad \dot{V} &= -l_2(\bar{e}_\alpha^2 + \bar{e}_\beta^2) \le 0\end{aligned} \tag{10.2.80}$$

implying that $\hat{e}_\alpha, \hat{e}_\beta$ tend to e_α, e_β asymptotically. Now, in the first two equations of (10.2.78), all terms except for $\bar{\omega}_e\hat{e}_\alpha$ and $\bar{\omega}_e\hat{e}_\beta$ are equal to zero. Since $\hat{e}_\alpha(t) \neq 0$ and $\hat{e}_\beta(t) \neq 0$, the speed estimation error $\bar{\omega}_e$ should be equal to zero as well. Thus the convergence of $\hat{\omega}_e$ to ω_e is proven.

For field-oriented control of PMSMs, transformation matrix $A_{d,q}^{\alpha,\beta}$ is needed, based on the electrical angular position θ_e for PMSMs. Denoting the estimated matrix as $\hat{A}_{d,q}^{\alpha,\beta}$, we have

$$\hat{A}_{d,q}^{\alpha,\beta} = \begin{bmatrix} \cos\hat{\theta}_e & \sin\hat{\theta}_e \\ -\sin\hat{\theta}_e & \cos\hat{\theta}_e \end{bmatrix} \tag{10.2.81}$$

where $\hat{\theta}_e$ is an estimate of θ_e. Considering relations

$$\begin{aligned} \hat{e}_\alpha &= -\lambda_0\hat{\omega}_e \sin\hat{\theta}_e \\ \hat{e}_\beta &= -\lambda_0\hat{\omega}_e \cos\hat{\theta}_e \end{aligned} \tag{10.2.82}$$

the matrix elements $\sin\hat{\theta}_e$ and $\cos\hat{\theta}_e$ can be obtained as

$$\begin{aligned} \sin\hat{\theta}_e &= -\frac{1}{\lambda_0}\frac{\hat{e}_\alpha}{\hat{\omega}_e} \\ \cos\hat{\theta}_e &= -\frac{1}{\lambda_0}\frac{\hat{e}_\beta}{\hat{\omega}_e} \end{aligned} \tag{10.2.83}$$

with $\sin\hat{\theta}_e \to \sin\theta_e$ and $\cos\hat{\theta}_e \to \cos\theta_e$.

The estimate of mechanical rotor speed, denoted by $\hat{\omega}_r$, can be obtained from (10.2.29) as

$$\hat{\omega}_r = \frac{\hat{\omega}_e}{N_r} \tag{10.2.84}$$

and $\hat{\omega}_r \to \omega_r$ as $t \to \infty$.

Remark 10.2

To develop the EMF model discussed in this section, we assumed that the motor speed varies slowly. Under this assumption, the observation problem can be simplified significantly. For the case of variable speed, it is also possible to design an observer; however, the convergence proof is a considerably more complicated problem. □

Figures 10.23 to 10.25 show the simulation results of the proposed observers for the speed sensorless control, i.e. from equations (10.2.72) to (10.2.76). As shown in Figure 10.23, the observed speed rapidly converges to the true speed. In order to show the convergence process, we select one set of initial conditions for the observed EMF components and a different set of initial conditions for the real EMF components: $\hat{e}_\alpha(0) = 0.002$ and $\hat{e}_\beta(0) = -0.002$, $e_\alpha(0) = 0$ and $e_\beta(0) = 0$. The observer gains in (10.2.67) and (10.2.76) are selected as $l_1 = 20\,000$ and $l_2 = 20\,000$. The lowpass filters used to obtain the equivalent controls in (10.2.74) are of *Butterworth* type, with a cutoff frequency of $10\,000$ rad s^{-1}.

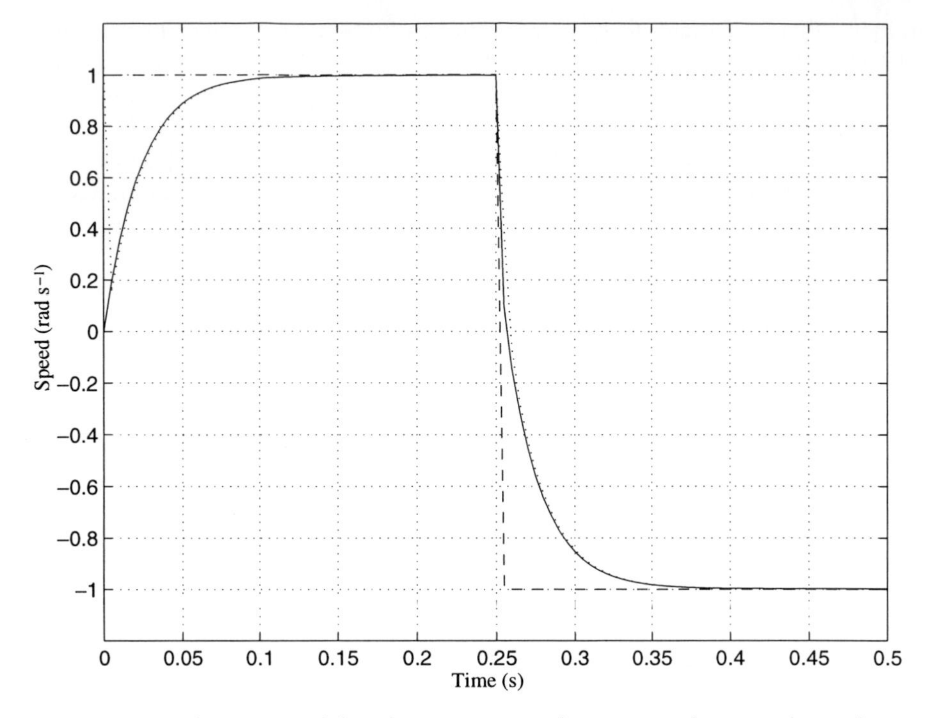

Figure 10.23 Speed response of the observer: (- - -) reference speed, (—) real speed, (- · -) observed speed.

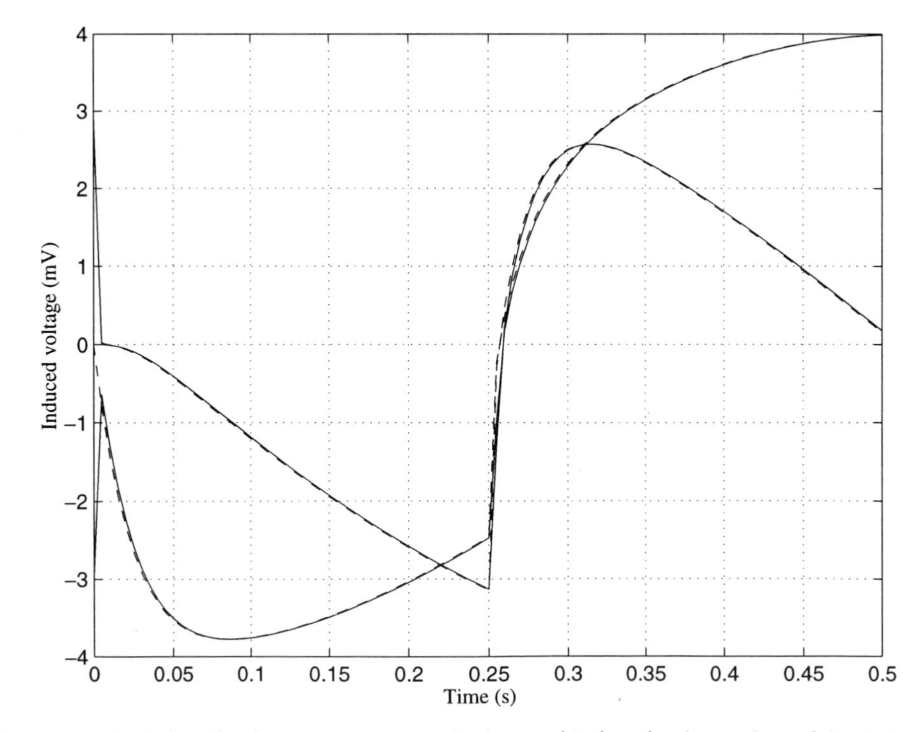

Figure 10.24 Induced voltage response: (- - -) observed induced voltages $\hat{e}_\alpha$ and $\hat{e}_\beta$, (—) real induced voltages $\hat{e}_\alpha$ and $\hat{e}_\beta$.

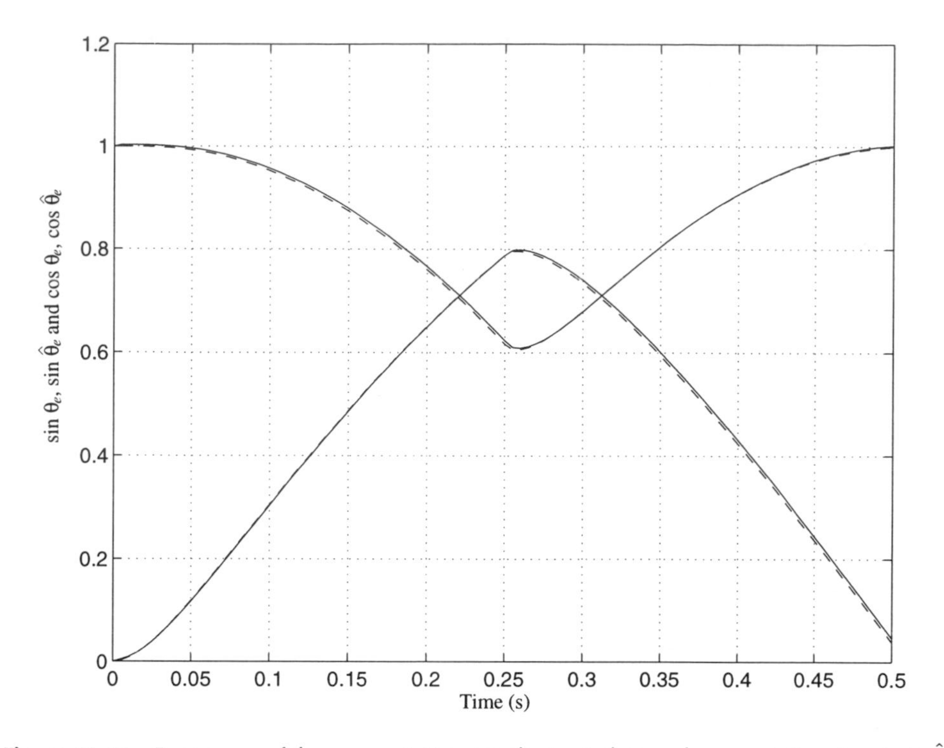

Figure 10.25 Responses of the commutation signals: (- - -) observed commutation signals $\sin\hat{\theta}_e$ and $\cos\hat{\theta}_e$, (—) real commutation signals $\sin\theta_e$ and $\cos\theta_e$.

10.2.7 Discussion

Dynamic models in each coordinate frame have been given as the starting point of this section. Two versions of current controllers have been presented. The first version is a standard method of sliding mode current control except the last part for deriving the switching patterns; instead of the standard lookup table approach, the so-called sliding mode PWM technique has been introduced.

The second version of the current control is actually an open-loop approach, i.e. the actual controller is implemented in phase coordinates with the reference currents being transformed from the field-oriented coordinates. If the link voltage is high enough, it is equivalent to the closed-loop approach (pure field-oriented approach). Moreover, the stability condition is given and the resulting control algorithm is very simple.

For speed control, we have not followed the traditional cascade control structure, because it has been discussed in many textbooks. Instead, the motor speed and acceleration are controlled simultaneously without an inner current control loop. The advantage is the fast response and the robustness with respect to the mechanical parameters, but the motor acceleration should be made available. The motor acceleration can be estimated based on the motor model and the speed measurements; this involves a larger calculation overhead.

To avoid employing a current sensor, thereby reducing the measurement noise, a current observer may be implemented in field-oriented coordinates. However, this observer is sensitive to the motor parameters and relies on the voltage vector which is usually not measured. Instead, it is normal to use the reference voltages feeding to the power converter, i.e. the outputs of the current controllers; these reference voltages are actually different from the real voltages applied to the motor windings.

The final topic was sensorless control based on a sliding mode observer combined with a conventional asymptotic observer. Any algorithm for a sensorless control problem will rely on model parameters. In the presence of modelling uncertainties, the control performance may not be as good as expected.

10.3 Induction motors

10.3.1 Introduction

From a control viewpoint, induction motors are nonlinear high-order dynamic systems of considerable complexity. They are very amenable to a formal mathematical analysis. However, it is not a trivial matter to comprehend the principles of their operation, e.g. under transient conditions, in an imaginative way. On the other hand, induction motors are widely used in practical systems due to their simple mechanical construction, low maintenance requirement and lower cost compared to other types of motors, such as brushless DC motors. Therefore, it is of great significance to investigate the dynamic control problems of these kinds of drive system.

As discussed in the previous section, for a permanent-magnet synchronous motor, the field-oriented coordinate frame coincides with the rotor coordinate frame, such that only three frames of reference are considered. For an induction motor, however, the field-oriented coordinate frame (normally the rotor flux frame) differs from the rotor coordinate frame, hence there are four different frames of reference to be considered.

For an induction motor, the angular difference between the rotor flux and the electrical angular position of the rotor is called *slip*. Its time derivative is called *slip frequency* and it is proportional to the developed motor torque. All state and control variables in the field-oriented coordinate frame are transformed based on the rotor flux angular position ρ, called the *rotor flux angle*, instead of the electrical rotor angle θ_e. To calculate the rotor flux angle, flux measurements are needed. However, rotor flux components are usually not determined by flux measurements, but rather by a *flux observer*. If we know the electrical angular speed of the rotor ω_e, the *flux observer* is simply the on-line simulation of the flux dynamics in the stator coordinate frame (the α, β coordinates); the rate of convergence depends on the rotor time constant. This result is proven later on. However, it is recognized that this time constant may vary slowly with changing ambient temperature. As a result, some on-line adaptation mechanisms based on physical criteria (e.g. the power balance condition) are often involved. Such additional adaptation algorithms, besides the *flux observer*, make the control problem of induction motors more complicated than the control of permanent-magnet synchronous motors. Moreover, if the rotor speed is not available, so-called sensorless control techniques have to be employed. Currently, sensorless control techniques are not mature enough for application in industrial systems.

Current research in the control of induction motors is characterized by a great variety of control methodologies with different control, observation and adaptation algorithms combined with different coordinate systems, different state variables and different notations. For the sliding mode control design described in this section, we mainly use the dynamic model given in the orthogonal stator coordinate frame, the (α, β) coordinates, with stator current components and rotor flux components as state variables, complemented by a mechanical equation.

10.3.2 Model of the induction motor

The structure of an induction motor drive system is shown in Figure 10.26. For sliding mode control design, it is convenient if the control inputs take values from the discrete set $\{-u_0, u_0\}$ instead of on-off signals from the discrete set $\{0, 1\}$. Let the six on-off signals be $\boldsymbol{s} = [s_{w1} \quad s_{w2} \quad s_{w3} \quad s_{w4} \quad s_{w5} \quad s_{w6}]^T$ with $s_{w4} = 1 - s_{w1}$, $s_{w5} = 1 - s_{w2}$, $s_{w6} = 1 - s_{w3}$, and let the control inputs for sliding mode control design be $\boldsymbol{U}_{gate} = [u_1 \quad u_2 \quad u_3]^T$, then the following relation holds:

$$\boldsymbol{U}_{gate} = u_0 \boldsymbol{L}_w \boldsymbol{s}_w \quad \text{with} \quad L_w = \begin{bmatrix} 1 & 0 & 0 & -1 & 0 & 0 \\ 0 & 1 & 0 & 0 & -1 & 0 \\ 0 & 0 & 1 & 0 & 0 & -1 \end{bmatrix} \tag{10.3.1}$$

The backward transformation can be obtained as

$$\begin{aligned} s_{w1} &= 0.5(1 + u_1/u_0) \qquad & s_{w4} &= 1 - s_{w1} \\ s_{w2} &= 0.5(1 + u_2/u_0) \qquad & s_{w5} &= 1 - s_{w2} \\ s_{w3} &= 0.5(1 + u_3/u_0) \qquad & s_{w6} &= 1 - s_{w3} \end{aligned} \tag{10.3.2}$$

Four frames of reference are normally used in describing the dynamic behaviour of the motor (Figure 10.27): the phase frame in (a, b, c) coordinates, the stator frame in (α, β) coordinates, the rotor frame in (x, y) coordinates and the field-oriented frame in (d, q) coordinates.

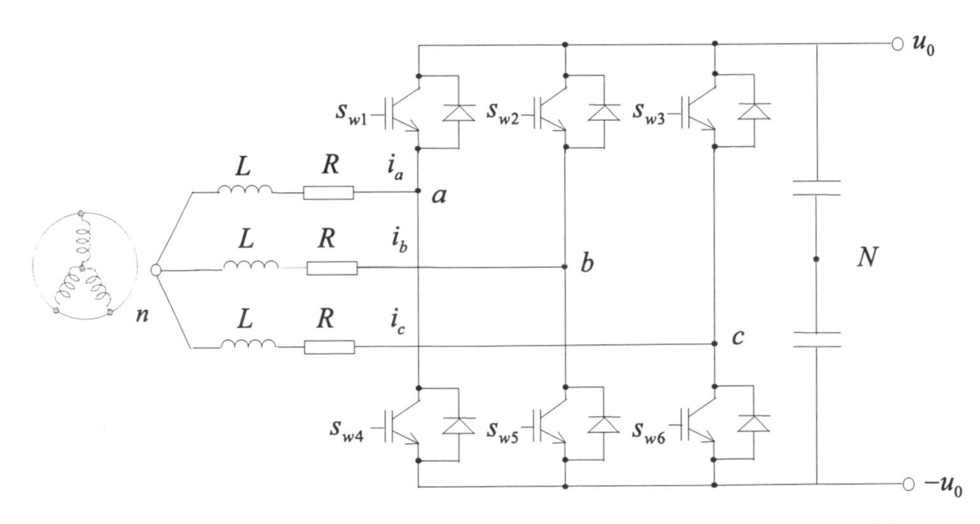

Figure 10.26 Structure of an induction motor drive system: n is the neutral point of the stator windings; u_a, u_b and u_c are the potential differences between points a, b, c and the neutral point n, respectively; i_a, i_b and i_c are the phase currents.

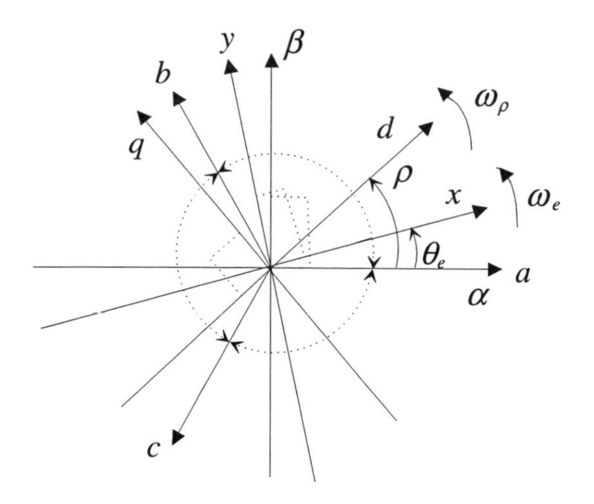

Figure 10.27 Coordinate systems of induction motor model: θ_e is the electrical rotor angular position and ω_e is the electrical rotor angular speed; ρ is the angular position of the rotor flux and ω_ρ is the angular speed of the rotor flux.

The relationship between the phase voltages u_a, u_b, u_c and the discontinuous controls u_1, u_2, u_3 can be given as follows:

$$[u_a \quad u_b \quad u_c]^T = G[u_1 \quad u_2 \quad u_3]^T \tag{10.3.3}$$

where matrix G is defined as

$$G = \frac{1}{3}\begin{bmatrix} 2 & -1 & -1 \\ -1 & 2 & -1 \\ -1 & -1 & 2 \end{bmatrix} \tag{10.3.4}$$

Note that matrix G is a singular matrix implying that the phase voltages u_a, u_b, u_c are not linearly independent. In fact, the sum of u_a, u_b, u_c is equal to zero due to the physical configuration given in Figure 10.26.

The motor model in (α, β) coordinates is important for our control design and can be written as

$$\begin{aligned}
\frac{di_\alpha}{dt} &= \beta\eta\lambda_\alpha + \beta\omega_e\lambda_\beta - \gamma i_\alpha + \frac{1}{\sigma L_s}u_\alpha \\
\frac{di_\beta}{dt} &= \beta\eta\lambda_\beta - \beta\omega_e\lambda_\alpha - \gamma i_\beta + \frac{1}{\sigma L_s}u_\beta \\
\frac{d\lambda_\alpha}{dt} &= -\eta\lambda_\alpha - \omega_e\lambda_\beta + \eta L_h i_\alpha \\
\frac{d\lambda_\beta}{dt} &= -\eta\lambda_\beta + \omega_e\lambda_\alpha + \eta L_h i_\beta \\
\tau &= \frac{3N_r}{2}\frac{L_h}{L_r}(i_\beta\lambda_\alpha - i_\alpha\lambda_\beta) \\
\frac{d\omega_r}{dt} &= \frac{1}{J}(\tau - \tau_l)
\end{aligned} \tag{10.3.5}$$

with

$$\eta = \frac{R_r}{L_r} \qquad \sigma = 1 - \frac{L_h^2}{L_s L_r} \qquad \beta = \frac{L_h}{\sigma L_s L_r} \qquad \gamma = \frac{1}{\sigma L_s}\left(R_s + \frac{L_h^2}{L_r^2} R_r\right) \tag{10.3.6}$$

where

$u_\alpha,\ u_\beta$ = Stator voltages in (α, β) coordinates
$i_\alpha,\ i_\beta$ = Stator currents in (α, β) coordinates
$\lambda_\alpha,\ \lambda_\beta$ = Rotor flux components in (α, β) coordinates
$L_r,\ L_s,\ L_h$ = Rotor, stator and manual inductance
$R_r,\ R_s$ = Rotor and state resistance
N_r = Number of pole pairs
ω_e = Electrical rotor speed
J = Moment of inertia
τ = Motor torque
τ_l = Load torque

The stator currents i_α, i_β and the stator voltages u_α, u_β are transformed from the phase currents i_a, i_b, i_c and the phase voltages u_a, u_b, u_c by

$$\begin{aligned} [i_\alpha \quad i_\beta]^T &= A_{\alpha,\beta}^{a,b,c}[i_a \quad i_b \quad i_c]^T \\ [u_\alpha \quad u_\beta]^T &= A_{\alpha,\beta}^{a,b,c}[u_a \quad u_b \quad u_c]^T \end{aligned} \tag{10.3.7}$$

where $A_{\alpha,\beta}^{a,b,c}$ denotes the transformation matrix

$$A_{\alpha,\beta}^{a,b,c} = \frac{2}{3}\begin{bmatrix} 1 & -1/2 & -1/2 \\ 0 & \sqrt{3}/2 & -\sqrt{3}/2 \end{bmatrix}. \tag{10.3.8}$$

Since the rank of this matrix is 2, the backward transformation has no unique solution. For the backward transformation it is possible to use the pseudoinverse concept. The pseudoinverse of matrix $A_{\alpha,\beta}^{a,b,c}$, denoted as $(A_{\alpha,\beta}^{a,b,c})^+$, is calculated as

$$(A_{\alpha,\beta}^{a,b,c})^+ = (A_{\alpha,\beta}^{a,b,c})^T(A_{\alpha,\beta}^{a,b,c}(A_{\alpha,\beta}^{a,b,c})^T)^{-1} = \frac{3}{2}(A_{\alpha,\beta}^{a,b,c})^T = \begin{bmatrix} 1 & -1/2 & -1/2 \\ 0 & \sqrt{3}/2 & -\sqrt{3}/2 \end{bmatrix}^T \tag{10.3.9}$$

It is easy to show that $A_{\alpha,\beta}^{a,b,c}(A_{\alpha,\beta}^{a,b,c})^+$ is a 2×2 identity matrix. As a result, the backward transformation of the stator currents and voltages to the phase currents and voltages can be given as

$$\begin{aligned} [i_a \quad i_b \quad i_c]^T &= (A_{\alpha,\beta}^{a,b,c})^+[i_\alpha \quad i_\beta]^T \\ [u_a \quad u_b \quad u_c]^T &= (A_{\alpha,\beta}^{a,b,c})^+[u_\alpha \quad u_\beta]^T \end{aligned} \tag{10.3.10}$$

The motor model in (d, q) coordinates, which fixes on the rotor flux vector, can also be obtained by transforming the motor model from (α, β) coordinates to (d, q) coordinates using the rotor flux angle $\rho = \tan^{-1}(\lambda_\beta/\lambda_\alpha)$:

$$\begin{aligned}
\frac{di_d}{dt} &= -\gamma i_d + \eta\beta\lambda_d + \omega_e i_q + \eta\frac{L_h i_q^2}{\lambda_d} + \frac{1}{\sigma L_2}u_d \\
\frac{di_q}{dt} &= -\gamma i_q - \omega_e\beta\lambda_d - \omega_e i_d + \eta\frac{L_h i_d i_q}{\lambda_d} + \frac{1}{\sigma L_s}u_q \\
\frac{d\lambda_d}{dt} &= -\eta\lambda_d + \eta L_h i_d \\
\frac{d\rho}{dt} &= \omega_\rho = \omega_e + \omega_s, \quad \text{with} \quad \omega_s = \eta\frac{L_h i_q}{\lambda_d} \\
\tau &= \frac{N_r L_h}{L_r}\lambda_d i_q \\
\frac{d\omega_r}{dt} &= \frac{1}{J}(\tau - \tau_l)
\end{aligned} \tag{10.3.11}$$

where i_d and i_q are the stator currents in (d, q) coordinates; u_d and u_q are the associated stator voltages; λ_d is the d-component of the rotor flux (the q-component of the rotor flux is equal to zero, i.e. $\lambda_q = 0$). Currents i_d, i_q and voltages u_d, u_q are transformed from (α, β) coordinates by

$$\begin{aligned}
[i_d \quad i_q]^T &= A_{d,q}^{\alpha,\beta}[i_\alpha \quad i_\beta]^T \\
[u_d \quad u_q]^T &= A_{d,q}^{\alpha,\beta}[u_\alpha \quad u_\beta]^T
\end{aligned} \tag{10.3.12}$$

in which matrix $A_{d,q}^{\alpha,\beta}$ is defined as

$$A_{d,q}^{\alpha,\beta} = \begin{bmatrix} \cos\rho & \sin\rho \\ -\sin\rho & \cos\rho \end{bmatrix} \tag{10.3.13}$$

This matrix is an orthogonal matrix whose inverse is equal to its transpose. As a result, the backward transformation can be written as

$$\begin{aligned}
[i_\alpha \quad i_\beta]^T &= (A_{d,q}^{\alpha,\beta})^T[i_d \quad i_q]^T \\
[u_\alpha \quad u_\beta]^T &= (A_{d,q}^{\alpha,\beta})^T[u_d \quad u_q]^T
\end{aligned} \tag{10.3.14}$$

The relationship between (d, q) coordinates and (a, b, c) coordinates can also be established. The phase currents i_a, i_b, i_c and the phase voltages u_a, u_b, u_c can be transformed into (d, q) coordinates using

$$\begin{aligned}
[i_d \quad i_q]^T &= A_{d,q}^{a,b,c}[i_a \quad i_b \quad i_c]^T \\
[u_d \quad u_q]^T &= A_{d,q}^{a,b,c}[u_a \quad u_b \quad u_c]^T
\end{aligned} \tag{10.3.15}$$

where matrix $A_{d,q}^{a,b,c}$ depends on the rotor flux angle ρ and is defined as

$$A_{d,q}^{a,b,c} = A_{d,q}^{\alpha,\beta}A_{\alpha,\beta}^{a,b,c} \tag{10.3.16}$$

Matrix $A_{d,q}^{a,b,c}$ is a 2×3 matrix, hence for the backward transformation we again need its pseudoinverse given as

$$(A_{d,q}^{a,b,c})^{+} = (A_{d,q}^{a,b,c})^{T}(A_{d,q}^{a,b,c}(A_{d,q}^{a,b,c})^{T})^{-1} = \frac{3}{2}(A_{d,q}^{a,b,c})^{T} \tag{10.3.17}$$

leading to the backward transformation

$$\begin{aligned} [i_a \quad i_b \quad i_c]^T &= (A_{d,q}^{a,b,c})^{+}[i_d \quad i_q]^T \\ [u_a \quad u_b \quad u_c]^T &= (A_{d,q}^{a,b,c})^{+}[u_d \quad u_q]^T \end{aligned} \tag{10.3.18}$$

The relationship between the control voltages u_d, u_q and the discontinuous controls u_1, u_2, u_3 can be established as

$$[u_d \quad u_q]^T = A_{d,q}^{1,2,3}[u_1 \quad u_2 \quad u_3]^T \tag{10.3.19}$$

where matrix $A_{d,q}^{1,2,3}$ is defined as

$$A_{d,q}^{1,2,3} = A_{d,q}^{\alpha,\beta}A_{d,q}^{a,b,c}A_{d,q}^{1,2,3} \tag{10.3.20}$$

Following the properties of matrices $A_{d,q}^{a,b,c}$ and $A_{d,q}^{1,2,3}$, we have $A_{\alpha,\beta}^{a,b,c}A_{a,b,c}^{1,2,3} = A_{\alpha,\beta}^{a,b,c}$ such that

$$A_{d,q}^{1,2,3} = A_{d,q}^{\alpha,\beta}A_{\alpha,\beta}^{a,b,c} = A_{d,q}^{a,b,c} \tag{10.3.21}$$

However, to maintain clarity, we prefer to use matrix $A_{d,q}^{1,2,3}$ to denote the transformation between the discontinuous controls u_1, u_2, u_3 and the control voltages u_d, u_q. Matrix $A_{d,q}^{1,2,3}$ is a 2×3 matrix, hence for the backward transformation, we need its pseudoinverse as well:

$$(A_{d,q}^{1,2,3})^{+} = (A_{d,q}^{1,2,3})^{T}(A_{d,q}^{1,2,3}(A_{d,q}^{1,2,3})^{T})^{-1} = \frac{3}{2}(A_{d,q}^{1,2,3})^{T} \tag{10.3.22}$$

leading to the backward transformation

$$[u_1 \quad u_2 \quad u_3]^T = (A_{d,q}^{1,2,3})^{+}[u_d \quad u_q]^T \tag{10.3.23}$$

Throughout this section the parameters of the induction motor used to verify the design principles are $L_r = 650 \times 10^{-6}\,\text{H}$, $L_s = 650 \times 10^{-6}\,\text{H}$, $L_h = 610 \times 10^{-6}\,\text{H}$, $R_r = 0.015\,\Omega$, $R_s = 0.035\,\Omega$, $N_r = 2$, $J = 4.33 \times 10^{-4}\,\text{kg}\,\text{m}^2$ and $\tau_l = B\omega_r$ where B is the coefficient of viscous friction $0.01\,\text{Nm}\,\text{s}\,\text{rad}^{-1}$. The supplied link voltage is $u_0 = 12\,\text{V}$.

Thus far, we have presented the induction motor model in the (α, β) coordinate system and the (d, q) coordinate system. Current and voltage transformations between the different coordinate systems have also been given. They are very important for the control design of induction motors.

Field-oriented current control of induction motors is very similar to field-oriented control of PMSMs (Section 10.2.3). The only difference lies in the angle used for the current and voltage transformations. Here the rotor flux angle ρ is used instead of the electrical rotor angle θ_e. It is recognized that for a high control performance, angle ρ should be available. For a sliding mode control design, the angle ρ is needed as well.

We begin with the rotor flux observation problem under known rotor angular speed, then we discuss speed servo control without explicitly involving the current control loop (the issue of sliding mode current control is omitted here due to its similarity with the PMSM case in Section 10.2.3). Finally, we discuss simultaneous observation of rotor flux and rotor speed, essential for sensorless control of an induction motor.

10.3.3 Rotor flux observer with known rotor speed

The original motion equation for the rotor flux can be used directly as the observer equation. The implementation of such a flux observer is just the on-line simulation of the controlled induction motor. As proven below, this observer is stable in the large with the rate of convergence depending on the rotor time constant. In this case the observer equation is given by

$$\begin{aligned} \frac{d\hat{\lambda}_\alpha}{dt} &= -\eta\hat{\lambda}_\alpha - \omega_e\hat{\lambda}_\beta + \eta L_h i_\alpha \\ \frac{d\hat{\lambda}_\beta}{dt} &= -\eta\hat{\lambda}_\beta + \omega_e\hat{\lambda}_\alpha + \eta L_h i_\beta \end{aligned} \tag{10.3.24}$$

where $\hat{\lambda}_\alpha$, $\hat{\lambda}_\beta$ are the estimates of the rotor flux components. Defining the estimation errors as $\bar{\lambda}_\alpha = \hat{\lambda}_\alpha - \lambda_\alpha$ and $\bar{\lambda}_\beta = \hat{\lambda}_\beta - \lambda_\beta$ results in the following error dynamics:

$$\begin{aligned} \frac{d\bar{\lambda}_\alpha}{dt} &= -\frac{R_r}{L_r}\bar{\lambda}_\alpha - \omega_e\bar{\lambda}_\beta \\ \frac{d\bar{\lambda}_\beta}{dt} &= -\frac{R_r}{L_r}\bar{\lambda}_\beta + \omega_e\bar{\lambda}_\alpha \end{aligned} \tag{10.3.25}$$

Select a Lyapunov function candidate as

$$V = \frac{1}{2}(\bar{\lambda}_\alpha^2 + \bar{\lambda}_\beta^2) > 0 \tag{10.3.26}$$

and calculate the time derivative of V along the solution of (10.3.25) to yield

$$\dot{V} = -\eta(\bar{\lambda}_\alpha^2 + \bar{\lambda}_\beta^2) = -2\eta V < 0 \tag{10.3.27}$$

This means that if we simply integrate equation (10.3.24), the mismatch between the real and estimated rotor flux tends to zero asymptotically and the rate of convergence depends on the rotor time constant $1/\eta$. The rate of convergence may be improved if properly designed observer gains are introduced, as is usually done for reduced-order observers.

With the estimates of the rotor flux components, the rotor flux angle $\hat{\rho}$ can be calculated as

$$\hat{\rho} = \tan^{-1}(\hat{\lambda}_\beta/\hat{\lambda}_\alpha). \tag{10.3.28}$$

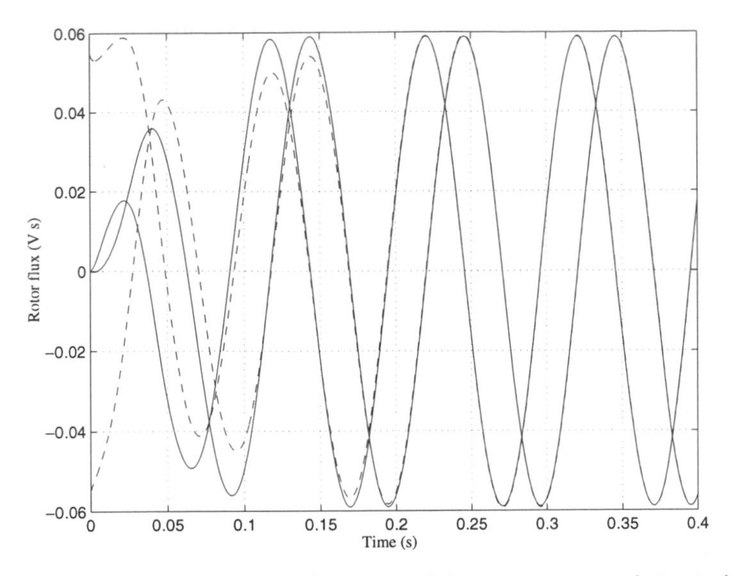

Figure 10.28 Responses of the flux observer with known rotor speed: (- - -) observed flux components $\hat{\lambda}_\alpha$ and $\hat{\lambda}_\beta$, (—) real flux components λ_α and λ_β

Figure 10.28 shows the simulation results of the rotor flux observer (10.3.24). To show the convergence process, select as initial conditions the observed flux components $\hat{\lambda}_\alpha(0) = -0.055$ and $\hat{\lambda}_\beta(0) = 0.055$, and the real flux components $\lambda_\alpha(0) = 0$ and $\lambda_\beta(0) = 0$. Since the flux observer is just the on-line simulation of the motor model, no observer gain is to be adjusted. From (10.3.27) the convergence time is equal to $1/2\eta = 0.022\,\text{s}$, which is confirmed by the simulation.

10.3.4 Speed control

Similar to PMSMs, speed control of induction motors can be realized in two different ways. The first method employs a current controller in an inner loop along with a speed and flux controller in an outer loop, which provide the reference currents i_d^* and i_q^* for the inner loop. Unlike PMSMs, the reference current i_d^* is designed as the output of the flux controller, where the magnitude of the rotor flux should be maintained at some constant level to ensure the desired motor torque. Like PMSMs, the reference current i_q^* is the output of the speed controller which directly corresponds to the motor torque. For the outer loop speed/flux controller, conventional control techniques like PI control are normally used.

The second method of speed/flux control directly uses the sliding mode technique to design discontinuous controls without explicitly involving current control. The advantage of this approach is its robustness with respect to the mechanical parameters. Let $\omega_r^*(t)$ be the reference rotor speed and λ_0 the reference value of the rotor flux magnitude given by the control designer; two switching functions are defined as follows:

$$\begin{aligned} s_\lambda &= c_1(\lambda_0 - \|\lambda(t)\|) + \frac{d}{dt}(\lambda_0 - \|\lambda(t)\|) \\ s_\omega &= c_2(\omega_r^* - \omega_r(t)) + \frac{d}{dt}(\omega_r^* - \omega_r(t)) \end{aligned} \tag{10.3.29}$$

where $\|\lambda(t)\| = \sqrt{\lambda_\alpha^2 + \lambda_\beta^2}$ is the magnitude of the rotor flux and c_1, c_2 are positive constants determining the motion performance in sliding mode. The motion projection in the subspace spanned by s_ω and s_λ has the form

$$\begin{bmatrix} \dot{s}_\lambda \\ \dot{s}_\omega \end{bmatrix} = \begin{bmatrix} F_1 \\ F_2 \end{bmatrix} - \begin{bmatrix} a_\lambda \cos\rho & a_\lambda \sin\rho \\ -a_\omega \sin\rho & a_\omega \cos\rho \end{bmatrix} \begin{bmatrix} u_\alpha \\ u_\beta \end{bmatrix} \tag{10.3.30}$$

where F_1, F_2 are functions of the state variables, the reference inputs and their time derivatives, but they do not depend on the controls u_α, u_β; $\rho = \tan^{-1}(\lambda_\beta/\lambda_\alpha)$ and a_λ, a_ω are defined by

$$a_\lambda = R_r \frac{L_h}{L_s L_r - L_h^2} \qquad a_\omega = \frac{3}{2} \frac{P\|\lambda(t)\|}{J} \frac{L_h}{L_s L_r - L_h^2} \tag{10.3.31}$$

with $L_s L_r - L_h^2 > 0$.

Rewriting (10.3.30) in matrix form yields

$$\dot{\boldsymbol{S}}_{\lambda\omega} = \boldsymbol{F}_{\lambda\omega} - A_{\lambda\omega} \begin{bmatrix} u_\alpha \\ u_\beta \end{bmatrix} \tag{10.3.32}$$

where $\boldsymbol{S}_{\lambda\omega} = [s_\lambda \quad s_\omega]^T \qquad \boldsymbol{F}_{\lambda\omega} = [F_1 \quad F_2]^T \qquad A_{\lambda\omega} = \begin{bmatrix} a_\lambda \cos\rho & a_\lambda \sin\rho \\ -a_\omega \sin\rho & a_\omega \cos\rho \end{bmatrix}$

In order to establish the dependence between $\dot{\boldsymbol{S}}_{\lambda\omega}$ and the switching controls $\boldsymbol{U}_{gate} = [u_1 \quad u_2 \quad u_3]^T$, relations (10.3.7) and (10.3.3) should be substituted into (10.3.32):

$$\dot{\boldsymbol{S}}_{\lambda\omega} = \boldsymbol{F}_{\lambda\omega} - D_{\lambda\omega} \boldsymbol{U}_{gate}, \tag{10.3.33}$$

with $D_{\lambda\omega} = A_{\lambda\omega} A_{\alpha,\beta}^{a,b,c} G = A_{\lambda\omega} A_{\alpha,\beta}^{a,b,c}$ (since $A_{\alpha,\beta}^{a,b,c} G = A_{\alpha,\beta}^{a,b,c}$).

Equation (10.3.33) is the starting point of the sliding mode control design. Since matrix $D_{\lambda\omega}$ is time-varying, the controls defined in $\boldsymbol{U}_{gate}$ cannot perform their switching depending on $\boldsymbol{S}_{\lambda\omega}$. In this case a transformed vector of new switching functions is required to achieve a time-invariant control matrix. Design the transformation as

$$\boldsymbol{S}^* = D_{\lambda\omega}^+ \boldsymbol{S}_{\lambda\omega} \tag{10.3.34}$$

where $\boldsymbol{S}^* = [s_1^* \quad s_2^* \quad s_3^*]^T$ is the vector representing the transformed switching functions; $D_{\lambda\omega}^+$ is the pseudoinverse of matrix $D_{\lambda\omega}$ given as

$$D_{\lambda\omega}^+ = D_{\lambda\omega}^T (D_{\lambda\omega} D_{\lambda\omega}^T)^{-1} = \frac{3}{2} (A_{\alpha,\beta}^{a,b,c})^T (A_{\lambda\omega})^T Q \tag{10.3.35}$$

with Q being defined by

$$Q = (A_{\lambda\omega}(A_{\lambda\omega})^T)^{-1} = \begin{bmatrix} 1/a_\lambda^2 & 0 \\ 0 & 1/a_\omega^2 \end{bmatrix} \tag{10.3.36}$$

Q is a positive definite time-varying matrix (since parameter a_ω is proportional to $\|\lambda(t)\|$), provided that $\|\lambda(t)\| \neq 0$. The controls u_1, u_2, u_3 are designed as

$$\boldsymbol{U}_{gate} = u_0 \ \mathrm{sign}(\boldsymbol{S}^*) \tag{10.3.37}$$

with $\mathrm{sign}(\boldsymbol{S}^*) = [\mathrm{sign}(s_1^*) \ \mathrm{sign}(s_2^*) \ \mathrm{sign}(s_3^*)]^T$.

Theorem 10.3

For high enough link voltage u_0, system (10.3.33) converges to the origin $s_\lambda = 0, s_\omega = 0$ in finite time under control (10.3.37) and transformation (10.3.34). □

Proof

For the Lyapunov function candidate

$$V = \frac{1}{2}(\boldsymbol{S}_{\lambda\omega})^T Q \boldsymbol{S}_{\lambda\omega} \tag{10.3.38}$$

its time derivative along the solutions of (10.3.33) will be

$$\begin{aligned} \dot{V} &= (\boldsymbol{S}_{\lambda\omega})^T Q \dot{\boldsymbol{S}}_{\lambda\omega} + \frac{1}{2}(\boldsymbol{S}_{\lambda\omega})^T \dot{Q} \boldsymbol{S}_{\lambda\omega} \\ &= (\boldsymbol{S}^*)^T (D_{\lambda\omega})^T Q \boldsymbol{F}_{\lambda\omega} - (\boldsymbol{S}^*)^T (D_{\lambda\omega}^T) Q D_{\lambda\omega} \boldsymbol{U}_{gate} + \frac{1}{2}(\boldsymbol{S}_{\lambda\omega})^T \dot{Q} \boldsymbol{S}_{\lambda\omega} \end{aligned} \tag{10.3.39}$$

Substituting control (10.3.37) into (10.3.39) defining $\boldsymbol{F}^* = [F_1^* \ \ F_2^* \ \ F_3^*] = (D_{\lambda\omega})^T Q \boldsymbol{F}_{\lambda\omega}$ and considering the definitions of matrices $D_{\lambda\omega}$ and Q, we obtain

$$\dot{V} = (\boldsymbol{S}^*)^T \boldsymbol{F}^* - (\boldsymbol{S}^*)^T (A_{\alpha,\beta}^{a,b,c})^T A_{\alpha,\beta}^{a,b,c} u_0 \ \mathrm{sign}(\boldsymbol{S}^*) + \frac{1}{2}(\boldsymbol{S}_{\lambda\omega})^T \dot{Q} \boldsymbol{S}_{\lambda\omega} \tag{10.3.40}$$

where the term $\frac{1}{2}(\boldsymbol{S}_{\lambda\omega})^T \dot{Q} \boldsymbol{S}_{\lambda\omega}$ can be expanded as

$$\frac{1}{2}(\boldsymbol{S}_{\lambda\omega})^T \dot{Q} \boldsymbol{S}_{\lambda\omega} = -f(\lambda_\alpha, \lambda_\beta, \dot{\lambda}_\alpha, \dot{\lambda}_\beta) s_\omega^2 \tag{10.3.41}$$

with $f(\lambda_\alpha, \lambda_\beta, \dot{\lambda}_\alpha, \dot{\lambda}_\beta)$ being a bounded function, given as

$$f(\lambda_\alpha, \lambda_\beta, \dot{\lambda}_\alpha, \dot{\lambda}_\beta) = \left(\frac{3}{2}\frac{J}{P}\frac{L_s L_r - L_h^2}{L_h^2}\right)^2 \frac{\lambda_\alpha \dot{\lambda}_\alpha + \lambda_\beta \dot{\lambda}_\beta}{(\lambda_\alpha^2 + \lambda_\beta^2)^2} \tag{10.3.42}$$

Note that $\dot{\lambda}_\alpha$ and $\dot{\lambda}_\beta$ do not depend on the control. Matrix $(A_{\alpha,\beta}^{a,b,c})^T A_{\alpha,\beta}^{a,b,c}$ is a *singular* matrix and can be calculated as

$$(A_{\alpha,\beta}^{a,b,c})^T A_{\alpha,\beta}^{a,b,c} = \frac{4}{9}\begin{bmatrix} 1 & -\frac{1}{2} & -\frac{1}{2} \\ -\frac{1}{2} & 1 & -\frac{1}{2} \\ -\frac{1}{2} & -\frac{1}{2} & 1 \end{bmatrix} \tag{10.3.43}$$

Depending on the signs of s_1^*, s_2^* and s_3^*, there are eight possible combinations of $\text{sign}(s_1^*), \text{sign}(s_2^*)$ and $\text{sign}(s_3^*)$. By evaluating (10.3.34), it is easy to verify that two of these combinations are not possible, i.e. $\text{sign}(s_1^*), \text{sign}(s_2^*)$ and $\text{sign}(s_3^*)$ cannot be all $+1$ or all -1 simultaneously. The remaining six combinations can be summarized as

$$\text{sign}(s_l^*) \neq \text{sign}(s_m^*) = \text{sign}(s_n^*), \quad \text{with} \quad l \neq m \neq n \text{ and } l, m, n \in \{1, 2, 3\}. \tag{10.3.44}$$

Starting from this notion, equation (10.3.40) can be expanded as

$$\dot{V} = (s_1^* F_1^* + s_2^* F_2^* + s_3^* F_3^*) - (2/3)^2 u_0 (2|s_l^*| + |s_m^*| + |s_n^*|) - f(\lambda_\alpha, \lambda_\beta, \dot{\lambda}_\alpha, \dot{\lambda}_\beta) s_\omega^2 \tag{10.3.45}$$

with $l \neq m \neq n$ and $l, m, n \in \{1, 2, 3\}$.

As long as $\|\boldsymbol{S}^*\| \neq 0$ and $\|\lambda(t)\| \neq 0$, there exists a DC link voltage u_0 high enough such that $\dot{V} < 0$ will be guaranteed for all possible combinations of l, m, n. This proof implies that $s_\lambda = 0, s_\omega = 0$ can be reached in finite time (Section 3.5) by directly switching the controls u_1, u_2 and u_3.

For case $\|\lambda(t)\| = 0$, $\|\boldsymbol{F}^*\| = \|(D_{\lambda\omega})^T Q \boldsymbol{F}_{\lambda\omega}\|$ and $f(\lambda_\alpha, \lambda_\beta, \dot{\lambda}_\alpha, \dot{\lambda}_\beta)$ will be unbounded, thus sliding mode cannot be provided. This problem also exists in the conventional field-oriented control approach. In critical applications the induction motor can be started by an open-loop control such that flux can be established. The closed-loop control can be applied to the motor thereafter. ∎

With high enough u_0, the proposed speed controller is robust with respect to any disturbances that are included in F_1^*, F_2^* and F_3^*. For example, a very important feature of the speed controller is the robustness with respect to the load torque variation, since $\dot{\tau}_l$ is contained in $\boldsymbol{F}^*$. From an implementation viewpoint, the exact values of F_1^*, F_2^* and F_3^* are not required. They depend on the system states, system parameters and reference inputs as well as their time derivatives. Their norms should not be unbounded, otherwise no DC link voltage can be found to enforce sliding mode.

To implement the control algorithm, matrix $D_{\lambda\omega}^+$ is needed for calculating the transformation (10.3.34). Matrix $D_{\lambda\omega}^+$ can be found as

$$D_{\lambda\omega}^+ = \frac{3}{2}(A_{\alpha,\beta}^{a,b,c})^T (A_{\lambda\omega})^T Q = \begin{bmatrix} \frac{1}{a_\lambda}\cos\rho_a & -\frac{1}{a_\omega}\sin\rho_a \\ \frac{1}{a_\lambda}\cos\rho_b & -\frac{1}{a_\omega}\sin\rho_b \\ \frac{1}{a_\lambda}\cos\rho_c & -\frac{1}{a_\omega}\sin\rho_c \end{bmatrix} \tag{10.3.46}$$

where $\rho_a = \rho$, $\rho_b = \rho - 2\pi/3$, $\rho_c = \rho + 2\pi/3$ and ρ is the rotor flux angle. Since the rotor flux angle is not available in most practical cases, the observer given in Section 10.3.3 can be used.

Again, for constructing the sliding surface s_ω, the acceleration signal of the rotor revolution is needed. Since the rotor flux components can be estimated using the flux observer of Section 10.3.3 and the stator current components can be measured, the developed motor torque can be calculated according to (10.3.5). Based on this torque signal, the mechanical equation in (10.3.5) and the assumption that the load torque

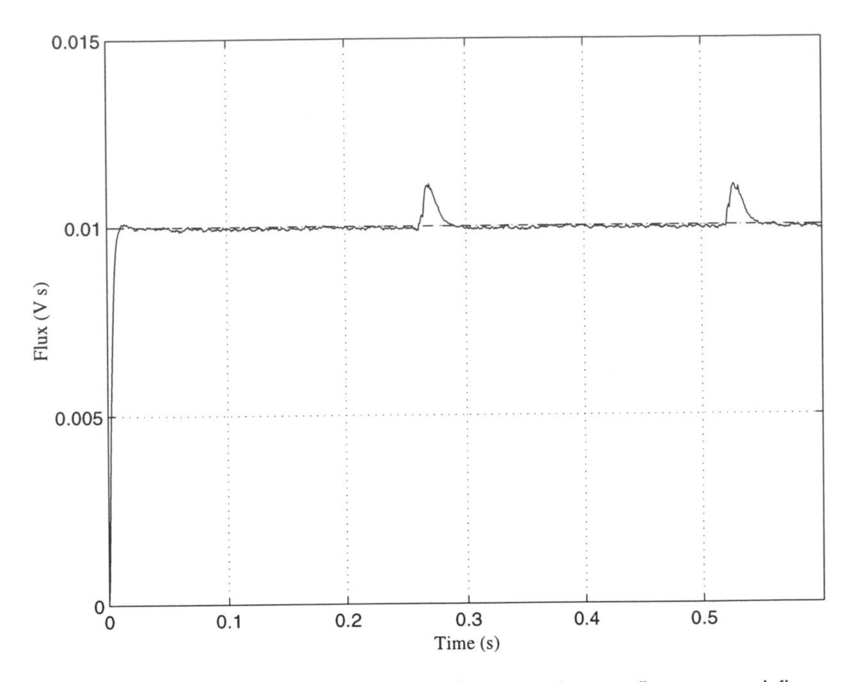

Figure 10.29 Response of the rotor flux control: (- - -) reference flux, (—) real flux.

varies slowly with respect to time, similar techniques as described for DC motor control can be applied to estimate the rotor acceleration (Section 10.1.6). Furthermore, to construct the sliding surface s_λ, the time derivative of the rotor flux amplitude is needed. Observer equation (10.3.24) can be used to calculate this signal. However, if there exist uncertainties in the motor parameters, both the resulting acceleration signal and the time derivative of the rotor flux amplitude will be corrupted by disturbances. Then, sensitivity analysis is needed for critical applications. On-line adaptation mechanisms may be used to update the motor parameters (e.g. rotor resistance).

Although the stability proof is rather involved, the implementation of the speed/flux control is actually very simple: only equations (10.3.29), (10.3.34) and (10.3.37) need to be calculated. Figures 10.29 and 10.30 show the simulation results of the proposed speed/flux controller. The control gains in (10.3.29) are designed as $c_1 = 200$ and $c_2 = 100$. From a practical viewpoint, an integral term of the speed control error may be added to the second equation of (10.3.29) to reduce the steady-state error which may appear due to modelling uncertainties or external disturbances. At the initial time instant, the amplitude of the rotor flux is zero, i.e. $\|\lambda(0)\| = 0$. Theoretically, the convergence may not be guaranteed. Nevertheless, the simulation results show a stable behaviour during the initial time interval.

10.3.5 Observer for rotor flux and rotor speed

Simultaneous observation of rotor flux and rotor speed has been a challenging problem in the control of induction motors. The speed/flux observer presented here follows the ideas of Utkin and Shi (1995), and is a modified version of Isozimov (1983).

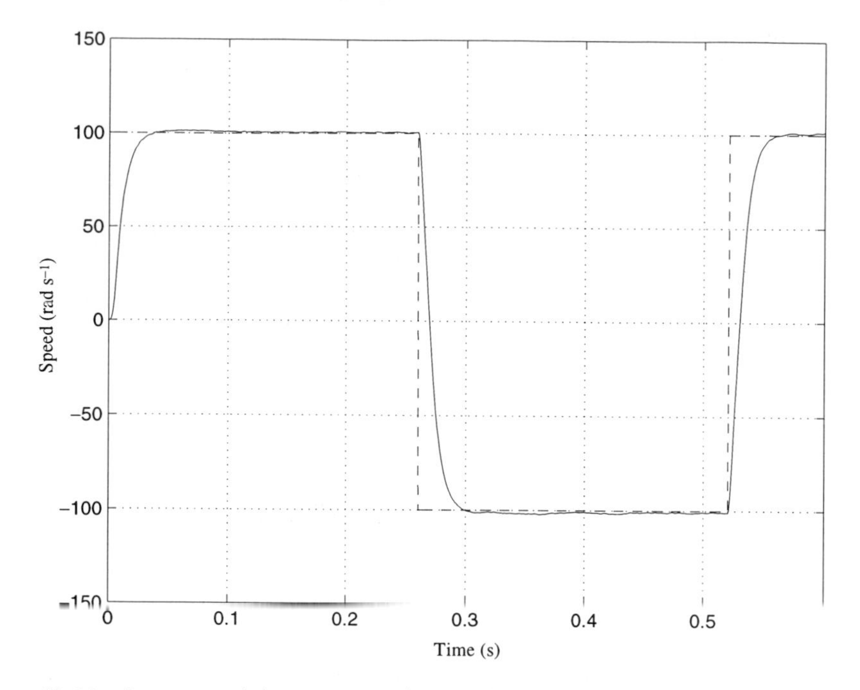

Figure 10.30 Response of the rotor speed control: (- - -) reference speed, (—) real speed.

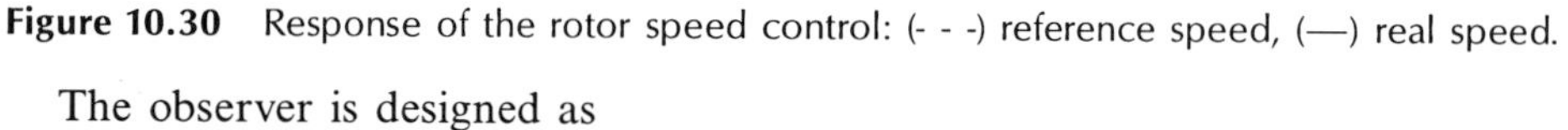

The observer is designed as

$$
\begin{aligned}
\frac{d\hat{i}_\alpha}{dt} &= \beta\eta\hat{\lambda}_\alpha + \beta\hat{\omega}_e\hat{\lambda}_\beta - \gamma\hat{i}_\alpha + \frac{1}{\sigma L_s}u_\alpha \\
\frac{d\hat{i}_\beta}{dt} &= \beta\eta\hat{\lambda}_\beta - \beta\hat{\omega}_e\hat{\lambda}_\alpha - \gamma\hat{i}_\beta + \frac{1}{\sigma L_s}u_\beta \\
\frac{d\hat{\lambda}_\alpha}{dt} &= -\eta\hat{\lambda}_\alpha - \hat{\omega}_e\hat{\lambda}_\beta + \eta L_h i_\alpha \\
\frac{d\hat{\lambda}_\beta}{dt} &= -\eta\hat{\lambda}_\beta + \hat{\omega}_e\hat{\lambda}_\alpha + \eta L_h i_\beta
\end{aligned}
\tag{10.3.47}
$$

in which $\hat{\lambda}_\alpha, \hat{\lambda}_\beta$ are the estimates for rotor flux components; $\hat{i}_\alpha, \hat{i}_\beta$ are estimated stator currents used for generating sliding mode; and $\hat{\omega}_e$ is the estimate for the electrical rotor speed, which is discontinuous and given by

$$\hat{\omega}_e = \omega_0 \operatorname{sign}(s_n) \tag{10.3.48}$$

where ω_0 is a positive control gain. In (10.3.48) the sliding surface s_n is defined as

$$s_n = (\hat{i}_\beta - i_\beta)\hat{\lambda}_\alpha - (\hat{i}_\alpha - i_\alpha)\hat{\lambda}_\beta \tag{10.3.49}$$

By selecting a large enough ω_0, sliding mode $s_n = 0$ will occur, implying that the estimated currents $\hat{i}_\alpha, \hat{i}_\beta$ will converge to their real values i_α, i_β. The equivalent value of $\hat{\omega}_e$, $\hat{\omega}_e^{eq}$, can be obtained by solving the algebraic equation $\dot{s}_n = 0$ (Section 2.3).

It may be interpreted as a smoothed estimate of the electrical rotor speed. The analytical value of $\hat{\omega}_e^{eq}$ can be obtained as

$$\hat{\omega}_e^{eq} = \omega_e \frac{\lambda^T \hat{\lambda}}{\|\hat{\lambda}\|^2} + k \frac{(\bar{\lambda}_\beta \hat{\lambda}_\alpha - \bar{\lambda}_\alpha \hat{\lambda}_\beta)}{\|\hat{\lambda}\|^2} \tag{10.3.50}$$

where $\lambda^T \hat{\lambda} = \lambda_\alpha \hat{\lambda}_\alpha + \lambda_\beta \hat{\lambda}_\beta$ and k is a constant depending on the motor parameters. Equation (10.3.50) implies that if the estimated rotor flux converges to the true flux, then the equivalent value of $\hat{\omega}_e$ is equal to the true rotor speed, i.e. $\lim_{\hat{\phi} \to \phi} \hat{\omega}_e^{eq} = \omega_e$.

In practice $\hat{\omega}_e^{eq}$ can be obtained through a lowpass filter with the discontinuous value $\hat{\omega}_e$ given in (10.3.48) as the input of the filter (Section 2.3), i.e.

$$\hat{\omega}_e^{eq} = \frac{1}{1 + \mu p} \hat{\omega}_e \tag{10.3.51}$$

where μ is the time constant of the filter, and p is the Laplace variable. The filter time constant μ should be selected such that the high-frequency component will be filtered out and the equivalent control $\hat{\omega}_e^{eq}$ should be preserved.

The proposed observer has been implemented with a Digital Signal Processor-based hardware system. Simulation and experimental results show that the estimated rotor flux components do indeed converge to the real flux components, as does the rotor speed (Utkin and Shi, 1995). Moreover, this observer was verified to be very reliable experimentally. While the observer equations were being calculated, the hardware system was switched off and then switched on again after several seconds, but the observer still converged (recognized by comparing the observed speed with the real speed obtained through a precise incremental encoder).

For the proof of flux convergence, the theoretical value of the equivalent rotor speed (10.3.50) should be substituted into the observer equation (10.3.47). However, at the time of writing, this proof has not been performed satisfactorily. Readers interested in a proof under certain assumptions should consult the technical reports by Utkin and Jin (1997) and by Zaremba (1995).

Remark 10.3

In a real implementation, control gain ω_0 may not be a constant, it may depend on $\hat{\omega}_e^{eq}$. Theoretically, ω_0 should be made large enough such that the robustness of the observer can be guaranteed (sliding mode can be enforced), especially in the case where the mechanical equation is not used due to the lack of knowledge about the load torque and the moment of inertia of the drive system. But practically, the integration error of the observer will become higher if ω_0 is getting larger, due to the limited sampling rate of the chosen microprocessor. In this case the observed rotor flux will substantially differ from the real rotor flux. A discrete-time version of the observer should provide more accurate results, and Chapter 9 gives general design principles for discrete-time sliding mode control systems. □

Figures 10.31 to 10.33 depict the simulation results of the proposed observer algorithms in (10.3.47) to (10.3.49). The motor is controlled in open loop. To show the convergence process, select as initial conditions the observed flux components $\hat{\lambda}_\alpha(0) = -0.025$ and

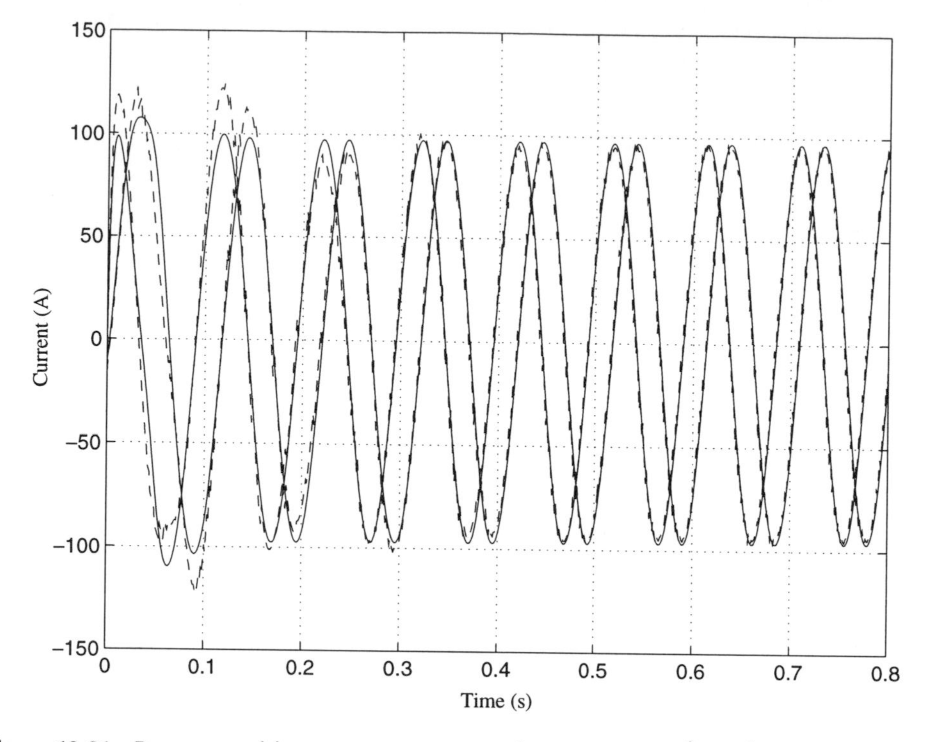

Figure 10.31 Responses of the stator currents: (- - -) observed currents $\hat{i}_\alpha$ and $\hat{i}_\beta$, (—) real currents i_α and i_β.

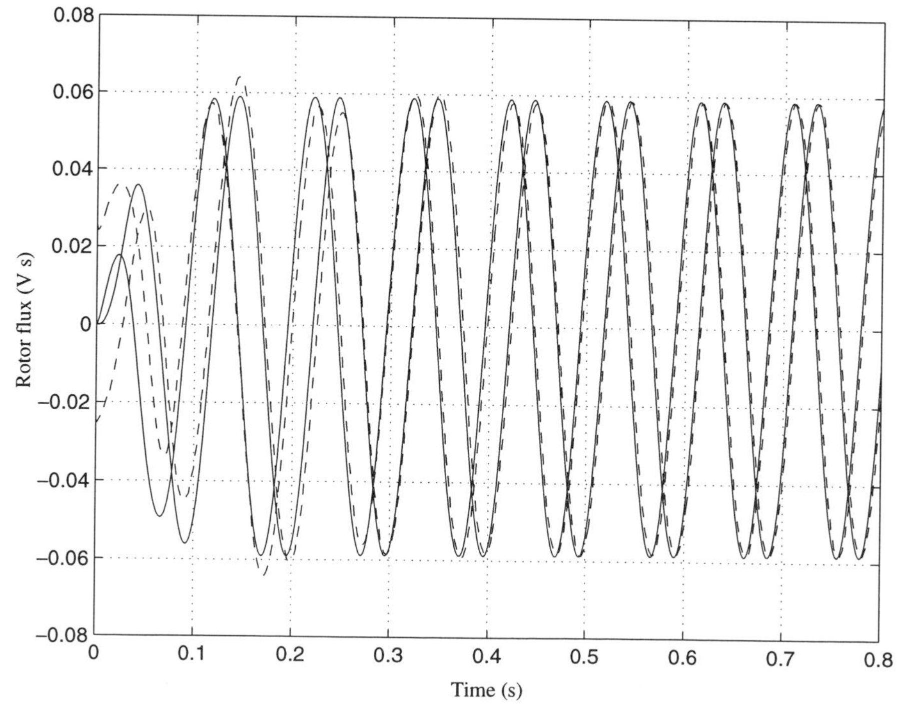

Figure 10.32 Responses of the rotor flux: (- - -) observed flux components $\hat{\lambda}_\alpha$ and $\hat{\lambda}_\beta$, (—) real flux components λ_α and λ_β.

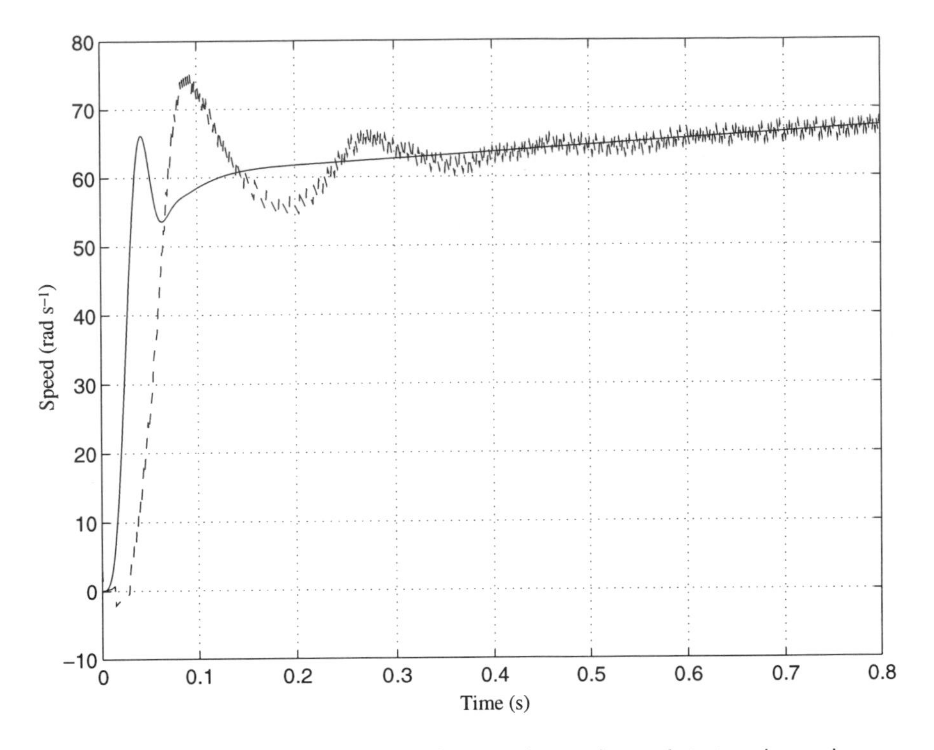

Figure 10.33 Response of the rotor speed: (- - -) observed speed, (—) real speed.

$\hat{\lambda}_\beta(0) = 0.025$, and the real flux components $\lambda_\alpha(0) = 0$ and $\lambda_\beta(0) = 0$. The discontinuous control gain in (10.3.48) is designed as $\omega_0 = 240$. The cutoff frequency of the lowpass filter in (10.3.51) is selected to be 15 Hz.

The observer design problem is essential for the control of induction motors, especially when the rotor speed is not known. If no information about any mechanical variable is available, the design of the flux observer as well as the speed observer strongly depends on the motor model and associated parameters. Moreover, it has been recognized that some of these parameters vary with the operating conditions, e.g. the rotor time constant changes with respect to the ambient temperature and with respect to the rotor position. Hence the design of such an observer is not a trivial matter.

The most difficult problem in sensorless control is the *zero-speed* problem. At speeds close to zero, equation (10.3.5) shows that the rotor flux components are strongly dependent on the motor parameters, i.e. $\eta = R_r/L_r$. Since these parameters may not be known exactly or they are time-varying, the resulting estimation of the rotor flux may substantially differ from the true rotor flux. Additional information is then required about the system. Different approaches use different physical properties of the induction motors:

- Lorenz (1995) uses structural modification on the motor rotor.
- Jiang and Holtz (1997) extract speed information from the harmonic components.
- Blaschke *et al.* (1995) use the test current method.
- Xu and Novotny (1993) employ stator flux estimation.

These methods were tested in practical systems and shown to yield a highly acceptable performance at very low rotor speed. Another trend in research on sensorless control is to use sophisticated control/estimation algorithms These methods include some adaptive/robust algorithms:

- Henneberger *et al.* (1991) use a Kalman filter (EKF) method.
- Tamai *et al.* (1987) use model reference adaptive control (MRAC).

10.3.6 Discussion

Dynamic models of induction motors in stator orthogonal coordinates and in field-oriented coordinates were the starting point of this section. We omitted the current control problem for induction motors since it is very similar to the problem for PMSMs. The only difference is the angle used for the field orientation. The rotor flux angle is estimated by an on-line simulation of the rotor flux model, resulting in a stable rotor flux observer. This observer depends on the rotor time constant (rotor resistance and rotor inductance) and on the speed measurements.

For speed control, we did not follow the traditional control structure, i.e. the cascaded control structure, because it has been discussed in many textbooks. Instead, we looked at simultaneous control of the speed, the acceleration, the magnitude of the rotor flux and its time derivative. The motor acceleration and the magnitude of the rotor flux (as well as its time derivative) are normally not measured. They should be estimated using the motor model and the speed measurements. The rotor flux angle is also necessary for the field orientation. We gave the stability condition for the proposed controller.

Finally, we discussed observer design for speed sensorless control based on the proposed sliding mode observer. Simulation results showed that the estimated rotor flux components do indeed converge to the true rotor flux components, as does the rotor speed.

10.4 Summary

This chapter has presented control algorithms for DC motors, permanent-magnet synchronous motors and induction motors using the sliding mode design principle. Beside the advantages provided by the sliding mode approach in the sense of control techniques, this unified design principle also provides a deeper understanding of the functionality and mechanism of an electric drive system. For AC motors, the ultimate object of field-oriented control is to enable a decoupling between the motor torque and the motor flux, resulting in a system similar to a separately excited DC motor. However, the traditional decoupling characteristic is sensitive to variations in the motor parameters (resistance and inductance). On the other hand, the design goals of sliding mode control are also given in field-oriented coordinates and the field orientation is performed by the switching surface transformation. Since no exact decoupling is required for sliding mode control, the resulting system is insensitive to the motor parameters and the solutions are quite simple. In this sense, sliding mode control may be considered as an extension of the traditional control techniques like hysteresis control, but with the following advantages:

- It achieves the performance of field-oriented control.

- It has strict stability conditions.
- It extends beyond current control to outer-loop control, e.g. speed and flux control.

Another impressive feature of sliding mode techniques is the combination with asymptotic observers and even the observer design itself. In these applications the concept of equivalent control plays a key roll. This concept originated from the observation of physical systems, providing an additional source of information to the control design and thus reducing the complexity of the overall system.

References

BLASCHKE, F., 1974, 'Das Verfahren der Feldorientierung zur Regelung der Drehfeldmaschine', PhD thesis, Technische Universität Braunschweig.

BLASCHKE, F., VANDENPUT, A. and VAN DER BURGT, J., 1995, Feldorientierung der geberlosen Drehfeldmaschine (in German), *Elektromotoren*, **21**, 14–23.

HENNEBERGER, G., BRUNSBACH, B. J. and KLEPSCH, T., 1991, Field-oriented control of synchronous and asynchronous drives without mechanical sensors using a Kalman filter, *Proceedings of the European Conference on Power Electronics and Applications*, **3**, 664–71.

ISOZIMOV, D. B., 1983, Sliding mode nonlinear state observer of an induction motor, in Meerov, M.V. and Kuznetsov, N. A. (Eds), *Control of Multiconnected Systems* (in Russian), pp. 133–39, Moscow: Nauka.

JIANG, J. and HOLTZ, J., 1997, High dynamic speed sensorless AC drive with on-line model parameter tuning for steady-state accuracy, *IEEE Transactions on Industrial Electronics*, **44**, 240–46.

KOKOTOVIC, P. V., O'MALLEY, R. B. and SANNUTI, P., 1997, Singular perturbations and order reduction in control theory, *Automatica*, **12**, 123–32.

LORENZ, R., 1995, Future trends in power electronic control of drives: robust, zero speed sensorless control and new standard approaches for field orientation, *Proceedings of the International Power Electronics Conference (IPEC)*, pp. 23–28.

SABANOVIC, A., SABANOVIC, N. and OHNISHI, K., 1993, Sliding mode in power converters and motion control systems, *International Journal of Control*, **57**, 1237–59.

SABANOVIC, A. and UTKIN, V. I., 1994, Sliding mode applications in switching converters and motion control systems, Materials Workshop at the International Conference on Industrial Electronics, Control, and Instrumentation (IECON), Bologna.

SHI, J. and LU, Y., 1996, Field-weakening operation of cylindrical permanent-magnet motors, *Proceedings of the IEEE Conference on Control Applications*, pp. 864–69..

TAMAI, S., SUGIMOTO, H. and YANO, M., 1987, Speed sensorless vector control of induction motor with model reference adaptive system, *Proceedings of the IEEE/IAS Annual Meeting*, pp. 189–95.

UTKIN, V. I., 1993, Sliding mode control design principle and applications to electric drives, *IEEE Transactions on Industrial Electronics*, **40**, 23–36.

UTKIN, V. I. and SHI, J., 1995, Torque control of an induction motor without mechanical variable sensors, Technical report of Ford Motor Company and Ohio State University.

UTKIN, V. I. and JIN, C., 1997, Sensorless sliding mode torque control of induction motor, Technical report of Ford Motor Company and Ohio State University.

XU, X. and NOVOTNY, D. W., 1993, Implementation of direct stator flux oriented control on a versatile DSP based system, *IEEE Transactions on Industrial Applications*, **29**, 344–48.

ZAREMBA, A., 1995, Reduced order sliding mode speed observer of induction motors, Technical report of Ford Motor Company.

CHAPTER ELEVEN

Power Converters

This chapter will present a sliding mode approach for the design of control systems for power converters. A cascaded control structure is chosen for ease of control realization and to exploit the motion separation property of power converters. For power converters, the fast motion is dominated by the dynamics of the winding currents whereas the slow motion stems from the dynamics of the output voltage. Since power converters inherently include switching devices, it is straightforward to design sliding mode control that yields a discontinuous control law. Detailed theoretical developments and numerical simulations will demonstrate the efficiency of sliding mode control in this field as a powerful alternative to the existing PWM techniques.

11.1 DC/DC converters

For circuits controlled by switching devices, where the control variable can assume only a discrete set of values, it is natural to consider sliding mode strategies to synthesize the switching policy, natural from a technological viewpoint and from a theoretical viewpoint. In the past, the method of state-space averaging has been widely used to analyze DC/DC converters. In the state-space averaging method, the linear circuit models and the state-space equation are identified for each of the possible switch positions of the converter during the switching period. These state-space equations are then averaged over the switching period, leading to a low-frequency equivalent model of the converter. The low-frequency model thus obtained may be linearized in order to apply linear control theory to design feedback compensators. In essence, state-space averaging provides a method of low-frequency characterization for converters, allowing the use of frequency domain design. Sliding mode techniques belong to the category of time domain design; they can be used to characterize the system under small-signal and large-signal conditions. Sliding mode control uses state feedback and directly sets up the desired closed-loop response in the time domain, or in terms of differential equations. The most important feature of the sliding mode approach is the low sensitivity to system parameter variations.

11.1.1 Bilinear systems

Commonly used DC/DC converters can be classified into *buck*, *boost*, *buck-boost* and *Cuk* converters. Some of these DC/DC converters can be summarized with a unified state-space formulation in the form of a bilinear system defined on $\Re^n$

$$\dot{x} = Ax + uBx \tag{11.1.1}$$

where $x \in \Re^n$ is the state vector; $A \in \Re^{n \times n}$ and $B \in \Re^{n \times n}$ are matrices with constant real entries; u is a scalar control taking values from the discrete set $U = \{0, 1\}$. For system (11.1.1), we may design a discontinuous control as

$$u = \frac{1}{2}(1 - \text{sign}(s)) \tag{11.1.2}$$

where $s(x)$ is a scalar switching function in the sense of sliding mode theory, defined by

$$s(x) = c^T x \tag{11.1.3}$$

with vector $c = [\partial s/\partial x]$ and $c \in \Re^n$ denoting the gradient of the scalar function s with respect to state-space vector x. The motion projection of system (11.1.1) onto the manifold s can be obtained as

$$\begin{aligned} \dot{s} &= c^T \dot{x} = c^T Ax + uc^T Bx \\ &= c^T Ax + \frac{1}{2} c^T Bx - \frac{1}{2} \text{ sign}(s) c^T Bx \end{aligned} \tag{11.1.4}$$

For sliding mode to exist in the manifold $s = 0$, system (11.1.4) needs to satisfy the convergence (to $s = 0$) and sliding condition $s\dot{s} < 0$, which implies that

$$s\dot{s} = s\left(c^T Ax + \frac{1}{2} c^T Bx\right) - \frac{1}{2} |s| c^T Bx < 0 \tag{11.1.5}$$

From this inequality we may obtain the necessary condition for sliding mode to exist.
If sliding mode exists, then in the vicinity of $s = 0$, the following relations hold:

$$\begin{aligned} \dot{s}_{s>0} &= C^T Ax < 0 \\ \dot{s}_{s<0} &= C^T Bx > -C^T Ax \end{aligned} \tag{11.1.6}$$

The motion equation of system (11.1.1) in sliding mode can be derived using the equivalent control method (Section 2.3). The equivalent control of the discontinuous control u is calculated by formally setting $\dot{s} = 0$ and solving equation (11.1.4) for u to yield

$$u_{eq} = -\frac{c^T Ax}{c^T Bx} \tag{11.1.7}$$

The motion equation of the sliding motion is governed by

$$\dot{x} = Ax + u_{eq}Bx \qquad (s(x) = 0) \tag{11.1.8}$$

Theorem 11.1

For sliding mode to exist locally on $s(x) = 0$, the corresponding equivalent control satisfies

$$0 < u_{eq} = -\frac{c^T Ax}{c^T Bx} < 1 \tag{11.1.9}$$

□

Proof

For sliding mode to exist, $s\dot{s} < 0$ holds. The second line of equation (11.1.4) states that

$$\frac{1}{2}c^T Bx > \left|\frac{1}{2}c^T Bx + c^T Ax\right| \tag{11.1.10}$$

Solving this inequality leads to

$$0 < u_{eq} = -\frac{c^T Ax}{c^T Bx} < 1 \tag{11.1.11}$$

■

Note that condition $s\dot{s} < 0$ defines an attraction domain of the sliding manifold. It is the task of the control designer to ensure this condition is always fulfilled, both in transient behaviour and in steady state. This may also involve careful choices of initial conditions, as will become clearer during the study of different types of power converters.

In the rest of this chapter, the control problems of buck and boost converters are investigated. The design approaches can be naturally extended to the buck-boost and Cuk converters. Boost converters are used for applications where the required output voltage is higher than the source voltage. Buck converters, on the other hand, are employed for applications with the output voltage being smaller than the source voltage. From a control design viewpoint, the boost converters are more difficult than the buck converters, since the standard design model of a boost converter is a *nonminimum phase* system (Venkataramanan *et al.*, 1985).

Traditionally, the control problems of the DC/DC converters are solved by using pulse-width modulation (PWM) techniques. Sira-Ramirez (1988) demonstrated the equivalence between sliding mode control and PWM control in the low-frequency range for a boost converter. Generally speaking, hardware implementation is much easier for sliding mode control than for PWM control. Since the maximum frequency of commercially available switching elements grows higher and higher, the sliding mode approach is expected to become increasingly popular in the field of power converter control. The following sections will discuss the problems of output voltage regulation for both types of DC/DC converter.

11.1.2 Direct sliding mode control

For DC/DC converters, the input inductor (the word 'inductor' is often used in the literature for DC/DC converters instead of 'inductance') current and the output capacitor voltage are normally selected as the state variables. For most converters

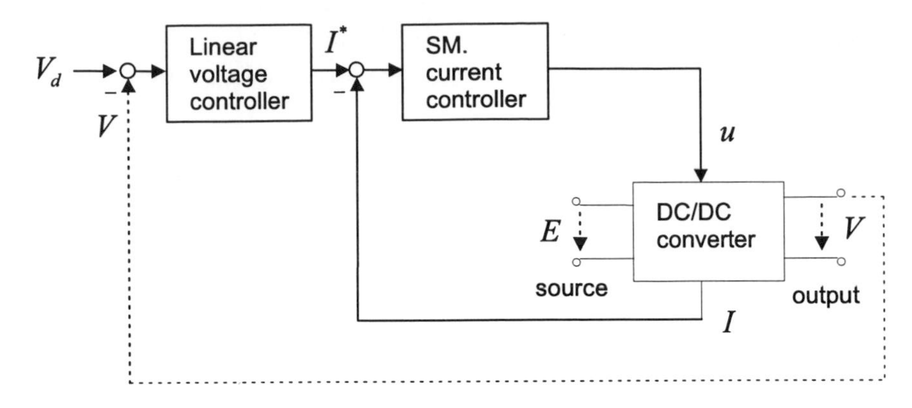

Figure 11.1 Cascaded control structure of DC/DC converters.

used in practice, the motion rate of the current is much faster than the motion rate of the output voltage. Calling upon the theory of *singular perturbations* (Kokotovic *et al.*, 1976), the control problem can be solved by using a cascaded control structure with two control loops: an inner current control loop and an outer voltage control loop. The voltage control is usually realized with standard linear control techniques, whereas the current control is implemented using either PWM or hysteresis control. Here we use the sliding mode approach for the control of inductor current. Figure 11.1 shows the general structure of the control system for DC/DC converters.

Buck-type DC/DC converter

The circuit structure of a buck DC/DC converter is shown in Figure 11.2, where the parameters and variables are defined as follows:

L = loop inductor
C = storage capacitor
R = load resistance
E = source voltage
I = input current
V = output voltage
u = switching signal

The switching signal u takes values from discrete set $U = \{0, 1\}$.

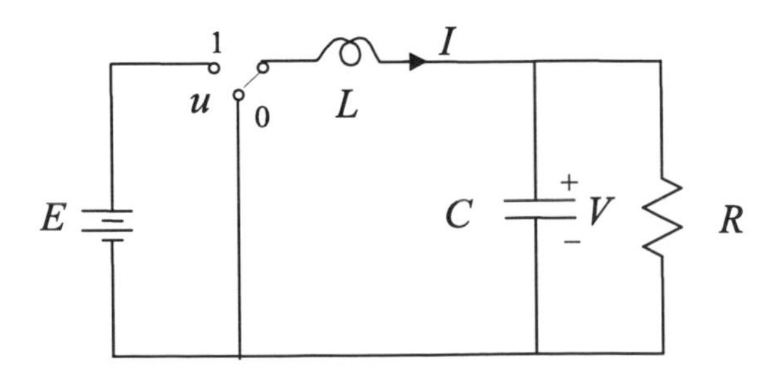

Figure 11.2 Buck DC/DC converter.

The dynamic model of the buck-type converter is given by

$$\begin{aligned}\dot{x}_1 &= -\frac{1}{L}x_2 + u\frac{E}{L}\\ \dot{x}_2 &= \frac{1}{C}x_1 - \frac{1}{RC}x_2\end{aligned} \tag{11.1.12}$$

with $x_1 = I$ and $x_2 = V$.

The goal of control is to achieve a constant output voltage denoted by V_d. In other words, the steady state behaviour of the buck converter (11.1.12) should be given by

$$\begin{aligned}x_2 &= V_d\\ \dot{x}_2 &= \dot{V}_d = 0\end{aligned} \tag{11.1.13}$$

The control design follows a two-step procedure known as integrator backstepping (Krstic *et al.*, 1995) or regular form control (Section 3.3). First, it is assumed that x_1 in the second equation of (11.1.12) can be handled as a control input. However, since x_1 is the output of the current loop in the first equation of (11.1.12), this first design step yields desired current x_1^*. The control goal (11.1.13) is substituted into the voltage loop, i.e. the second equation of (11.1.12), to yield the desired current

$$x_1^* = V_d/R \tag{11.1.14}$$

The task of the second design step is to ensure the actual current x_1 tracks the desired current (11.1.14) exactly. Due to its exact tracking properties, the sliding mode approach is an ideal tool for this task. If sliding mode is enforced in

$$s = x_1 - x_1^* = 0 \tag{11.1.15}$$

then $x_1 = V_d/R$. In order to enforce sliding mode in the manifold $s = 0$ in (11.1.15), control u taking only two values, 0 or 1, in the first equation of (11.1.12) is defined as

$$u = \frac{1}{2}(1 - \text{sign}(s)) \tag{11.1.16}$$

The condition for sliding mode to exist is derived from $s\dot{s} < 0$. In compliance with the derivations in Section 11.1.1 sliding mode exists if

$$0 < x_2 < E \tag{11.1.17}$$

This condition defines an attraction domain of the sliding manifold. Since the control (11.1.16) contains no control gain to be adjusted, the domain of attraction (11.1.17) is predetermined by the system architecture. In steady state, condition (11.1.17) is fulfilled by the definition of a buck converter: the output voltage is smaller than the source voltage.

After the state of the inner current loop has reached the sliding manifold, i.e. converged to $s = 0$ at time $t = t_h$, $x_1 = x_1^* = V_d/R$ holds for $t > t_h$. The outer voltage loop is governed by

$$\dot{x}_2 = -\frac{1}{RC}x_2 + \frac{1}{RC}V_d \tag{11.1.18}$$

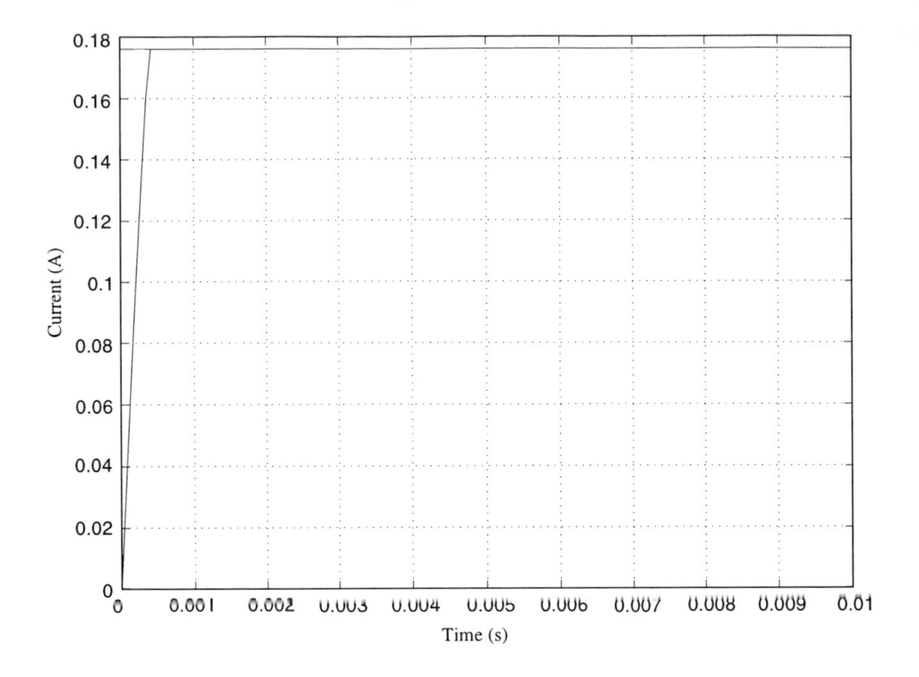

Figure 11.3 Current response of a sliding mode buck DC/DC converter.

The solution of this system is

$$x_2(t) = V_d + (x_2(t_h) - V_d)e^{-(t-t_h)/RC} \tag{11.1.19}$$

tends to V_d exponentially. Hence the design goal of control is achieved.

Figures 11.3 and 11.4 show the simulation results of the proposed control algorithm for the buck DC/DC converter. The converter parameters are selected as $E = 20\,\text{V}$, $C = 4\,\mu\text{F}$, $R = 40\,\Omega$ and $L = 40\,m\text{H}$. The desired output voltage is $V_d = 7\,\text{V}$. Notice that the inductor current and the output capacitor voltage converge rapidly to their reference values.

Boost-type DC/DC converter

Figure 11.5 shows the principle of a boost-type converter, where the parameters and variables are defined as follows:

L = loop inductor
C = storage capacitor
R = load resistance
E = source voltage
I = input current
V = output voltage
u = switching signal

The switching signal u takes values from discrete set $U = \{0, 1\}$.

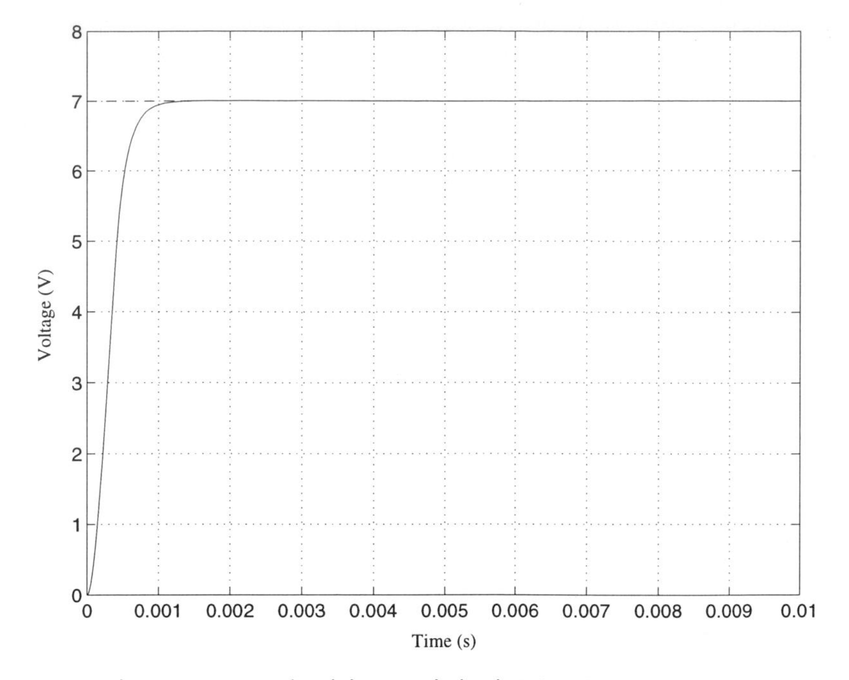

Figure 11.4 Voltage response of a sliding mode buck DC/DC converter.

The main difference between the boost converter (Figure 11.5) and the buck converter (Figure 11.2) is the location of inductor L. The dynamic model of the boost converter is given as

$$\begin{aligned} \dot{x}_1 &= -(1-u)\frac{1}{L}x_2 + \frac{E}{L} \\ \dot{x}_2 &= (1-u)\frac{1}{C}x_1 - \frac{1}{RC}x_2 \end{aligned} \tag{11.1.20}$$

with $x_1 = I$ and $x_2 = V$.

The topological modification of locating the inductor before the switching element rather than after it, as in the buck converter, enables a higher output voltage than the source voltage. However, the boost converter is more difficult to control than the buck converter. This is because the control u appears in both current and voltage equations, each time in bilinear fashion. Such a configuration implies a highly nonlinear system and a difficult control design.

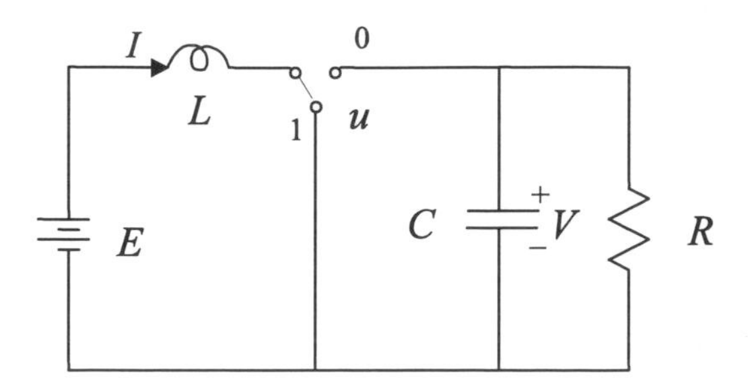

Figure 11.5 Boost DC/DC converter.

Just like buck converters, boost converters satisfy the *motion separation* principle, which originates from singular perturbation theory. This means the motion rate of the current is much faster than the motion rate of the output voltage. Consequently, the control problem can again be solved by using two cascaded control loops: an inner current control loop and an outer voltage control loop. We present a design for the current control loop based on sliding mode techniques.

Similar to the control design for the buck converter in the previous section, a desired current is obtained from the outer voltage loop as

$$x_1^* = V_d^2/RE \tag{11.1.21}$$

where V_d is the desired output capacitor voltage. The switching function for the inner current control is defined as

$$s = x_1 - x_1^* \tag{11.1.22}$$

in order to enforce the current x_1 to track the desired current x_1^*. Control u can be designed as

$$u = \frac{1}{2}(1 - \text{sign}(s)) \tag{11.1.23}$$

Under this control scheme, the *equivalent control* of u is derived by formally solving $\dot{s} = \dot{x}_1 = 0$ in (11.1.20) for the control input u:

$$u_{eq} = 1 - \frac{E}{x_2} \tag{11.1.24}$$

where x_2 is the output voltage of the slow voltage loop. The motion equation of the outer voltage loop during sliding mode in the inner current loop is obtained by substituting the equivalent control (11.1.24) into the second equation of (11.1.20):

$$\dot{x}_2 = -\frac{1}{RC}\left(x_2 - \frac{V_d^2}{x_2}\right) \tag{11.1.25}$$

This equation can be solved explicitly as

$$x_2(t) = \left(V_d^2 + (x_2^2(t_h) - V_d^2)e^{-2(t-t_h)/RC}\right)^{1/2}, \tag{11.1.26}$$

where t_h stands for the reaching instant of the sliding manifold $s = 0$ and $x_2(t_h)$ is the output voltage at time t_h. Apparently, x_2 tends to V_d asymptotically as t goes to infinity.

The attraction domain of the sliding manifold $s = 0$ is found by applying the convergence condition $s\dot{s} < 0$ to the system (11.1.20), yielding

$$x_2 > E \quad \text{or} \quad 0 < u_{eq} = 1 - \frac{E}{x_2} < 1 \tag{11.1.27}$$

Condition (11.1.27) implies that as long as the output voltage is higher than the source voltage, sliding mode can be enforced. This requirement is essential for a boost-type DC/DC converter and careful consideration of the initial conditions is required to guarantee convergence to $s = 0$.

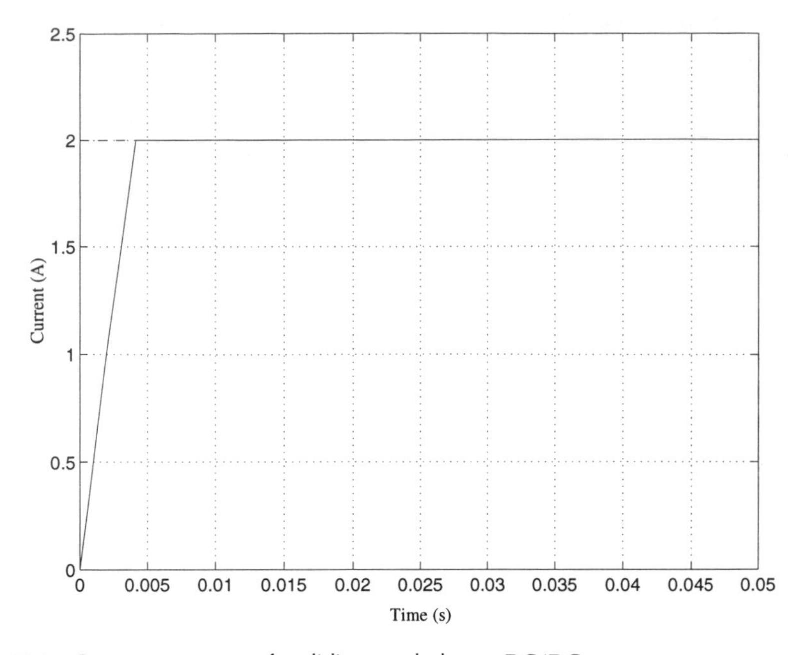

Figure 11.6 Current response of a sliding mode boost DC/DC converter.

Figures 11.6 and 11.7 show the simulation results of the proposed control algorithm for the boost DC/DC converter. The converter parameters are selected as $E = 20\,\text{V}$, $c = 4\mu\,\text{F}$, $R = 40\,\Omega$ and $L = 40\,\text{mH}$. The desired output voltage is $V_d = 40\,\text{V}$. Notice that the inductor current and the output capacitor voltage converge rapidly to their reference values.

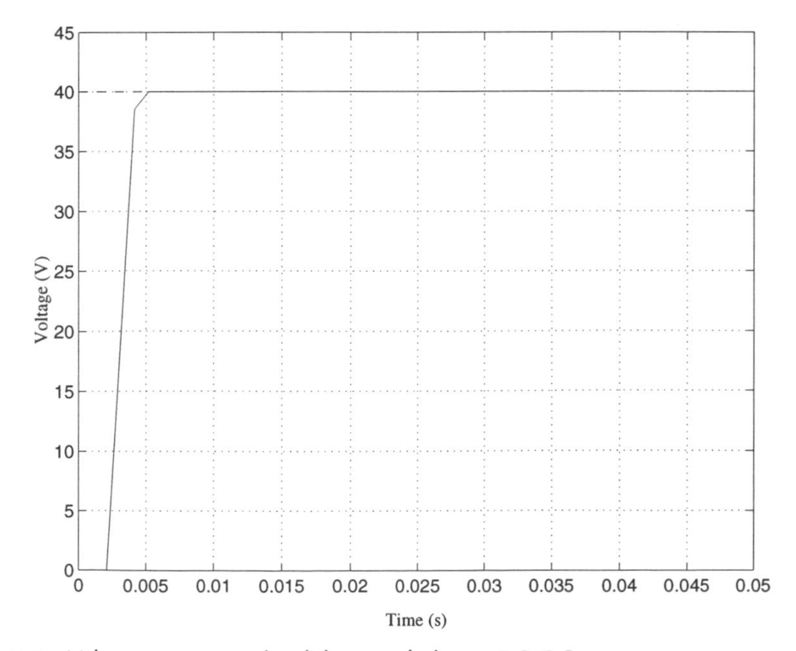

Figure 11.7 Voltage response of a sliding mode boost DC/DC converter.

11.1.3 Observer-based control

Recently, research on sliding mode control theory has revealed great advantages of introducing certain dynamics into a sliding mode controller (Chapters 6, 7 and 8). These approaches fall into the category of dynamic feedback control. For observer-based sliding mode control, an ideal model is simulated in the controller in parallel with the real plant. For the sliding mode control itself, measurements of real plant states are substituted by observer states, reducing the number of plant states to be measured. The mismatch between the measured states and the observer states is the 'bridge' to keep both systems operating 'closely'. This mismatch has been used in different ways to improve the control performance. The following control approaches are similar in this sense:

- global sliding mode control
- full-order sliding mode control
- integral sliding mode control
- model-following sliding mode control
- sliding mode control based on asymptotic observers
- frequency-shaped sliding mode control

These advanced control approaches aim to increase the control accuracy and robustness, while reducing or even alleviating any *chattering* effects. But an on-line simulated plant model is essential for special control problems such as sensorless control. Almost all of these control approaches need a high gain to generate sliding mode, but the high gain can often be placed outside the main feedback loop.

However, with power converter control, no control gain can be adjusted; the attraction domain of the sliding manifold $s = 0$ is always bounded. The remaining degrees of freedom for the control design are the initial conditions of the auxiliary system (the simulated plant model). It has been shown that by proper selection of the initial conditions of the auxiliary system, the performance of an observer-based sliding mode controller may be improved significantly (Sira-Ramirez *et al.*, 1996a,b).

To put this in perspective, consider how sliding mode control theory has evolved over the years. As a general control design methodology for a large class of nonlinear systems, sliding mode control theory originally fell into the field of nonlinear control with order reduction and decoupling capabilities. The order reduction and decoupling issues were some of the most difficult problems in the control of high-order nonlinear dynamic systems. Applications of sliding mode techniques in the fields of robotics and electric drives have confirmed the validity of this compact, uniform and straight-forward design methodology. In recent years, researchers have started to incorporate the dynamic feedback into the sliding mode controller to achieve higher control performance. Dynamic feedback increases the system order in the controller space and establishes coupling between the plant system space and the controller space. As a result, the complexity of the overall system will be increased.

Similar to the direct sliding mode control of DC/DC converters, the following sections investigate inductor current control and subsequently output capacitor voltage regulation based on observed states. Figure 11.8 shows the structure of the observer-based control system for DC/DC converters. The price for observer-based

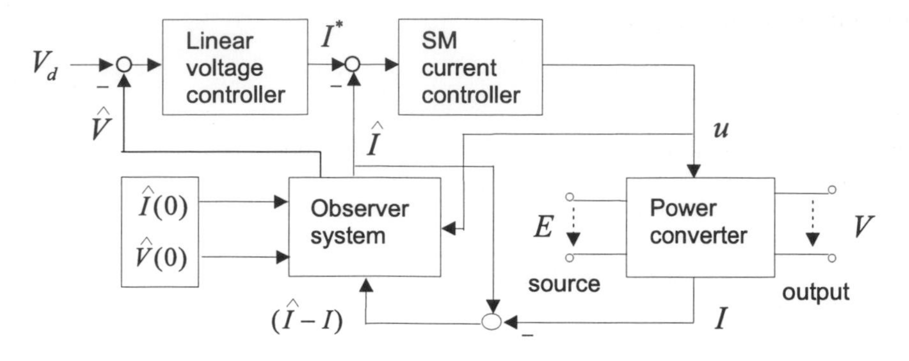

Figure 11.8 Observer-based control structure.

control is a more complicated control design and an increasingly complex stability analysis. A general design strategy proceeds as follows:

1. The observer dynamics are derived with a similar structure as the plant model.
2. The stability of the observer system is examined to ensure the observer states converge to the states of the real system (asymptotic stability).
3. A sliding mode current controller is designed based on the observed current rather than the measured current. Supposing sliding mode exists, the observed current tracks the desired reference current and associated equivalent control can be obtained.
4. A reduced-order system – consisting of the real plant model and the output voltage equation of the observer system – can be derived by substitution of the equivalent control.
5. Under equivalent control, the reduced-order system provides that (a) the real current converges to the desired reference current and (b) the real output voltage converges to the observed output voltage.
6. The observed output voltage converges to the desired output voltage, and so does the real output voltage.
7. The existence of sliding mode is established by proper selection of the initial conditions for the observer dynamics.

Observer-based control of buck converters

The observer equations, or the auxiliary system, are designed as

$$\begin{aligned}\dot{\hat{x}}_1 &= -\frac{1}{L}\hat{x}_2 + u\frac{E}{L} - l(\hat{x}_1 - x_1)\\ \dot{\hat{x}}_2 &= \frac{1}{C}\hat{x}_1 - \frac{1}{RC}\hat{x}_2\end{aligned} \tag{11.1.28}$$

where l is a positive scalar observer gain and $\hat{x}_1, \hat{x}_2$ are the observed inductor current and output voltage, i.e. the outputs of the observer. Note that only measurement of the inductor current x_1 is required; there is no need to measure the output voltage x_2.

Theorem 11.2

Observer (11.1.28) is an asymptotic observer whose outputs $\hat{x}_1, \hat{x}_2$ converge to the real states x_1, x_2 asymptotically. □

Proof

Defining the observer errors as $\bar{x}_1 = \hat{x}_1 - x_1$ and $\bar{x}_2 = \hat{x}_2 - x_2$, the error dynamics can be derived by subtracting (11.1.12) from (11.1.28)

$$\begin{aligned} \dot{\bar{x}}_1 &= -\frac{1}{L}\bar{x}_2 - l\bar{x}_1 \\ \dot{\bar{x}}_2 &= \frac{1}{C}\bar{x}_1 - \frac{1}{RC}\bar{x}_2 \end{aligned} \tag{11.1.29}$$

Since control u is applied to both the observer system and the real plant model, it is cancelled out in the error dynamics. The characteristic polynomial of this linear system is

$$p^2 + \left(l + \frac{1}{RC}\right)p + \left(\frac{l}{RC} + \frac{1}{LC}\right) = 0 \tag{11.1.30}$$

Since all coefficients in (11.1.30) are positive, system (11.1.29) is stable implying that observer errors $\bar{x}_1$ and $\bar{x}_2$ tend to zero asymptotically. ■

The switching function for the sliding mode current control will be designed based on observed current $\hat{x}_1$ instead of measured current x_1 in (11.1.15):

$$\hat{s} = \hat{x}_1 - V_d/R \tag{11.1.31}$$

The control u, applied to both the real plant and the observer system, is of the same form as the control scheme without an observer (11.1.16):

$$u = \frac{1}{2}(1 - \text{sign}(\hat{s})) \tag{11.1.32}$$

Suppose that sliding mode can be enforced in the vicinity of the sliding manifold $\hat{s} = 0$, which results in

$$\hat{x}_1(t) \equiv V_d/R \qquad (\forall t > t_h) \tag{11.1.33}$$

where t_h denotes the reaching instant of the sliding manifold $\hat{s} = 0$. The *equivalent control* of u can be obtained by solving $\dot{\hat{s}} = 0$:

$$u_{eq} = \left(\frac{1}{L}\hat{x}_2 + l(\hat{x}_1 - x_1)\right)\frac{L}{E} \tag{11.1.34}$$

Substituting u_{eq} into the real plant model (11.1.12) and considering the observer model (11.1.28), the motion in sliding mode can be represented as a reduced-order system comprising the motion of the real plant and the slow dynamics (output voltage loop) of the observer:

$$\begin{aligned}
\dot{x}_1 &= -\frac{1}{L}x_2 + \left(\frac{1}{L}\hat{x}_2 + l\left(\frac{V_d}{R} - x_1\right)\right) \\
\dot{x}_2 &= \frac{1}{C}x_1 - \frac{1}{RC}x_2 \\
\dot{\hat{x}}_2 &= \frac{1}{C}\frac{V_d}{R} - \frac{1}{RC}\hat{x}_2
\end{aligned} \tag{11.1.35}$$

By defining errors $\bar{x}_1^* = V_d/R - x_1$ and $\bar{x}_2 = \hat{x}_2 - x_2$, these equations can be transformed into a second-order error system:

$$\begin{aligned}
\dot{\bar{x}}_1^* &= -\frac{1}{L}\bar{x}_2 - l\bar{x}_1^* \\
\dot{\bar{x}}_2 &= \frac{1}{C}\bar{x}_1^* - \frac{1}{RC}\bar{x}_2
\end{aligned} \tag{11.1.36}$$

where $\bar{x}_2$ has the same meaning as defined for (11.1.29), but $\bar{x}_1^*$ represents the difference between the desired reference current and the real current rather than the difference between the observed current and the real current, denoted by $\bar{x}_1$.

Formally, equation system (11.1.36) is the same as the observer error system (11.1.29) which is proven to be stable, implying that the real current converges to the desired reference current asymptotically.

So far, we have proven that after sliding mode occurs, i.e. for $t \geq t_h$, the observed current is equal to the desired reference current, and the real current will also tend to the desired reference current. It means that

$$\lim_{t\to\infty} x_1(t) = \hat{x}_1(t)|_{t\geq t_h} = \frac{V_d}{R} \tag{11.1.37}$$

Next we should prove that the observed output voltage converges to the desired output voltage, and so does the real output voltage. Substitute (11.1.33) into the second equation of (11.1.28) and solve the resulting equation with a similar procedure as used for (11.1.18); this gives

$$\lim_{t\to\infty} \hat{x}_2(t) = V_d \tag{11.1.38}$$

Following Theorem 11.2, the real output voltage and the observed output voltage are identical as $t \to \infty$. Finally, we achieve

$$\lim_{t\to\infty} x_2(t) = \lim_{t\to\infty} \hat{x}_2(t) = V_d \tag{11.1.39}$$

The remaining task is to find the condition under which the occurrence of the sliding mode can be guaranteed. Applying the existence condition of sliding mode to the first equation of (11.1.28) with the substitution of (11.1.32) and (11.1.31) yields

$$-Ll(\hat{x}_1 - x_1) < \hat{x}_2 < E - Ll(\hat{x}_1 - x_1) \tag{11.1.40}$$

This condition is consistent with Theorem 11.1, i.e.

$$0 < u_{eq} = \left(\frac{1}{L}\hat{x}_2 + l(\hat{x}_1 - x_1)\right)\frac{L}{E} < 1 \tag{11.1.41}$$

Since x_1 is measured and $\hat{x}_1$, $\hat{x}_2$ are state variables in the controller space, i.e. variables in the control algorithm, the initial conditions of the observer $\hat{x}_1(0)$ and $\hat{x}_2(0)$ can be designed such that the occurrence of the sliding mode can always be guaranteed.

As an important specification of DC/DC converters, the so-called *stored error energy* has been defined in the literature (Sira-Ramircz *et al.*, 1996a). For the buck converter, this quantity is defined as

$$H(t) = \frac{1}{2}\left(L\left(\frac{V_d}{R} - x_1(t)\right)^2 + C(V_d - x_2(t))^2\right) \tag{11.1.42}$$

where $H(t)$ represents the energy difference between the desired value and the real value provided to the load. For a well-controlled DC/DC converter, this energy difference should converge to zero smoothly.

Figures 11.9, 11.10 and 11.11 show ω_0 simulation results of the observer-based control algorithm for the buck DC/DC converter. The converter parameters are selected as $E = 20$V, $C = 4\mu$F, $R = 40\,\Omega$ and $L = 40$ mH. The desired output voltage is $V_d = 7$ V. The observer gain is designed as $l = 200$. The initial conditions of the observer are selected as $\hat{x}_1(0) = 0.12$ A, $\hat{x}_2(0) = 5.0$V and $\hat{x}_1(0) = 0.07$ A, $\hat{x}_2(0) = 2.5$ V. Notice that both the inductor current and the output capacitor voltage converge rapidly to their reference values, and the system response can be influenced by the design of the observer initial conditions.

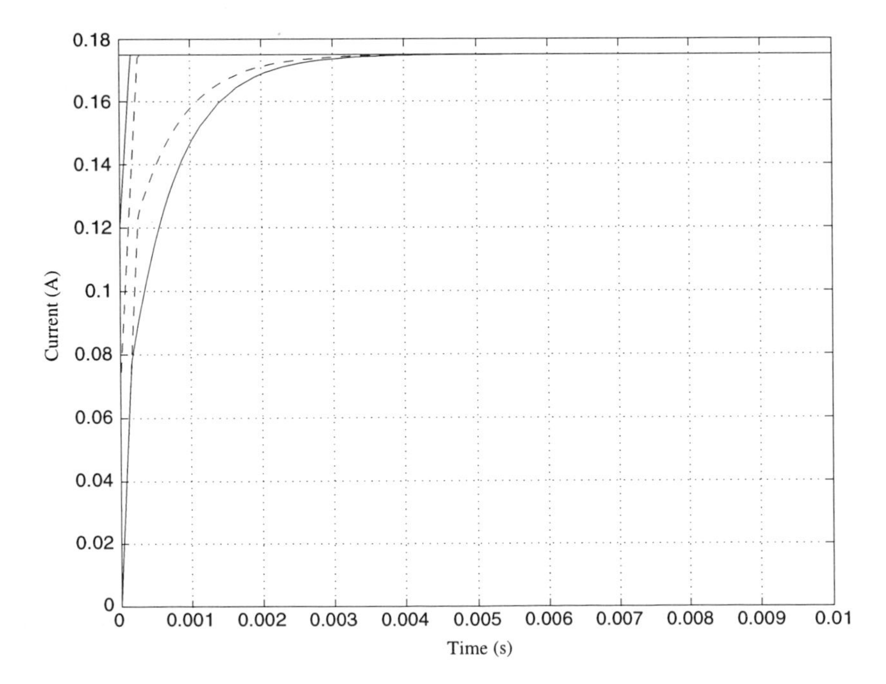

Figure 11.9 Current response: (—) $\hat{x}_1(0) = 0.12$, $\hat{x}_2(0) = 5.0$; (- - -) $\hat{x}_1(0) = 0.07$, $\hat{x}_2(0) = 2.5$.

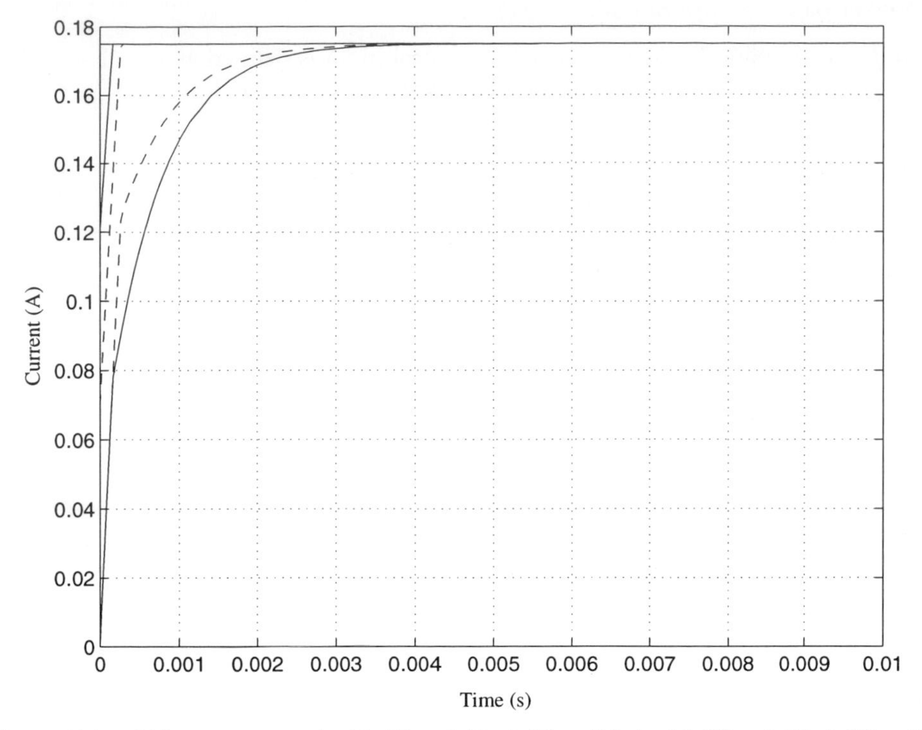

Figure 11.10 Voltage response: (—) $\hat{x}_1(0) = 0.12,\ \hat{x}_2(0) = 5.0$; (- - -) $\hat{x}_1(0) = 0.07,\ \hat{x}_2(0) = 2.5$.

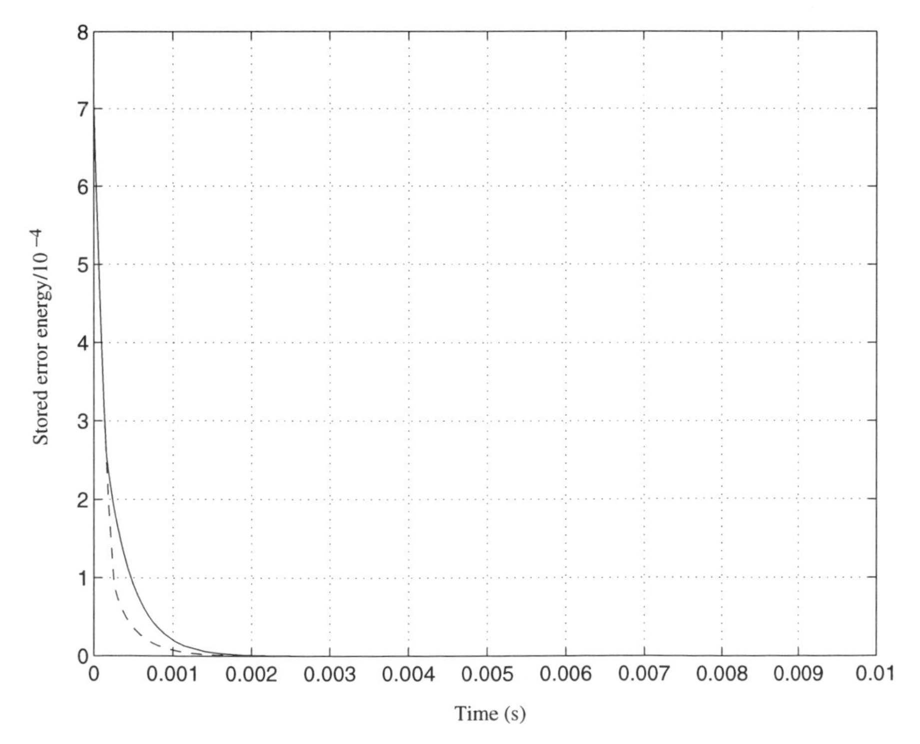

Figure 11.11 Stored error energy: (—) $\hat{x}_1(0) = 0.12,\ \hat{x}_2(0) = 5.0$; (- - -) $\hat{x}_1(0) = 0.07,\ \hat{x}_2(0) = 2.5$.

Observer-based control of boost converters

In order to simplify the derivation, a new control input is defined as $v = (1 - u)$. The observer dynamics designed for a boost converter are governed by

$$\begin{aligned}\dot{\hat{x}}_1 &= -v\frac{1}{L}\hat{x}_2 + \frac{E}{L} - l(\hat{x}_1 - x_1)\\ \dot{\hat{x}}_2 &= v\frac{1}{C}\hat{x}_1 - \frac{1}{RC}\hat{x}_2\end{aligned} \tag{11.1.43}$$

where $\hat{x}_1, \hat{x}_2$ are the observed inductor current and output voltage, i.e. the outputs of the observer, and l is a positive scalar observer gain.

Theorem 11.3

Observer (11.1.43) is an asymptotic observer whose outputs $\hat{x}_1, \hat{x}_2$ converge to the real states x_1, x_2 asymptotically. □

Proof

Defining the observer errors as $\bar{x}_1 = \hat{x}_1 - x_1$ and $\bar{x}_2 = \hat{x}_2 - x_2$, the error dynamics can be derived by subtracting (11.1.20) from (11.1.43):

$$\begin{aligned}\dot{\bar{x}}_1 &= -v\frac{1}{L}\bar{x}_2 - l\bar{x}_1\\ \dot{\bar{x}}_2 &= v\frac{1}{C}\bar{x}_1 - \frac{1}{RC}\bar{x}_2\end{aligned} \tag{11.1.44}$$

System (11.1.44) is a nonlinear system, since the system states are multiplied by the control input v. For the convergence proof, we design a Lyapunov function candidate as

$$V = \frac{1}{2}(L\bar{x}_1^2 + C\bar{x}_2^2) > 0 \tag{11.1.45}$$

Its time derivative along the solutions of (11.1.44) can be found as

$$\dot{V} = -Ll\bar{x}_1^2 - \frac{1}{R}\bar{x}_2^2 < 0 \tag{11.1.46}$$

therefore system (11.1.44) is stable for any $l > 0$. As a result, observer errors $\bar{x}_1$ and $\bar{x}_2$ tend to zero asymptotically. The convergence rate for the inductor current estimation can be adjusted by the observer gain l. ■

The switching function for the sliding mode current control will be designed based on observed current $\hat{x}_1$ instead of measured current x_1 as in (11.1.22):

$$\hat{s} = \hat{x}_1 - \frac{V_d^2}{RE} \tag{11.1.47}$$

The control u applied to both the real plant and the observer is of the same form as the control scheme without an observer:

$$u = \frac{1}{2}(1 - \text{sign}(\hat{s})) \tag{11.1.48}$$

In terms of the new control input $v = (1 - u)$, we have

$$v = \frac{1}{2}(1 + \text{sign}(\hat{s})) \tag{11.1.49}$$

Suppose that sliding mode is enforced in the manifold $\hat{s} = 0$, then according to (11.1.47)

$$\hat{x}_1 \equiv \frac{V_d^2}{RE} \qquad (\forall t > t_h) \tag{11.1.50}$$

where t_h denotes the reaching instant of the sliding manifold $\hat{s} = 0$. The *equivalent control* of v can be obtained by solving $\dot{\hat{s}} = 0$:

$$v_{eq} = \frac{E - Ll(V_d^2/RE - x_1)}{\hat{x}_2} \tag{11.1.51}$$

The motion of the system in sliding mode is of reduced order and comprises the motion of the real plant and the slow dynamics (output voltage loop) of the observer:

$$\begin{aligned} \dot{x}_1 &= -v_{eq}\frac{1}{L}x_2 + \frac{E}{L} \\ \dot{x}_2 &= v_{eq}\frac{1}{C}x_1 - \frac{1}{RC}x_2 \\ \dot{\hat{x}}_2 &= v_{eq}\frac{1}{C}\frac{V_d^2}{RE} - \frac{1}{RC}\hat{x}_2 \end{aligned} \tag{11.1.52}$$

By defining errors $\bar{x}_1^* = V_d^2/RE - x_1$ and $\bar{x}_2 = \hat{x}_2 - x_2$, with the substitution of (11.1.51), these equations can be transformed into a second-order error system:

$$\begin{aligned} \dot{\bar{x}}_1^* &= -\frac{x_2}{\hat{x}_2}l\bar{x}_1^* + \frac{E}{L}\frac{x_2}{\hat{x}_2} - \frac{E}{L} \\ \dot{\bar{x}}_2 &= \frac{E}{C}\frac{1}{\hat{x}_2}\bar{x}_1^* - \frac{Ll}{C\hat{x}_2}(\bar{x}_1^*)^2 - \frac{1}{RC}\bar{x}_2 \end{aligned} \tag{11.1.53}$$

where $\bar{x}_2$ has the same meaning as defined for (11.1.44), but $\bar{x}_1^*$ represents the difference between the desired reference current and the real current rather than the difference between the observed current and the real current, denoted by $\bar{x}_1$.

Substituting $x_2 = \hat{x}_2 - \bar{x}_2$ into (11.1.53) further simplifies the equations to

$$\begin{aligned} \dot{\bar{x}}_1^* &= -l\bar{x}_1^* + \frac{l}{\hat{x}_2}\bar{x}_1^*\bar{x}_2 - \frac{E\,\bar{x}_2}{L\,\hat{x}_2} \\ \dot{\bar{x}}_2 &= -\frac{1}{RC}\bar{x}_2 - \frac{Ll}{C\hat{x}_2}(\bar{x}_1^*)^2 + \frac{E}{C}\frac{1}{\hat{x}_2}\bar{x}_1^* \end{aligned} \tag{11.1.54}$$

For the convergence proof, we design a Lyapunov function candidate as

$$V = \frac{1}{2}(L(\bar{x}_1^*)^2 + C\bar{x}_2^2) > 0 \tag{11.1.55}$$

Its time derivative along the solutions of (11.1.54) can be found as

$$\dot{V} = -Ll(\bar{x}_1^*)^2 - \frac{1}{R}\bar{x}_2^2 < 0 \tag{11.1.56}$$

therefore the system (11.1.54) is stable for any $l > 0$. As a result, errors $\bar{x}_1^*$ and $\bar{x}_2$ tend to zero asymptotically, showing that the real current converges to the desired reference current. So far we have proven that after sliding mode occurs, i.e. for $t \geq t_h$, the observed current is equal to the desired reference current and the real current will also tend to the desired reference current. Mathematically, this statement can be expressed as

$$\lim_{t\to\infty} x_1(t) = \hat{x}_1(t)|_{t \geq t_h} = \frac{V_d^2}{RE} \tag{11.1.57}$$

The next task is to prove that $\hat{x}_2$ converges to the desired voltage V_d. Substitution of (11.1.51) transforms the third equation of (11.1.52) into

$$\dot{y} = -\frac{2}{RC}y - \frac{2}{RC}\frac{V_d^2}{E}Ll\bar{x}_1^* \quad \text{with} \quad y = \hat{x}_2^2 - V_d^2 \tag{11.1.58}$$

This equation is a linear asymptotically stable system with the input x_1^* tending to zero, hence its output y tends to zero as well. Consequently, $\hat{x}_2 = \pm V_d$. Since $\hat{x}_2 > 0$ is required for generating sliding mode (proven below), the unique steady-state solution is $\hat{x}_2 = V_d$.

Following Theorem 11.3, the real output voltage and the observed output voltage are identical as $t \to \infty$. Finally, we achieve

$$\lim_{t\to\infty} x_2(t) = \lim_{t\to\infty} \hat{x}_2(t) = V_d \tag{11.1.59}$$

So far we have proven that the real inductor current x_1 tends to the desired reference current V_d^2/RE, if sliding mode can be enforced in the manifold $\hat{s} = \hat{x}_1 - V_d^2/RE = 0$. As a consequence, both the observed output voltage $\hat{x}_2$ and the real output voltage x_2 converge to the desired output voltage V_d.

The remaining task is to find the condition under which the occurrence of the sliding mode can be guaranteed. Applying the existence condition of sliding mode to (11.1.47) with the substitution of (11.1.49) and the first equation of (11.1.43) yields

$$0 < E - Ll(\hat{x}_1 - x_1) < \hat{x}_2 \tag{11.1.60}$$

This condition is consistent with Theorem 11.1, i.e.

$$0 < v_{eq} = \frac{E - Ll(\hat{x}_1 - x_1)}{\hat{x}_2} < 1 \tag{11.1.61}$$

Bearing in mind that $v = 1 - u$, inequality (11.1.61) also gives us

$$0 < u_{eq} < 1 \tag{11.1.62}$$

Since x_1 is measured and $\hat{x}_1$, $\hat{x}_2$ are state variables in the controller space, i.e. variables in the control algorithm, the initial conditions of the observer, $\hat{x}_1(0)$ and $\hat{x}_2(0)$, can be designed such that the occurrence of sliding mode can always be guaranteed.

The stored error energy for boost DC/DC converters is defined as

$$H(t) = \frac{1}{2}\left(L\left(\frac{V_d^2}{RE} - x_1(t)\right)^2 + C(V_d - x_2(t))^2\right) \tag{11.1.63}$$

where $H(t)$ represents the energy difference between the desired value and the real value provided to the load. For a well-controlled DC/DC converter, this energy difference should converge to zero smoothly.

Figures 11.12, 11.13 and 11.14 show ω_0 simulation results of the observer-based control algorithm for the boost DC/DC converter. The converter parameters are selected as $E = 20\,\text{V}$, $C = 4\,\mu\text{F}$, $R = 40\,\Omega$ and $L = 40\,\text{mH}$. The desired output voltage is $V_d = 40\,\text{V}$. The observer gain is designed as $l = 200$. The initial conditions of the observer are selected as $\hat{x}_1(0) = 0$, $\hat{x}_2(0) = 0$ and $\hat{x}_1(0) = 1.95\,\text{A}$, $\hat{x}_2(0) = 38.5\text{V}$.

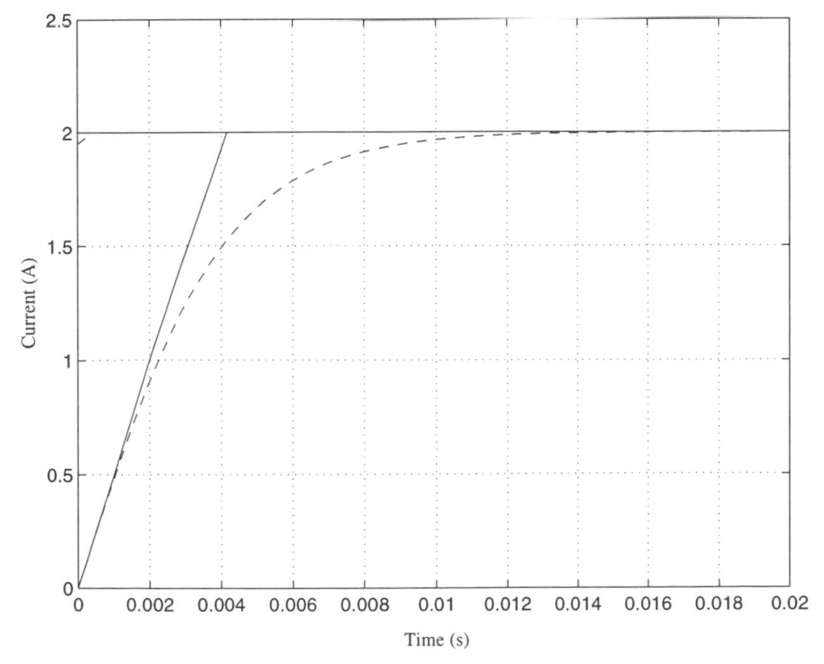

Figure 11.12 Current response: (—) $\hat{x}_1(0) = 0$, $\hat{x}_2(0) = 0$; (- - -) $\hat{x}_1(0) = 1.95$, $\hat{x}_2(0) = 38.5$.

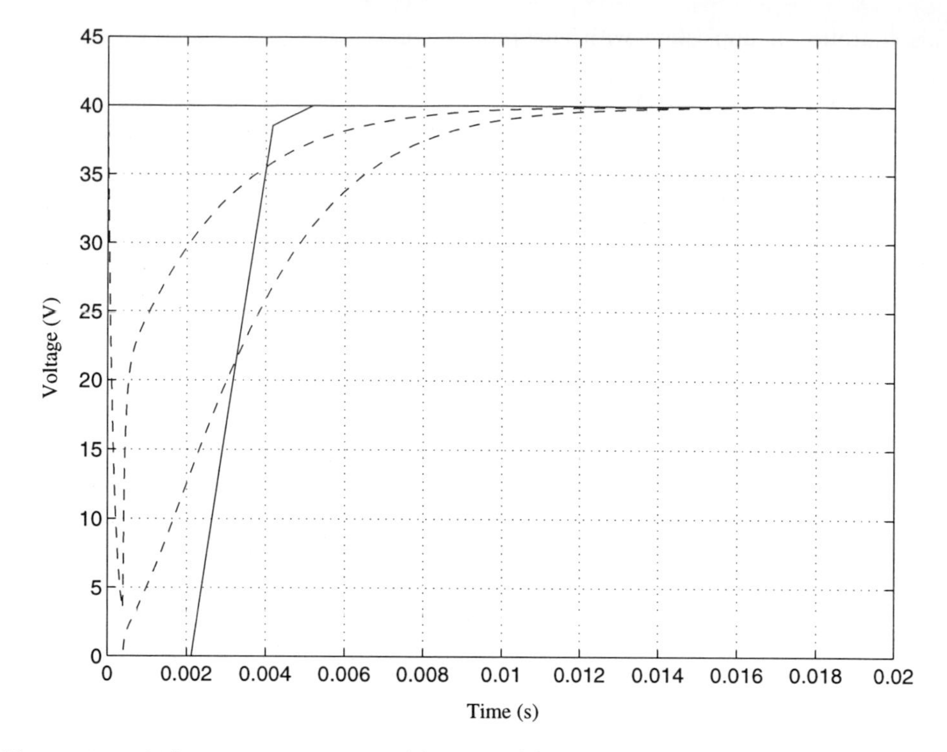

Figure 11.13 Voltage response: (—) $\hat{x}_1(0) = 0$, $\hat{x}_2(0) = 0$; (- - -) $\hat{x}_1(0) = 1.95$, $\hat{x}_2(0) = 38.5$.

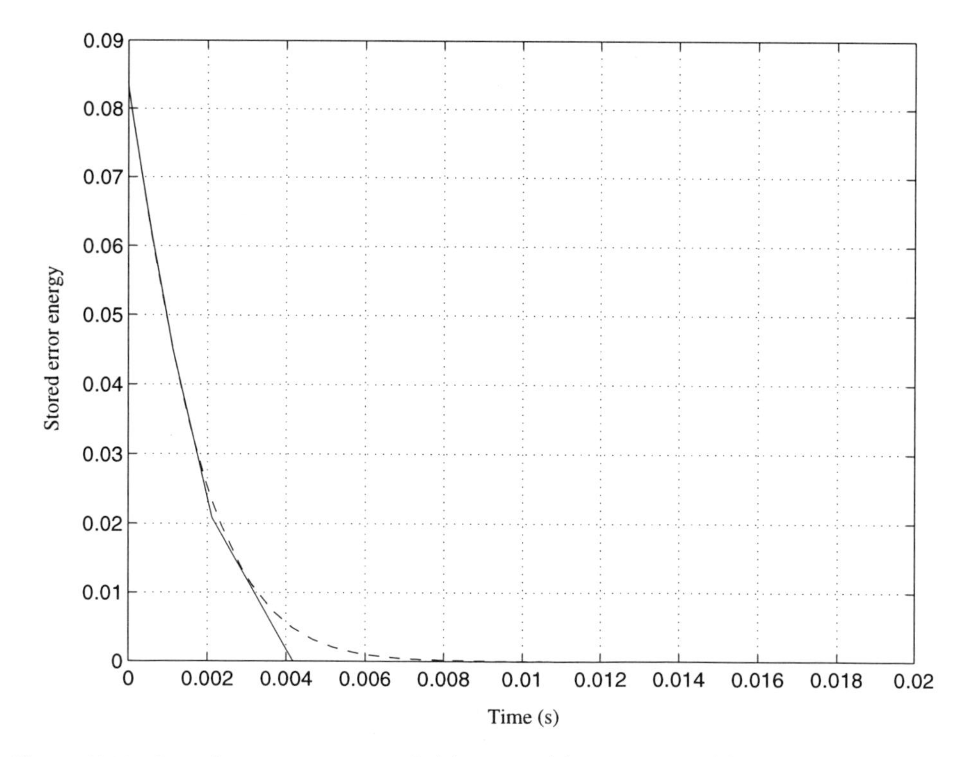

Figure 11.14 Stored error energy: (—) $\hat{x}_1(0) = 0$, $\hat{x}_2(0) = 0$; (- - -) $\hat{x}_1(0) = 1.95$, $\hat{x}_2(0) = 38.5$.

Notice that the inductor current and the output capacitor voltage converge rapidly to their reference values, and the system response can be influenced by selection of the observer initial conditions.

For zero initial conditions, i.e. $\hat{x}_1(0) = 0$ and $\hat{x}_2(0) = 0$, the observer-based control system converges to the nonobserver-based control system. In this case the stored error energy is not smooth (Figure 11.14); the derivative of $H(t)$ has a discontinuity point. However, if the initial conditions of the observer are designed properly, the stored error energy decreases to zero smoothly (Figure 11.14). This important improvement to the boost DC/DC converter is produced by the observer-based control design.

11.2 Boost-type AC/DC converters

Nowadays, semiconductors using high-frequency switching devices such as MOSFET, IGBT and MCT are commonly used for drive systems. Voltage modulation is one of the main control techniques to be employed. Phase currents and output voltages as well as switching losses and dynamic responses not only depend on the choice of power semiconductor, but also on the choice of pulse-width modulation. This explains why pulse-width modulation (PWM) techniques have been the subject of intensive research during the last few decades. A large variety of methods, different in concept and performance, have been developed.

These modulation approaches can be classified into two categories depending on the control techniques employed: feedforward PWM based on the so-called space-vector method and feedback modulation based on sliding mode or bang-bang control methods. The feedforward PWM technique is characterized by formation of the output voltage via an open-loop control structure. In this case the drive system does not exhibit high dynamic performance and the effects of disturbances, which always exist in a drive system, are not automatically reduced. On the other hand, it is very simple to realize a minimum-loss PWM strategy. The feedback modulation approach allows the switching pattern to be realized on-line. It ensures that the frequency and the pulse width are generated together automatically, solving the control task. Feedback control systems possess good dynamic performance since they use all available control resources to reduce the error. The influence of disturbances to the system is thus minimized. Many researchers have demonstrated the effectiveness of the sliding mode control scheme, e.g. Sabanovic *et al.* (1993) and Vilathgamuwa *et al.* (1996).

Compared to conventional PWM techniques, the sliding mode approach has no fixed switching frequency, resulting in higher switching losses and higher acoustic noise. Another development is sliding mode control under constant switching frequencies and with minimized switching losses for three-phase converters (Sergy and Izosimov, 1997).

Three distinguishing characteristics may be used to classify multiple-phase converters. First, there are two types of tasks for multiple-phase converters, *rectification* and *inversion*. Rectification transforms AC power into DC power, whereas inversion transforms DC power into AC power. Second, the power supply on the DC side may be in the form of a *voltage source* or a *current source*. Third, similar to DC/DC converters, multiple-phase converters can also be classified into *buck* type and *boost* type depending on the circuit topology. As a result, there are eight different types of multiple-phase power converter (Table 11.1). An exhaustive study of all these types of converter goes beyond the scope of this text. In this section we give an example of sliding mode control for a three-phase boost AC/DC converter, a typical converter used in industrial applications.

Table 11.1 Characteristics used to classify multiple-phase converters[a]

Type of transform	Power supply DC side	Circuit topology
Rectification (AC to DC)	Voltage source	Buck
Inversion (DC to AC)	Current source	Boost

[a]By choosing one characteristic from each column, it is possible to create eight different converters.

11.2.1 Model of the boost-type AC/DC converter

Figure 11.15 shows the structure of a boost-type AC/DC converter, where the parameters and variables are defined as follows:

$$\begin{aligned} L &= \text{phase inductance} \\ R_w &= \text{phase resistance} \\ C &= \text{storage capacitor} \\ i_1, i_2, i_3 &= \text{phase currents} \\ i_{link} &= \text{link current} \\ i_l &= \text{load current} \\ u_{g1}, u_{g2}, u_{g3} &= \text{source voltages} \\ u_0 &= \text{output voltage} \end{aligned}$$

For sliding mode control design, symmetric control inputs from set $\{-1, 1\}$ are more convenient than the on-off signals from set $\{0, 1\}$. Let the six on-off signals of an AC/DC converter be denoted by

$$s_w = [s_{w1} \quad s_{w2} \quad s_{w3} \quad s_{w4} \quad s_{w5} \quad s_{w6}]^T \tag{11.2.1}$$

The control inputs, as they appear in the converter model, are defined as $\boldsymbol{U}_{gate} = [u_1 \quad u_2 \quad u_3]^T$. Note that $\boldsymbol{U}_{gate}$ does not represent some control voltages,

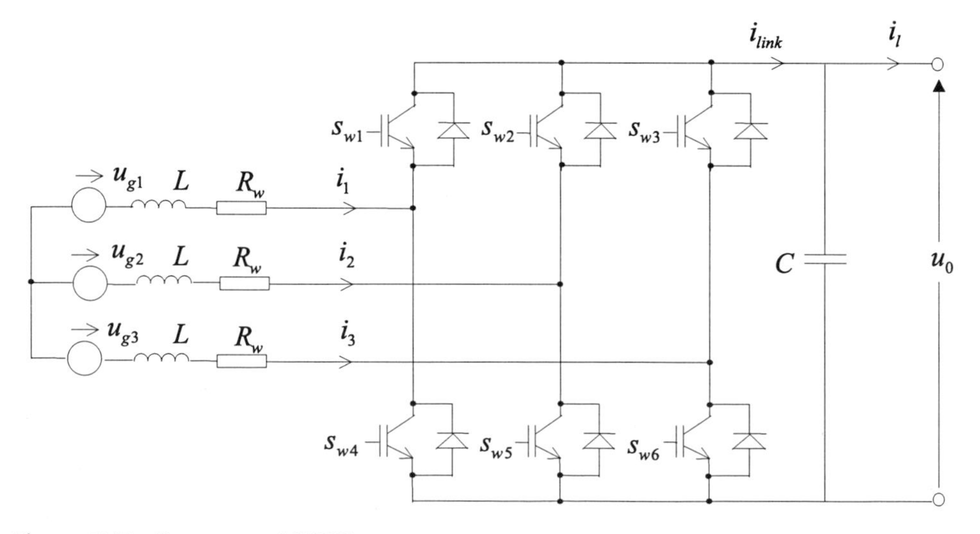

Figure 11.15 Boost type AC/DC converter.

but rather a set of transformed control inputs taking values from $\{-1, 1\}$ instead of $\{0, 1\}$ as done by s_w. Thus the following relation holds

$$U_{gate} = G_w s_w \quad \text{with } G_w = \begin{bmatrix} 1 & 0 & 0 & -1 & 0 & 0 \\ 0 & 1 & 0 & 0 & -1 & 0 \\ 0 & 0 & 1 & 0 & 0 & -1 \end{bmatrix} \tag{11.2.2}$$

In the following, the dynamic model of the boost AC/DC converter will be given both in the phase coordinate frame and the field-oriented coordinate frame. The models in these two coordinate frames are useful for our sliding mode controller and observer design.

Model in phase coordinate frame

Starting from the notation given in (11.2.2), the dynamic model of a boost AC/DC converter in phase coordinate frame can be obtained using the theory of switched electric circuits:

$$\begin{aligned}
\frac{di_1}{dt} &= -\frac{R_w}{L} i_1 - \frac{u_0}{6L}\left(2u_1 - \sum_{j=2}^{3} u_j\right) + \frac{1}{L} u_{g1} \\
\frac{di_2}{dt} &= -\frac{R_w}{L} i_2 - \frac{u_0}{6L}\left(2u_2 - \sum_{j=1, j\neq 2}^{3} u_j\right) + \frac{1}{L} u_{g2} \\
\frac{di_3}{dt} &= -\frac{R_w}{L} i_3 - \frac{u_0}{6L}\left(2u_3 - \sum_{j=1}^{2} u_j\right) + \frac{1}{L} u_{g3} \\
\frac{du_0}{dt} &= -\frac{i_l}{C} + \frac{1}{2C} \sum_{i=1}^{3} u_k i_k
\end{aligned} \tag{11.2.3}$$

Model in (d, q) coordinate frame

The control design of an AC motor is often performed in a field-oriented coordinate system, usually called (d, q) coordinates. For the control of an AC synchronous motor, the field-oriented coordinate system is simply the rotor coordinates. Similarly, for the control of three-phase AC/DC converters, it is convenient to design the control in the rotating reference frame synchronized with the supply frequency, i.e. the (d, q) coordinate system. In this case, all state variables should be transformed into the (d, q) coordinate system using the following relations:

$$\begin{aligned}
x_\alpha + jx_\beta &= \sqrt{2/3}(x_1 + e^{j2\pi/3} x_2 + e^{j4\pi/3} x_3) \\
x_d + jx_q &= (x_\alpha + jx_\beta)e^{-j\theta} \qquad \theta = \int \omega(t)\, dt
\end{aligned} \tag{11.2.4}$$

where $\omega(t)$ is the supply frequency; x_1, x_2, x_3 denote the state variables in phase coordinates, e.g. the phase currents (i_1, i_2, i_3), the source voltages (u_{g1}, u_{g2}, u_{g3}) and the switching controls (u_1, u_2, u_3); x_α, x_β represent the transformed state variables in a fixed orthogonal coordinate system with the α-axis aligned with the axis of phase 1, i.e. i_α, i_β and $u_{g\alpha}, u_{g\beta}$ and u_α, u_β; finally x_d, x_q are the transformed state variables in the (d, q) coordinate frame, i.e. i_d, i_q and u_{gd}, u_{gq} and u_d, u_q.

Remark 11.1

Here u_d, u_q are the transformed switching controls u_1, u_2, u_3, rather than the transformed phase voltages as in the case of AC motors. □

Equation (11.2.4) can be rewritten in matrix form as

$$\begin{bmatrix} x_d \\ x_q \end{bmatrix} = A_{d,q}^{\alpha,\beta} A_{\alpha,\beta}^{1,2,3} \begin{bmatrix} x_1 \\ x_2 \\ x_3 \end{bmatrix} \tag{11.2.5}$$

where the transformation matrices are defined as

$$A_{\alpha,\beta}^{1,2,3} = \frac{2}{3}\begin{bmatrix} 1 & -1/2 & -1/2 \\ 0 & \sqrt{3}/2 & -\sqrt{3}/2 \end{bmatrix} \qquad A_{d,q}^{\alpha,\beta} = \begin{bmatrix} \cos\theta & \sin\theta \\ -\sin\theta & \cos\theta \end{bmatrix} \tag{11.2.6}$$

The inverse transformation is given by

$$\begin{bmatrix} x_1 \\ x_2 \\ x_3 \end{bmatrix} = (A_{\alpha,\beta}^{1,2,3})^T (A_{d,q}^{\alpha,\beta})^T \begin{bmatrix} x_d \\ x_q \end{bmatrix} \tag{11.2.7}$$

Using these relations, the dynamic model (11.2.3) can be transformed into the (d, q) coordinate frame

$$\begin{aligned} \frac{du_{0d}}{dt} &= -\frac{i_l}{C} + \frac{i_d u_d + i_q u_q}{2C} \\ \frac{di_d}{dt} &= -\frac{R_w}{L} i_d + \frac{u_{gd}}{L} + \omega i_q - \frac{u_{od}}{2L} u_d \\ \frac{di_q}{dt} &= -\frac{R_w}{L} i_q + \frac{u_{gq}}{L} - \omega i_d - \frac{u_{od}}{2L} u_q \end{aligned} \tag{11.2.8}$$

where $u_{0d} = u_0$ ($u_{0q} = 0$ due to the field orientation).

11.2.2 Control problems

A well-controlled three-phase AC/DC power converter should have the following characteristics:

- unity power factor
- sinusoidal input currents
- regenerative capability
- ripple-free output voltage

To minimize electromagnetic interference, a *fixed switching frequency* is also an important characteristic, and PWM control will always have a fixed switching frequency. But a fixed switching frequency cannot be guaranteed for sliding mode control since the switching action occurs according to the value of a sliding function and is not synchronized with a frequency source. In fact, varying switching frequencies are a distinct feature of sliding mode control systems.

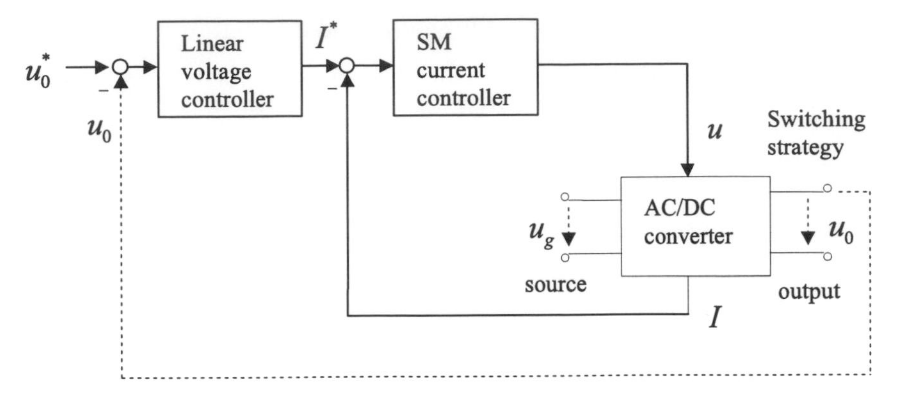

Figure 11.16 Cascaded control structure of AC/DC converters.

In industrial power converter systems, one of the major design challenges is an unknown *varying load* that requires sufficient control robustness. Moreover, not all state variables which are necessary for the control purpose are measured, or are measurable. As a result, a control engineer has to face the problem of designing a *robust* control system with observed state variables as well as achieving a high performance of the closed-loop system.

Control design of power converters is usually performed in two steps: current control in an inner loop and voltage control in an outer loop. In the frame of this cascaded control structure, sliding mode control is usually applied to the current control, whereas the outer-loop control, i.e. the output voltage regulation, is designed using linear control techniques. This holds for DC/DC converters and for AC/DC converters (Figure 11.16).

Sliding mode current control

Similar to the control design for electric motors, the current control of a boost-type AC/DC converter can be designed either in phase coordinates or in the (d, q) coordinate frame. Since the control criteria (as listed in the performance characteristics) are normally given in the (d, q) coordinate frame, it is more convenient to design the current control in the (d, q) coordinate frame than in phase coordinates.

Rewrite the current equations in (11.2.8) as follows:

$$\begin{aligned} \frac{di_d}{dt} &= -\frac{R_w}{L} i_d + \frac{u_{gd}}{L} + \omega i_d - \frac{u_{od}}{2L} u_d \\ \frac{di_q}{dt} &= -\frac{R_w}{L} i_q + \frac{u_{gq}}{L} - \omega i_d - \frac{u_{od}}{2L} u_q \end{aligned} \tag{11.2.9}$$

These equations are the starting point for the current control. To simplify the derivations and to use the results developed for the electric motors, the equations are represented in a generalized matrix form:

$$\dot{\boldsymbol{I}}_{dq} = \boldsymbol{f}_{dq}(\boldsymbol{I}_{dq}, \boldsymbol{U}_g, \omega) - b\boldsymbol{U}_{dq} \tag{11.2.10}$$

where $b = u_{0d}/2L, \boldsymbol{I}_{dq} = [i_d \quad i_q]^T, \boldsymbol{U}_{dq} = [u_d \quad u_q]^T$ and

$$\boldsymbol{f}_{dq}(\boldsymbol{I}_{dq}, \boldsymbol{U}_g, \omega) = \begin{bmatrix} -\frac{R_w}{L} i_d + \frac{u_{gd}}{L} + \omega i_q \\ -\frac{R_w}{L} i_q + \frac{u_{gq}}{L} - \omega i_d \end{bmatrix} \tag{11.2.11}$$

The switching functions for the current control will be designed as

$$\begin{aligned} s_d &= i_d^* - i_d \\ s_q &= i_q^* - i_q \end{aligned} \tag{11.2.12}$$

where i_d^* and i_q^* are the desired values of the currents in the (d, q) coordinate frame to be determined by the outer control loop for the output voltage regulation.

The next task is to find the condition under which sliding mode can be enforced. As for the AC/DC converters, no control gain can be adjusted. The solution is to find a domain in the system space from which any state trajectory converges to the sliding manifold defined by $s_d = 0,\ s_q = 0$. Defining $\boldsymbol{S}_{dq} = [s_d \quad s_q]^T$ and taking the time derivative of $\boldsymbol{S}_{dq}$ yields

$$\dot{\boldsymbol{S}}_{dq} = \dot{\boldsymbol{I}}_{dq}^* - \boldsymbol{f}_{dq}(\boldsymbol{I}_{dq}, \boldsymbol{U}_g, \omega) + b\boldsymbol{U}_{dq} = \boldsymbol{F}_{dq} + D\boldsymbol{U}_{gate} \tag{11.2.13}$$

in which $\boldsymbol{I}_{dq}^* = [i_d^* \quad i_q^*]^T$, $\boldsymbol{F}_{dq} = \dot{\boldsymbol{I}}_{dq}^* - \boldsymbol{f}_{dq}(\boldsymbol{I}_{dq}, \boldsymbol{U}_g, \omega)$ and

$$D = bA_{d,q}^{\alpha,\beta} A_{\alpha,\beta}^{1,2,3} \tag{11.2.14}$$

Design controls $\boldsymbol{U}_{gate} = [u_1 \quad u_2 \quad u_3]^T$ as follows

$$\boldsymbol{U}_{gate} = -\text{sign}(\boldsymbol{S}^*) \tag{11.2.15}$$

where $\boldsymbol{S}^* = [s_1^* \quad s_2^* \quad s_3^*]^T$ is a vector of transformed switching functions to be determined later, and

$$\text{sign}(\boldsymbol{S}^*) = [\text{sign}(s_1^*) \quad \text{sign}(s_2^*) \quad \text{sign}(s_3^*)]^T \tag{11.2.16}$$

Apparently, controls u_1, u_2, u_3 take values from the discrete set $\{-1, +1\}$. The transformed vector $\boldsymbol{S}^*$ should be designed such that under controls (11.2.15), s_d and s_q vanish in finite time. Vector $\boldsymbol{S}^*$ is selected as

$$\boldsymbol{S}^* = \frac{3}{2b^2} D^T \boldsymbol{S}_{dq} \tag{11.2.17}$$

Note that $\boldsymbol{S}_{dq} = D\boldsymbol{S}^* = (3/2b^2)DD^T\boldsymbol{S}_{dq} = \boldsymbol{S}_{dq}$.

Theorem 11.4

Under control (11.2.15) and transformation (11.2.17) there exist a domain of $s_d(0)$ and $s_q(0)$ in which the state vector of system (11.2.13) converges to the origin $s_d = 0, s_q = 0$ in finite time. □

Proof

Design a Lyapunov function candidate

$$V = \frac{1}{2} \boldsymbol{S}_{dq}^T \boldsymbol{S}_{dq} \tag{11.2.18}$$

Its time derivative along the solutions of (11.2.13) is of form

$$\dot{V} = (\boldsymbol{S}^*)^T \boldsymbol{F}^* + (\boldsymbol{S}^*)^T D^T D \boldsymbol{U}_{gate} \tag{11.2.19}$$

where $\boldsymbol{F}^* = [F_1^* \quad F_2^* \quad F_3^*]^T = D^T \boldsymbol{F}_{dq}$. Substituting controls (11.2.15) into (11.2.19) results in

$$\dot{V} = (\boldsymbol{S}^*)^T \boldsymbol{F}^* - (\boldsymbol{S}^*)^T D^T D \, \text{sign}(\boldsymbol{S}^*) \tag{11.2.20}$$

where matrix $D^T D$ is a *singular* matrix and can be calculated as

$$D^T D = b^2 \frac{4}{9} \begin{bmatrix} 1 & -\frac{1}{2} & -\frac{1}{2} \\ -\frac{1}{2} & 1 & -\frac{1}{2} \\ -\frac{1}{2} & -\frac{1}{2} & 1 \end{bmatrix} \tag{11.2.21}$$

Similar to the case of electric motor control (Chapter 10), depending on the signs of s_1^*, s_2^* and s_3^*, there are eight possible combinations of values for sign(s_1^*), sign(s_2^*) and sign(s_3^*). Evaluation of (11.2.17) shows that two of these combinations are not possible, i.e. sign(s_1^*), sign(s_2^*) and sign(s_3^*) are never all +1 or all −1. The remaining six combinations can be summarized as

$$\text{sign}(s_l^*) \neq \text{sign}(s_m^*) = \text{sign}(s_n^*) \quad \text{with} \quad l \neq m \neq n \text{ and } l, m, n \in \{1, 2, 3\} \tag{11.2.22}$$

Starting from this notation, (11.2.20) can be expanded as

$$\dot{V} = (s_1^* F_1^* + s_2^* F_2^* + s_3^* F_3^*) - (2/3)^2 b^2 (2|s_l^*| + |s_m^*| + |s_n^*|) \tag{11.2.23}$$

with $l \neq m \neq n$ and $l, m, n \in \{1, 2, 3\}$. This equation can be further represented as

$$\dot{V} = (s_1^* F_1^* + s_2^* F_2^* + s_3^* F_3^*) - (2/3)^2 b^2 (|s_l^*| + |s_m^*| + |s_n^*|) - (2/3)^2 b^2 |s_l^*| \tag{11.2.24}$$

Apparently, inequality

$$(2/3)^2 b^2 \geq \max(|F_1^*|, |F_2^*|, |F_3^*|) \tag{11.2.25}$$

is the sufficient condition for $\dot{V} < 0$. ■

Inequality (11.2.25) defines a subspace in the system space in which the state trajectories converge to the sliding manifold $\boldsymbol{S}_{dq} = \boldsymbol{0}$ in finite time. This is to show that the attraction domain of the sliding manifold is bounded in the state space. Note that parameter $b = u_{0d}/2L$ should be high enough at the initial time instant. Since $u_{0d} = u_0$ the output voltage should not be zero at the initial time instant. In critical applications this can be achieved by starting the converter operation with an open-loop control. In fact, as discussed for the DC/DC converters, an observer-based control scheme can also be applied here such that sliding mode occurs starting from the initial time instant. However, the associated convergence proof is rather complicated and thus goes beyond this text.

As the last step of the control design, the resulting controls u_1, u_2 and u_3 should be mapped into the switching patterns applying to the power converter using the following relations

$$\begin{aligned} s_{w1} &= \frac{1}{2}(1+u_1) \qquad s_{w4} = 1 - s_{w1} \\ s_{w2} &= \frac{1}{2}(1+u_2) \qquad s_{w5} = 1 - s_{w2} \\ s_{w3} &= \frac{1}{2}(1+u_3) \qquad s_{w6} = 1 - s_{w3} \end{aligned} \tag{11.2.26}$$

For implementation of the proposed current control, matrix D^T is needed for transformation (11.2.17) (exact values of $\boldsymbol{F}_{dq}$ are not required for the implementation). Matrix D^T can be found as

$$D^T = b\begin{bmatrix} \cos\theta_a & -\sin\theta_a \\ \cos\theta_b & -\sin\theta_b \\ \cos\theta_c & -\sin\theta_c \end{bmatrix} \tag{11.2.27}$$

where $\theta_a = \theta$, $\theta_b = \theta - 2\pi/3$ and $\theta_c = \theta + 2\pi/3$.

Output voltage regulation

In the following, the reference currents feeding to the current controller, i_d^* and i_q^*, will be determined to ensure asymptotic stability of the output voltage regulation in the outer loop. Neglecting the voltage drop over the phase resistance R_w, the system model in the (d, q) reference frame can be simplified to

$$\begin{aligned} \frac{du_{0d}}{dt} &= -\frac{i_l}{C} + \frac{i_d u_d + i_q u_q}{2C} \\ L\frac{di_d}{dt} &= u_{gd} + \omega L i_q - \frac{u_{0d}}{2} u_d \\ L\frac{di_q}{dt} &= u_{gq} - \omega L i_d - \frac{u_{0d}}{2} u_q \end{aligned} \tag{11.2.28}$$

Normally, the value of the inductance satisfies $L \ll 1$, and the right-hand sides in (11.2.28) have values of the same order. Hence $di_d/dt, di_q/dt \gg du_{0d}/dt$, implying that the dynamics of i_d and i_q are much faster than those of u_{0d}. Provided that the fast dynamics are stable, the outer-loop control can be simplified considerably. Using singular perturbation theory, we can formally let the left-hand sides of the second and third equation in (11.2.28) be equal to zero, and then solve the algebraic equations for u_d and u_q. As a result, the following system is valid for control design of the slow manifold:

$$\begin{aligned} u_d &= 2(u_{gd} + \omega L i_q^*)/u_{0d} \\ u_q &= 2(u_{gq} - \omega L i_d^*)/u_{0d} \\ \frac{du_{0d}}{dt} &= -\frac{i_l}{C} + \frac{i_d^* u_d + i_q^* u_q}{2C} \end{aligned} \tag{11.2.29}$$

where i_d^* and i_q^* are the reference values of i_d and i_q, respectively. Note that we replaced the real currents with their reference values, since we assume that the inner current control loop is in sliding mode with $s_d = i_d^* - i_d = 0$, $s_q = i_q^* - i_q = 0$. Based on (11.2.29) these reference currents will be determined depending on the desired system performance.

The design goals have been given in the performance characteristics at the beginning of Section 11.2.2. The demand of *sinusoidal input currents* has been fulfilled automatically by involving the (d, q) transformation. The following characteristics will cover the major requirements of a well controlled boost AC/DC converter:

1. The output voltage should converge to its reference value u_0^*.
2. The input current phase angle should trace its reference value $\rho^* = \tan^{-1}(i_q^*/i_d^*)$.
3. The power balance condition should be satisfied, i.e. $u_{gd} i_d^* + u_{gq} i_q^* = u_0^* i_l = u_{0d}^* i_l$.

The reference currents i_d^* and i_q^* will be calculated satisfying these requirements. Substitution of the first and second equation of (11.2.29) into the third equation yields

$$\frac{du_{0d}}{dt} = -\frac{i_l}{C} + \frac{u_{gd} i_d^* + u_{gq} i_q^*}{C u_{0d}} \tag{11.2.30}$$

Considering the power balance condition, this equation can be simplified to

$$\frac{du_{0d}}{dt} = -\frac{i_l}{C} + \frac{u_{0d}^* i_l}{C u_{0d}} \tag{11.2.31}$$

For a system with a pure resistance load R_l, the last term in (11.2.31) contains $i_l = u_{0d}/R_l$. Consequently, linear dynamics for output voltage u_{od} can be obtained as

$$\frac{du_{0d}}{dt} = \frac{u_{0d}^* - u_{0d}}{R_l C} \tag{11.2.32}$$

Defining the voltage regulation error as $\bar{u}_{0d} = u_{0d}^* - u_{0d}$ with a constant desired voltage $\dot{u}_{0d}^* = 0$, we have

$$\bar{u}_{0d} + R_l C \dot{\bar{u}}_{0d} = 0 \tag{11.2.33}$$

Obviously, the voltage error tends to zero asymptotically with the time constant $R_l C$. The above derivations mean that if the power balance condition is fulfilled, the output voltage converges to its reference value automatically. Solving the equations for requirements 2 and 3 with respect to i_d^* and i_q^* yields

$$
\begin{aligned}
i_d^* &= \frac{u_0^* i_l}{u_{gd} + u_{gq} \tan \rho^*} \\
i_q^* &= \frac{u_0^* i_l \tan \rho^*}{u_{gd} + u_{gq} \tan \rho^*}
\end{aligned}
\tag{11.2.34}
$$

The input current phase angle ρ^* defined in the second requirement is usually determined by the control designer.

Simulation results

The proposed sliding mode current controller and output voltage regulator are validated by the following simulation results. Parameters of the AC/DC converter are preset as $R_w = 0.45\,\Omega$, $L = 7.5\,\text{mH}$, $C = 820\,\mu\text{F}$ and load resistance $R_l = 182\,\Omega$. The amplitude of the supply voltage is selected as $E = 50/\sqrt{3}\,\text{V}$. The frequency of the supply voltage ω is $377\,\text{rad}\,\text{s}^{-1}$. The simulation results are shown in Figures 11.17 to 11.20.

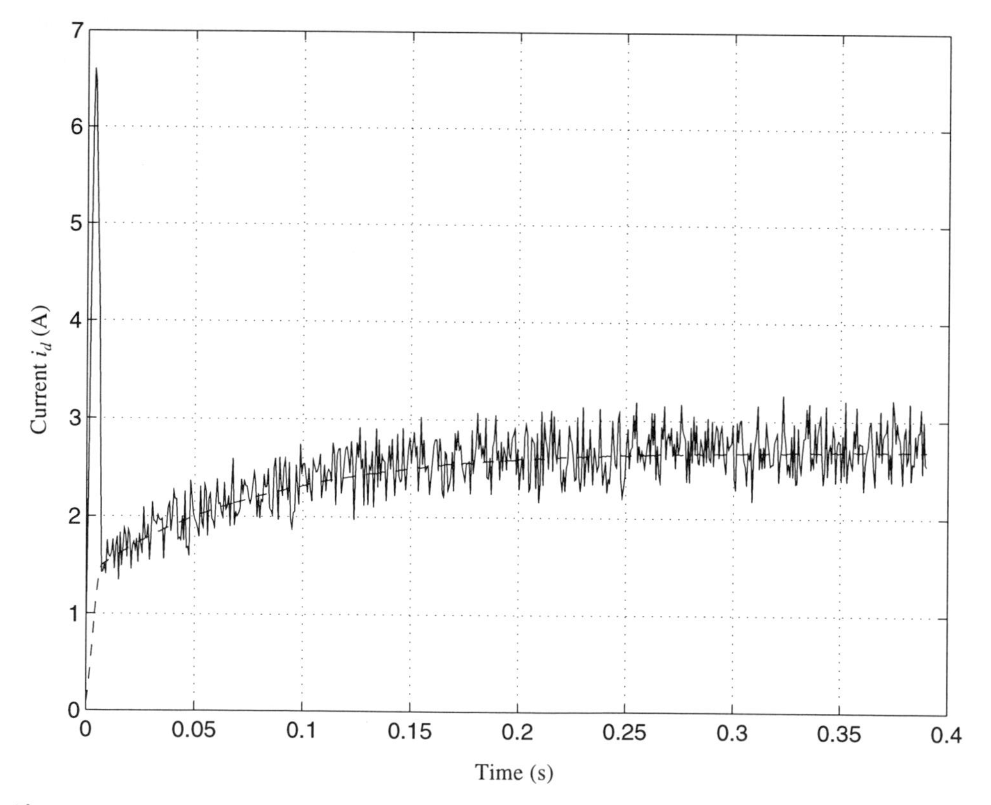

Figure 11.17 Current component i_d.

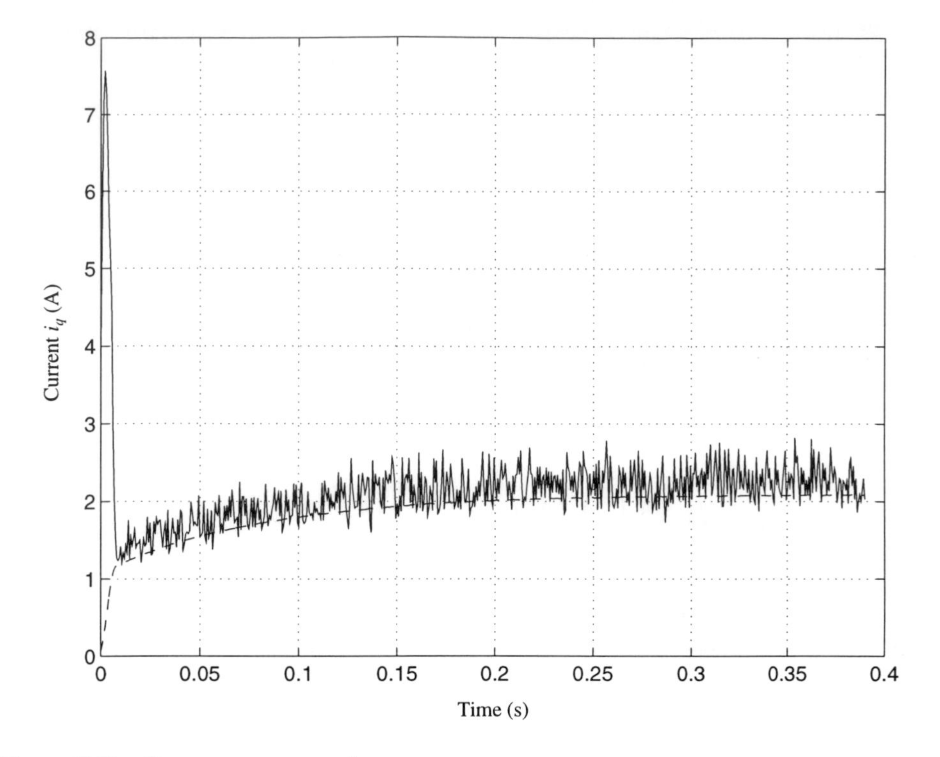

Figure 11.18 Current component i_q.

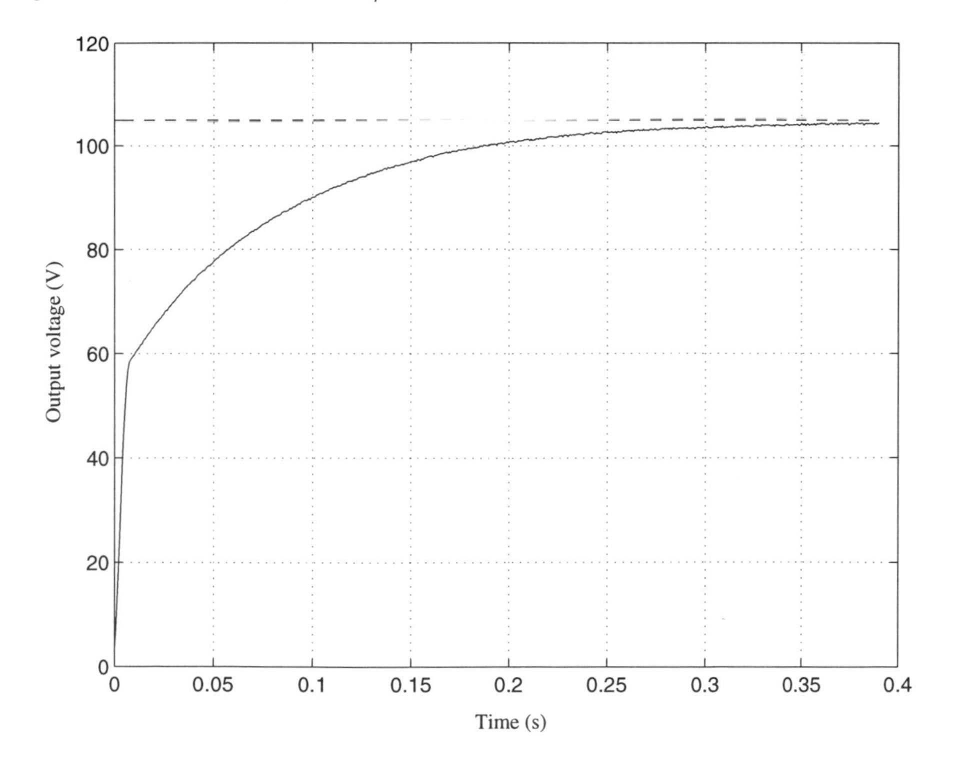

Figure 11.19 Output voltage u_0.

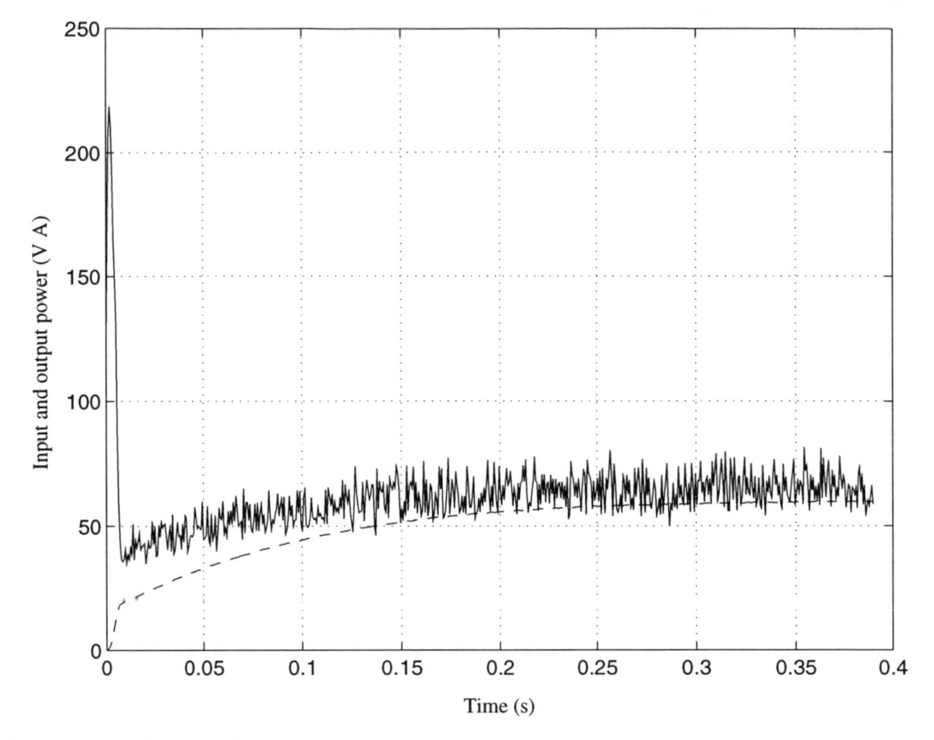

Figure 11.20 Input and output power.

11.2.3 Observer for sensorless control

In the previous sections we developed a cascaded control structure to control the phase currents, output voltage, power factor and the input current phase angle of boost-type AC/DC converters. Sliding mode current control ensures fast convergence of the real currents to their reference counterparts. However, this control structure is based on the assumption that the following information is available:

- phase currents i_1, i_2, i_3
- source voltages u_{g1}, u_{g2}, u_{g3}
- supply frequency ω
- load current i_l

In industrial systems the source voltages, assumed to have sinusoidal form, may come from a synchronous generator; transducers are needed for sensing their amplitude and frequency. The same holds for the phase currents and the load current. To minimize the number of sensors and hence the maintenance costs, a link current sensor may be integrated nearby the storage capacitor (Figure 11.15). Based on this link current sensor, all state variables needed for the proposed control structure can be estimated using a sliding mode observer combined with the design techniques of a conventional observer.

We have made the assumption that the source voltages are of sinusoidal form, but their frequency may not be constant. This is the case for source voltages coming from a synchronous generator which might be started, stopped, accelerated or decelerated.

Suppose the frequency of the source voltages changes approximately linearly with respect to time, i.e.

$$\omega(t) = \alpha t + \beta \tag{11.2.35}$$

where α, β are constant values. The phase of the source voltages is thus given by

$$\theta(t) = \frac{1}{2}\alpha t^2 + \beta t + \gamma \tag{11.2.36}$$

with γ also being a constant value. The model of the source voltage of phase $i = 1$ can then be obtained as

$$u_{g1} = \frac{d}{dt}(E\cos\theta(t)) = -E(\alpha t + \beta)\sin(\tfrac{1}{2}\alpha t^2 + \beta t + \gamma) \tag{11.2.37}$$

where E is a constant. Let us define new state variables as $x_1 = u_{g1}$ and $x_2 = \dot{x}_1$. It may be checked that u_{g1} in (11.2.3) is a solution to the system

$$\begin{aligned} \dot{x}_1 &= x_2 \\ \dot{x}_2 &= -\left(\omega^2 + \frac{3\alpha^2}{\omega^2}\right)x_1 + \frac{3\alpha}{\omega}x_2 \end{aligned} \tag{11.2.38}$$

for any E and α. In other words, we have derived the model of a source phase voltage.

In the following, we propose to design observers for the state variables of phase $i = 1$. The state variables associated with phase $i = 2$ and $i = 3$ can be obtained in a similar manner. Note that state variable x_2 stands for the time derivative of $x_1 = u_{g1}$ and should not be confused with the phase $i = 2$.

The link current is assumed to be measurable as a function of i_i and u_i for $i = 1, 2, 3$:

$$i_{link} = f(i_1, i_2, i_3, u_1, u_2, u_3) \tag{11.2.39}$$

where function f is defined as

$$i_{link} = \begin{cases} i_1 & \text{if } u_1 \neq u_2 = u_3 \\ i_2 & \text{if } u_2 \neq u_1 = u_3 \\ i_3 & \text{if } u_3 \neq u_1 - u_2 \end{cases} \tag{11.2.40}$$

Therefore, the link current consists of sequential 'windows' during which the phase currents can be observed sequentially. In other words, the link current is equal to one of the phase currents in certain windows, which are determined by the combination of the switching signals u_1, u_2, u_3. The lengths of the windows depend on the switching policy employed.

Current observer for source phase voltage

Design a sliding mode current observer as

$$\dot{\hat{i}}_1 = -\frac{R_w}{L}\hat{i}_1 - \frac{u_0}{6L}(2u_1 - u_2 - u_3) + \frac{1}{L}M\,\text{sign}(u_1 i_{link} - \hat{i}_1) \tag{11.2.41}$$

where M is a constant observer gain. To obtain the mismatch dynamics, assume the parameters R_w and L are identical with those in model (11.2.3) and subtract (11.2.41) from (11.2.3):

$$\dot{\bar{i}}_1 = -\frac{R_w}{L}\bar{i}_1 + \frac{1}{L}x_1 - \frac{1}{L}M \operatorname{sign}(\bar{i}_1) \tag{11.2.42}$$

where $\bar{i}_1 = i_1 - \hat{i}_1$. For a sufficiently large control gain M, sliding mode can be enforced in (11.2.42) with $\bar{i}_1 = 0$. Employing the equivalent control method for $\bar{i}_1 = 0$ and $\dot{\bar{i}}_1 = 0$ leads to

$$(M \operatorname{sign}(\bar{i}_1))_{eq} = x_1 \tag{11.2.43}$$

In order to extract the equivalent control from (11.2.42), we use a first-order linear filter (Section 2.3) with filter output z and filter time constant μ:

$$\mu\dot{z} + z = M \operatorname{sign}(\bar{i}_1) \tag{11.2.44}$$

implying that $\lim_{\mu\to 0} z = x_1 = u_{g1}$ asymptotically. Hence the source voltage of phase $i = 1$ can be reconstructed by using the sliding mode observer and exploiting the equivalent control method.

Observer for source voltage

The source voltage $x_1 = u_{g1}$ obtained above is often corrupted by high-frequency noise since the time constant of the lowpass filter is required to be small enough. Otherwise, the associated phase lag and time delay would destroy the information on u_{g1} equal to the average value of $M \operatorname{sign}(\bar{i}_1)$. In addition, the voltage u_{g1} is estimated only within the windows when the link current is equal to the current of phase $i = 1$.

Known supply frequency First, let us discuss a simple case to gain some theoretical insight of the EMF observer, i.e. the source voltage observer. Assuming the frequency ω and the acceleration α of the EMFs are known and assuming the link current can be measured continuously without the restriction of the watching windows, design the EMF observer as follows:

$$\begin{aligned} \dot{\hat{x}}_1 &= \hat{x}_2 - L_1\bar{x}_1 \\ \dot{\hat{x}}_2 &= -\left(\omega^2 + \frac{3\alpha^2}{\omega^2}\right)\hat{x}_1 + \frac{3\alpha}{\omega}\hat{x}_2 - L_2\bar{x}_1 \end{aligned} \tag{11.2.45}$$

where $\bar{x}_1 = \hat{x}_1 - z$ (z denotes the output of the lowpass filter (11.2.44) and we assume that $z = x_1$); L_1 and L_2 are the observer gains. The stability of the observer (11.2.45) can be proven by a proper choice of Lyapunov function.

Subtracting (11.2.38) from (11.2.45) yields

$$\dot{\bar{x}}_1 = \bar{x}_2 - L_1\bar{x}_1$$
$$\dot{\bar{x}}_2 = -\left(\omega^2 + \frac{3\alpha^2}{\omega^2}\right)\bar{x}_1 + \frac{3\alpha}{\omega}\bar{x}_2 - L_2\bar{x}_1 \tag{11.2.46}$$

with $\bar{x}_2 = \hat{x}_2 - x_2$. Design the Lyapunov function candidate as

$$V = \frac{1}{2}\omega_0^2\bar{x}_1^2 + \frac{1}{2}(\bar{x}_2 - L_1\bar{x}_1)^2 \quad \text{with} \quad V = 0 \text{ at } \bar{x}_1 = \bar{x}_2 = 0 \tag{11.2.47}$$

where ω_0 is constant. The time derivative of (11.2.47) along the solutions of (11.2.46) is given by

$$\dot{V} = -\left(L_1 - \frac{3\alpha}{\omega}\right)(\bar{x}_2 - L_1\bar{x}_1)^2 + \left[\omega_0^2 - \left(\omega^2 + \frac{3\alpha^2}{\omega^2}\right) + \frac{3\alpha}{\omega}L_1 - L_2\right]\bar{x}_1(\bar{x}_2 - L_1\bar{x}_1) \tag{11.2.48}$$

We assume the magnitude of $\omega(t)$ is lower bounded, then $\dot{V}(t)$ is negative semidefinite if

$$L_1 > 3\left|\frac{\alpha}{\omega}\right| \quad \text{and} \quad L_2(t) = \omega_0^2 - \left(\omega^2 + \frac{3\alpha^2}{\omega^2}\right) + \frac{3\alpha}{\omega}L_1 \tag{11.2.49}$$

The surface $\bar{x}_2 - L_1\bar{x}_1 = 0$ does not contain whole trajectories. Therefore, the EMF observer with the time-varying gain $L_2(t)$ is shown to be asymptotically stable.

If a crude measurement of the supply frequency is available, the acceleration α may be found using a linear observer

$$\dot{\hat{\omega}} = \hat{\alpha} - L_3(\hat{\omega} - \omega)$$
$$\dot{\hat{\alpha}} = -L_4(\hat{\omega} - \omega) \tag{11.2.50}$$

For constant positive parameters L_3 and L_4, both $\hat{\omega} \to \omega$ and $\hat{\alpha} \to \alpha$ as $t \to \infty$. It is preferable to use the estimated frequency $\hat{\omega}$ instead of the measured value ω if the crude measurement of the supply frequency is corrupted by noise.

Unknown supply frequency In real applications the supply frequency and the acceleration signals may not be measured. To meet this requirement, we need to simultaneously estimate the amplitude and the frequency of the EMFs. The convergence proof of this kind of observer becomes complicated, but simulations and experimental results have shown that the observer converges.

The following nonlinear observer for the EMFs and their frequency is proposed:

$$\begin{aligned}\dot{\hat{x}}_1 &= \hat{x}_2 - L_1 W_1 \bar{x}_1 \\ \dot{\hat{x}}_2 &= -\left(\hat{\omega}^2 + \frac{3\hat{\alpha}^2}{\hat{\omega}^2}\right)\hat{x}_1 + \frac{3\hat{\alpha}}{\hat{\omega}}\hat{x}_2 - L_2 W_1 \bar{x}_1 \\ \dot{\hat{\omega}} &= \hat{\alpha} + L_3 W_1 \bar{x}_1 \hat{x}_1 \\ \dot{\hat{\alpha}} &= -L_4 W_1 \bar{x}_1 \hat{x}_2\end{aligned} \tag{11.2.51}$$

where $L_i(i = 1, 2, 3, 4)$ are the observer gains and W_1 is the window signal defined by

$$W_1 = \begin{cases} 1 & \text{if } u_1 \neq u_2 = u_3 \\ 0 & \text{otherwise} \end{cases} \tag{11.2.52}$$

The problem of global convergence of the observer (11.2.51) is still open at the time of writing. Readers interested in a proof under certain assumptions should consult a technical report by Utkin and Drakunov (1995) and a thesis by Chen (1998).

The state variables associated with phase $i = 2$ and phase $i = 3$ can be obtained in a similar fashion. Several factors will deteriorate the performance of the proposed observers:

1. There exist parameter uncertainties, e.g. in parameter R_w or L.
2. The sampling rate of a digital implementation is too low, in this case we need a discrete-time version of the observer design (Utkin *et al.*, 1997).
3. The time constant of the lowpass filter μ (11.2.44) is selected inappropriately.

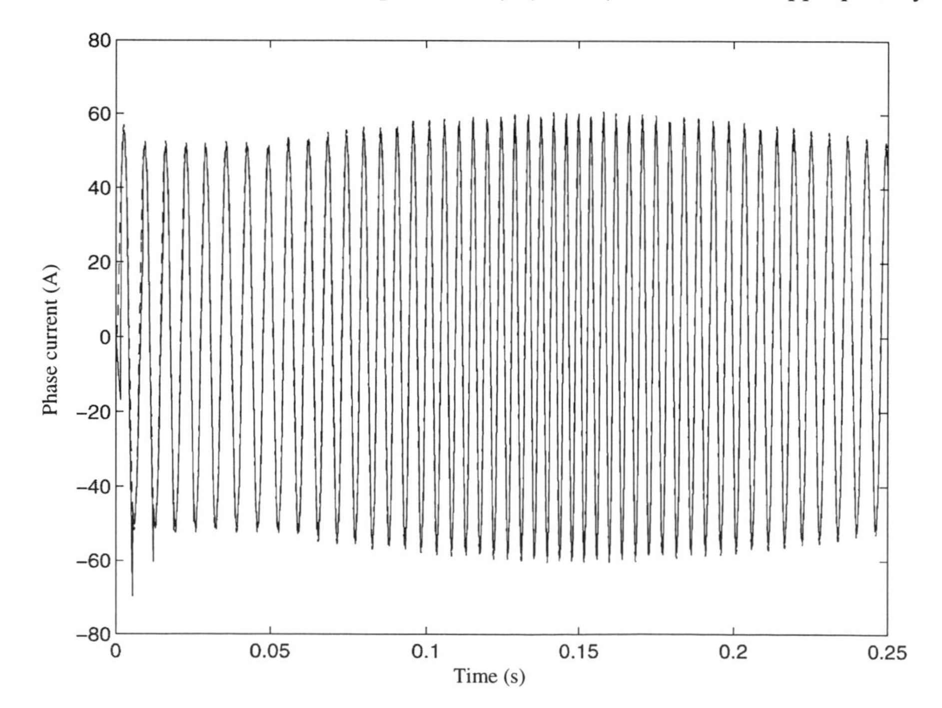

Figure 11.21 Estimation of the phase current $\hat{i}_1$.

Simulation results

The proposed sliding mode observer for the case of varying and unknown supply frequency is validated by the following simulation results. Parameters of the AC/DC converter are preset as $R_w = 0.033\,\Omega$, $L = 1.55 \times 10^{-4}$ H, $C = 40$ F and load resistance $R_l = 0.12\,\Omega$. The amplitude of the supply voltage is selected as $E = 10$ V. The frequency of the supply voltage ω is assumed to vary in the range 150–250 Hz and the rate of frequency change, i.e. the acceleration α, may change within the range $\pm 1000\,\mathrm{Hz\,s^{-1}}$. Results of the estimated signal are shown in Figures 11.21 to 11.24.

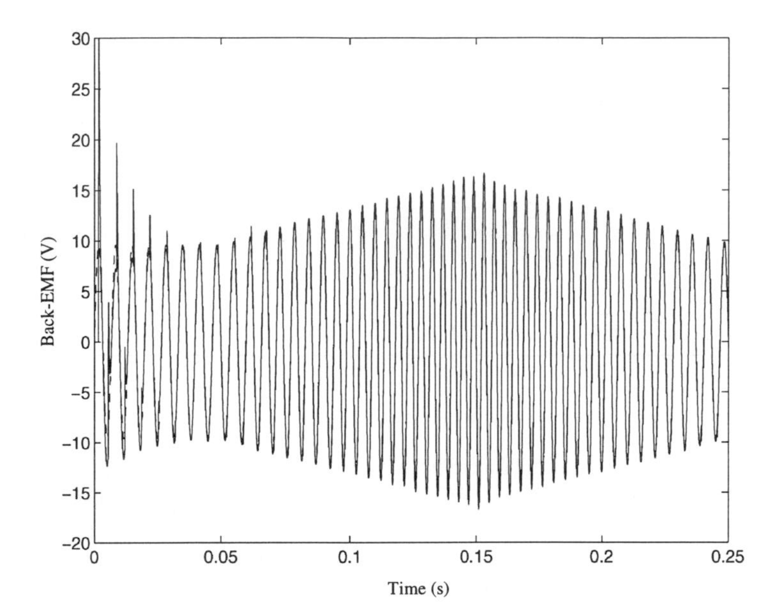

Figure 11.22 Estimation of the supply voltage $\hat{x}_1$.

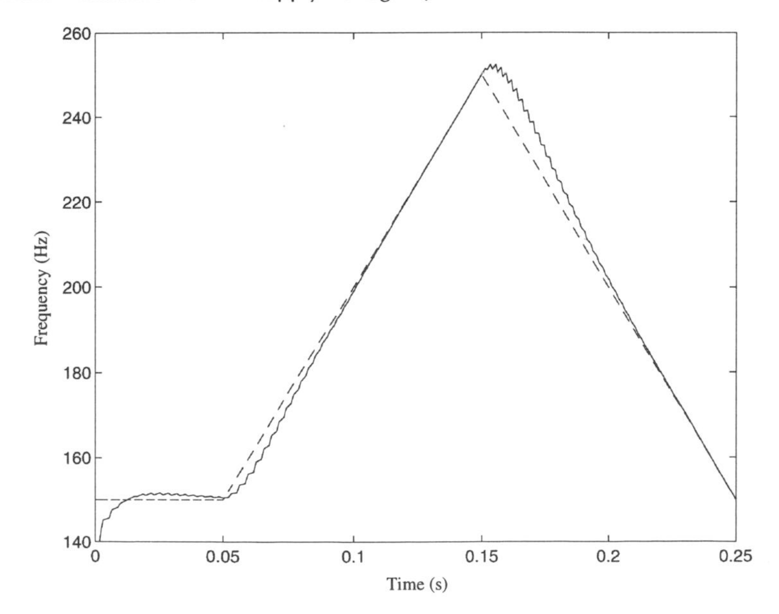

Figure 11.23 Estimation of the supply frequency $\hat{\omega}$.

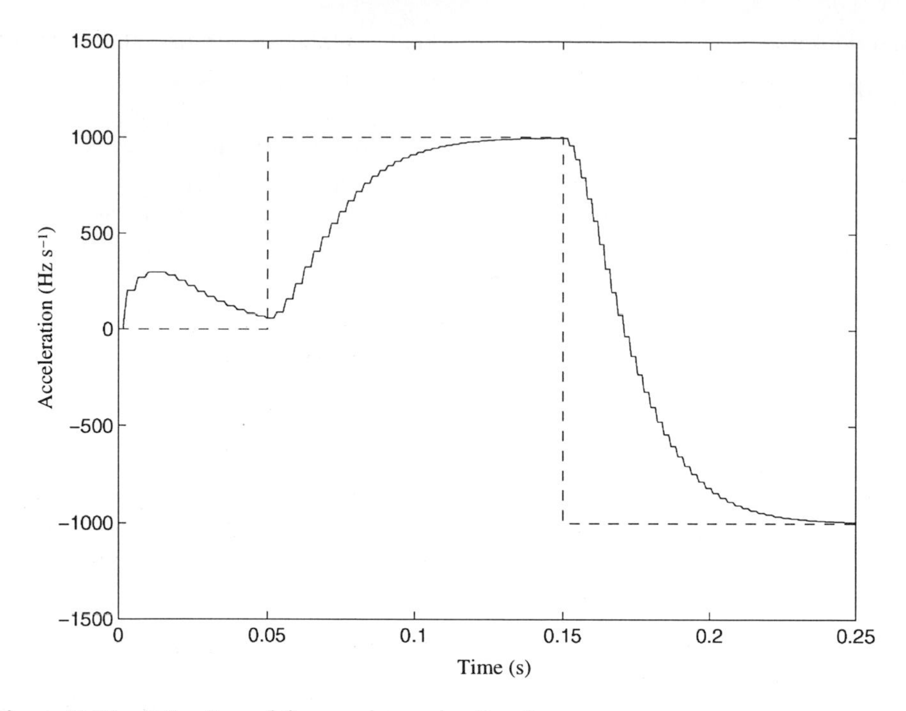

Figure 11.24 Estimation of the supply acceleration $\hat{\alpha}$.

The desired rate of convergence can be obtained by the choice of observer gains. In the simulation, the observer gains are designed as $L_1 = 4000$, $L_2 = 4\,000\,000$, $L_3 = 500$ and $L_4 = 20\,000$.

For all simulations, the information on the phase current is available only within the time windows where the link current is equal to the phase current. Beyond the windows, the observer operates in the open-loop model.

11.3 Summary

This chapter has discussed sliding mode approaches to switching power converters. The proper selection of the switching manifolds and a cascaded control structure was shown to have interesting features:

1. Sliding motion is performed in the inner current loop. The structure of this loop is similar for all types of power converters.
2. Because switching is the only way for controlling power converters, the sliding mode approach does not introduce any additional complexity or chattering.
3. In all systems the reference currents are calculated to be continuous variables. There are a few application-dependent parameters such that the controller can easily be realized in industrial systems.
4. The solutions for all types of converters are global, implying that no linearization procedure is necessary.

A very important aspect of this chapter are the observer-based control approaches it presented, either asymptotic observers or sliding mode observers. It was shown that an observer-based control system may achieve a higher control performance than a nonobserver-based control system. To reduce the number of sensors, sliding mode observers play an important role in the control design. The information is extracted through the concept of equivalent control, thus no high-order time derivatives of the internal state are necessary. Simulation results confirmed the effectiveness of the proposed control approaches.

References

CHEN, D. S., 1998, Discrete time sliding mode observer for automotive alternator, PhD thesis, Department of Mechanical Engineering, Ohio State University, Columbus, OH.

KOKOTOVIC, P. V., O'MALLEY, R. B. and SANNUTI, P., 1976, Singular perturbations and order reduction in control theory, *Automatica*, **12**, 123–32.

KRSTIC, M., KANELLAKOPOULOS, I. and KOKOTOVIC, P., 1995, *Nonlinear and Adaptive Control Design*, New York: J. Wiley.

SABANOVIC, A., SABANOVIC, N. and OHNISHI, K., 1993, Sliding mode in power converters and motion control systems, *International Journal of Control*, **57**, 1237–59.

SERGY, R. and IZOSIMOV, D., 1997, Novel switching losses optimal sliding mode control technique for three-phase voltage source, *Proceedings of the IEEE Symposium on Industrial Electronics*, pp. 288–93.

SIRA-RAMIREZ, H., 1988, Sliding mode control on slow manifolds of DC/DC power converters, *International Journal of Control*, **47**, 1323–40.

SIRA-RAMIREZ, H., ESCOBAR, G. and ORTEGA, R., 1996a, On passivity-based sliding mode control of switched DC-to-DC power converters, *Technical report of the Universidad de Los Andes, Mexico and the Université de Compiègne, France.*

SIRA-RAMIREZ, H., ESCOBAR, G. and ORTEGA, R., 1996b, On passivity-based sliding mode control of switched DC-to-DC power converters, *Proceedings of the 35th IEEE Conference on Decision and Control*, Kobe Japan, pp. 2525–26.

UTKIN, V. I. and DRAKUNOV, S., 1995, Sliding mode observer for automotive applications, Technical report of Ford Motor Company.

UTKIN, V. I., CHEN, D. S., S. ZAREI, S. and MILLER, J., 1997, Discrete time sliding mode observer for automotive alternator, *Proceedings of the European Control Conference*, paper WE-E K2, Conference ID 762.

VENKATARAMANAN, R., SABANOVIC, A. and SLOBODAN, C., 1985, Sliding mode control of DC-to-DC Converters, *Proceedings of the IEEE Conference on Industrial Electronics, Control, and Instrumentation (IECON)*, pp. 251–58.

VILATHGAMUWA, D. M., WALL, S. R. and JACKSON, R. D., 1996, Variable structure control of voltage-sourced reversible rectifiers, *IEE Proceedings on Electrical Power Applications*, **143**, 18–24.

CHAPTER TWELVE

Advanced Robotics

The control of robots, mobile robots and manipulator arms has fascinated control engineers for several decades. Robots are complex mechanical systems with highly nonlinear dynamics. Hence high-performance operation requires nonlinear control designs to fully exploit a robot's capabilities. After describing the dynamic models for robots, this chapter first discusses four basic sliding mode control design alternatives for the classic trajectory tracking problem, in which the robot is asked to follow a prescribed trajectory. Next, advanced robot control is studied using the example of gradient tracking control, where the robot motion is guided on-line by the gradient of an artificial potential field to avoid collisions with obstacles in its workspace. The chapter concludes with four practical examples of sliding mode control in advanced robotics.

12.1 Dynamic modelling

A large number of control problems for mechanical systems are based on controlling the position or location of a mass using a force or a torque as the input variable. Instead of the pure regulation problem of driving the output location to a specified value, the position of the mass is often required to follow a prescribed trajectory. Levels of complexity may be added by introducing sets of masses with coupled dynamics, to be controlled by sets of force/torque inputs. The standard 'fully actuated' case then features one control force/torque input associated with each primary mass and additional forces/torques arising from static and dynamic coupling between the different masses. A typical example is a robotic arm or robot manipulator with n links connected by n joints with force/torque generating actuators. Usually, an end effector tool is mounted at the tip of the last link for manipulating objects according to the specific robot application. The case of less control inputs than primary masses is called underactuation and requires extra consideration. Examples are given in Chapter 4.

The input forces/torques are the outputs of actuators, often electrical actuators, with their own complex dynamics. These actuator dynamics are usually neglected in the first step of control design for the electromechanical system, assuming they are stable and considerably faster than the inertial dynamics of the masses. Due to the large variety of actuators, we consider the treatment of actuator control in Section 12.2. Also, other dynamics such as structural flexibilities are often neglected when deriving a basic model for the mechanical system. In practice this leads to the chattering problem described in Chapter 8 and one of the solutions discussed there should be employed on top of the basic control designs outlined in this chapter.

Before designing control strategies for a mechanical system, a dynamic model describing the principle physical behaviour should be derived. In this section we consider holonomic mechanical systems in unconstrained motion and planar mobile robots with nonholonomic motion constraints. Several methods have been developed to obtain a dynamic model based on the physical properties of the system. A popular methodology is the Euler–Lagrange formulation for an energy-conserving system:

$$\frac{d}{dt}\frac{\partial L}{\partial \dot{q}} - \frac{\partial L}{\partial q} = \tau \tag{12.1.1}$$

where $q \in \Re^{n\times 1}$ is a vector of generalized configuration coordinates, $\tau \in \Re^{n\times 1}$ is a vector of generalized external (input) forces/torques (excluding gravity), and the *Lagrangian* $L = K - P$ is the difference between the total kinetic energy (K) and potential (P) of the system. For details and alternative formulations, please refer to textbooks on the dynamics of mechanical systems; for robotic systems, see Craig (1986) or Spong and Vidyasagar (1989).

12.1.1 Generic inertial dynamics

For the purpose of general control design considerations in the first part of this chapter, consider a continuous-time model of a generic, fully actuated n-dimensional robotic system with inertial dynamics of the form

$$M(q)\ddot{q} + N(q, \dot{q}) = \tau \tag{12.1.2}$$

where $q \in \Re^{n\times 1}$ is a vector of generalized configuration variables (translational or rotational), $M(q) \in \Re^{n\times n}$ denotes an inertial mass matrix, $N(q, \dot{q}) \in \Re^{n\times 1}$ comprises coupling forces/torques between the masses as well as gravity and friction, and $\tau \in \Re^{n\times 1}$ are the generalized input forces/torques. Equation (12.1.2) describes the principal relationship between inertial motion of the system masses, internal forces/torques $N(q, \dot{q})$ and external input forces/torques τ, hence it is well suited for control design.

Traditionally, robots have been categorized into *robot manipulators* and *mobile robots*. Robot manipulators usually have a fixed base and consist of a number of rigid links, connected by translatory or rotatory joints. A set of $q \in \Re^{n\times 1}$ configuration variables of the n joints prescribes a robot configuration, also called robot posture. The set of all possible configurations within the physical joint limitations defines the robot configuration space. The robot *kinematics* provide a mapping between joint coordinates and world coordinates. The associated locations of a given point of the manipulator, e.g. the tip of its end effector in a world coordinate system, define the robot workspace. Note that multiple configurations may result in similar end effector positions. Consequently, the inverse kinematic mapping between end effector location in world coordinates and the configuration vector $q \in \Re^{n\times 1}$ is not unique, so we omit the treatment of *inverse kinematics* and concentrate on control design in configuration space. Again, the interested reader is referred to textbooks on robotics for a detailed treatment of robot kinematics and inverse kinematic mappings.

Most manipulator arms have serial links, but there are also designs with parallel linkages. Mobile robots, on the other hand, possess wheels or other means to move about. Their workspace is defined by the set of points reachable via their means of mobility. Position and possibly orientation variables with respect to a workspace-fixed coordinate system define the configurations of a mobile robot. Recently, robot manipulators and mobile robots have been combined to form mobile manipulators, e.g. with three degrees of freedom for mobility in the plane and six degrees of freedom for manipulation.

For control design, we distinguish *holonomic* and *nonholonomic* robots. The motion of a holonomic robot is usually unconstrained. All joints may move arbitrarily within their physical limitations and the constraints of the robot workspace, i.e. only limits of the position variables exist. This class of robots, described in Section 12.1.2, incorporates both manipulators and so-called omni-directional mobile robots. Special cases include interaction between a robot and components of its workspace, and cooperative action of two or more robots, requiring special treatment beyond the scope of this text. Nonholonomic robots face additional constraints of the time derivatives $\dot{q} \in \Re^{n \times 1}$ of their position variables, i.e. constraints on the velocity variables. Section 12.1.3 describes a kinematic and dynamic model for nonholonomic robots.

12.1.2 Holonomic robot model

A well-known example of highly nonlinear, fully actuated mechanical systems with coupled dynamics is a robot manipulator with rigid links. For a large class of holonomic robot systems, the generic dynamics in (12.1.2) can be rewritten in configuration space as

$$M(q)\ddot{q} + V_m(q, \dot{q})\dot{q} + F(\dot{q}) + G(q) = \tau \tag{12.1.3}$$

where $q \in \Re^{n \times 1}$ denotes the joint configurations (translational or rotational) of the n robot links, $M(q)$ stands for the inertial mass matrix, $V_m(q, \dot{q}) \in \Re^{n \times n}$ comprises Coriolis and centripetal forces, vector $F(\dot{q}) \in \Re^{n \times 1}$ describes viscous friction and vector $G(q) \in \Re^{n \times 1}$ contains the gravity terms. The formulation (12.1.3) follows directly from the Euler–Lagrange equations of motion and encompasses robot manipulators operating freely without motion constraints. Craig (1988) revealed the following properties.

Mass matrix

The square mass matrix $M(q)$ is symmetric, positive definite, and can be written as

$$M(q) = \begin{bmatrix} m_{11}(q) & \cdots & m_{1n}(q) \\ \vdots & \ddots & \vdots \\ m_{1n}(q) & \cdots & m_{nn}(q) \end{bmatrix} \tag{12.1.4}$$

with bounded parameters $m_{ij}^{-} \leq m_{ij}(q) \leq m_{ij}^{+}, 1 \leq i, j \leq n$. Hence $M(q)$ can be bounded by

$$M^{-} \leq \|M(q)\|_2 \leq M^{+} \tag{12.1.5}$$

where any induced matrix norm may be used to define two known scalars $0 < M^- \leq M^+$ as bounds. The known scalars M^- and M^+ also bound the inverse of $M(q)$:

$$\frac{1}{M^+} \leq \|M^{-1}(q)\|_2 \leq \frac{1}{M^-}. \tag{12.1.6}$$

In (12.1.5) and in the sequel, the induced 2-norm will be used as an example of bounding norms, but many other norms may be employed instead. The induced 2-norm of matrix $M(q)$ is defined as

$$\|M(q)\|_2 = \sqrt{\max\{\lambda(M^T M)\}} \tag{12.1.7}$$

where $\lambda(M^T M)$ denotes the eigenvalues of matrix $M^T M$.

Skew symmetry

The time derivative of the mass matrix, $\dot{M}(q) = dM(q)/dt = (\partial M(q)/\partial q)\dot{q}$, and the Coriolis/centripetal matrix, $V_m(q, \dot{q})\dot{q}$, are skew symmetric, i.e.

$$y^T(\dot{M}(q) - 2V_m(q, \dot{q}))y = 0 \tag{12.1.8}$$

holds for any nonzero vector $y \in \Re^{n\times 1}$.

Boundedness of dynamic terms

The Coriolis/centripetal vector $V_m(q, \dot{q})\dot{q}$ is bounded by

$$\|V_m(q, \dot{q})\dot{q}\|_2 \leq V^+\|\dot{q}\|_2 \tag{12.1.9}$$

where V_m^+ is a positive scalar. Viscous friction may be bounded by positive scalars F^+ and F_0:

$$\|F(\dot{q})\|_2 \leq F^+\|\dot{q}\|_2 + F_0 \tag{12.1.10}$$

The gravity vector likewise is bounded by a positive scalar G^+:

$$\|G(q)\|_2 \leq G^+ \tag{12.1.11}$$

Example 12.1 Holonomic model of a two-link manipulator

A planar, two-link manipulator with revolute joints will be used as an example throughout the control development in Section 12.2. The manipulator and the associated variables are depicted in Figure 12.1. The geometry in Figure 12.1 reveals the *forward kinematics* of the two-link manipulator. The end effector position (x_w, y_w), i.e. the location of mass M_2 in world coordinate frame (x, y), is given by

$$\begin{aligned} x_w &= L_1 \cos q_1 + L_2 \cos(q_1 + q_2) \\ y_w &= L_1 \sin q_1 + L_2 \sin(q_1 + q_2) \end{aligned} \tag{12.1.12}$$

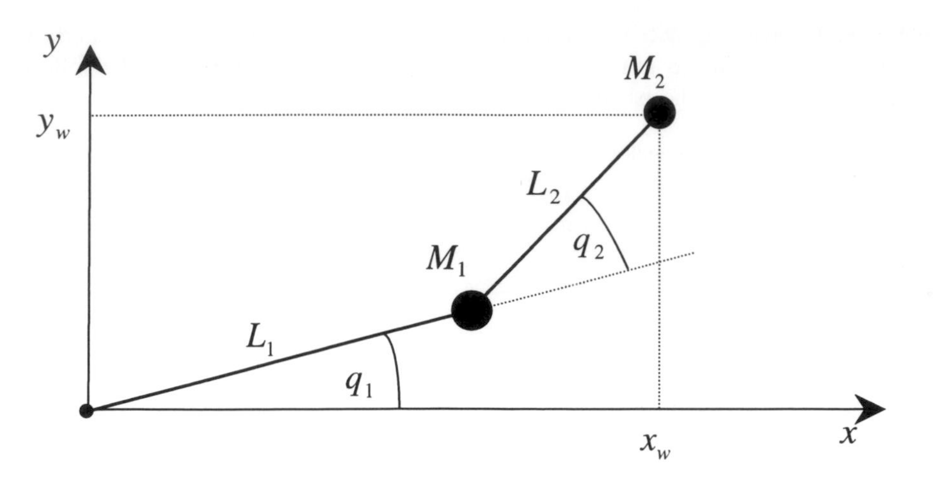

Figure 12.1 Two-link manipulator with link lengths L_1 and L_2, and concentrated link masses M_1 and M_2. The manipulator is shown in joint configuration (q_1, q_2), which leads to end effector position (x_w, y_w) in world coordinates. The manipulator is operated in the plane, i.e. gravity acts along the z-axis.

where q_1, q_2 denote the joint displacements and L_1, L_2 are the link lengths. Solving (12.1.12) for the joint displacements as a function of the end effector position (x_w, y_w) yields the *inverse kinematics* as

$$\begin{aligned} q_2 &= \operatorname{atan} 2(D, C) \quad \text{with} \quad C = \frac{x_w^2 + y_w^2 - L_1^2 - L_2^2}{2L_1^2 L_2^2}, D = \pm\sqrt{1 - C^2} \\ q_1 &= \operatorname{atan} 2(y_w, x_w) - \operatorname{atan} 2(L_2 \sin q_2, L_1 + L_2 \cos q_2) \end{aligned} \tag{12.1.13}$$

which is obviously not unique due to the two sign options of the square root in variable D. The function atan2(.) describes the arctan function normalized to the range $\pm 180°$.

Applying a standard modelling technique such as the Euler–Lagrange equations yields the dynamic model according to (12.1.2) as

$$\begin{bmatrix} \tau_1 \\ \tau_2 \end{bmatrix} = \begin{bmatrix} m_{11} & m_{12} \\ m_{21} & m_{22} \end{bmatrix} \begin{bmatrix} \ddot{q}_1 \\ \ddot{q}_2 \end{bmatrix} + \begin{bmatrix} n_1 \\ n_2 \end{bmatrix} \tag{12.1.14}$$

with

$$\begin{aligned} m_{22} &= L_2^2 M_2 \\ m_{12} &= m_{21} = m_{22} + L_1 L_2 M_2 \cos q_2 \qquad & M(q) &= \begin{bmatrix} m_{11} & m_{12} \\ m_{21} & m_{22} \end{bmatrix} \\ m_{11} &= L_1^2(M_1 + M_2) + 2m_{12} - m_{22} \\ n_2 &= L_1 L_2 M_2 \dot{q}_1^2 \sin q_2 \qquad & N(q, \dot{q}) &= \begin{bmatrix} n_1 \\ n_2 \end{bmatrix} \\ n_1 &= -L_1 L_2 M_2 (2\dot{q}_1 \dot{q}_2 - \dot{q}_2^2) \sin q_2 \end{aligned} \tag{12.1.15}$$

Note the absence of gravity terms in (12.1.15) since the manipulator is operated in the plane, perpendicular to gravity. For the control design examples in the following sections, we will use the parameters in Table 12.1.

Table 12.1 Two-link manipulator parameters

$M_1 = 10$ kg	$M_2 = 1$ kg
$L_1 = 1$ m	$L_2 = 1$ m

To examine the skew symmetry property in equation (12.1.8), take the derivative of mass matrix $M(q)$ in (12.1.15) to yield

$$\dot{M}(q, \dot{q}) = \begin{bmatrix} 2\dot{m}_{12} & \dot{m}_{12} \\ \dot{m}_{12} & 0 \end{bmatrix}$$
$$\dot{m}_{12}(q, \dot{q}) = L_1 L_2 M_2 \dot{q}_2 \sin q_2 \tag{12.1.16}$$

Then separate matrix $N(q, \dot{q})$ into its components according to (12.1.3). Due to the assumptions of planar operation and no friction, the gravity and friction terms are equal to zero and we obtain

$$N(q, \dot{q}) = V(q, \dot{q}) = V_m(q)\dot{q}$$
$$\text{with} \quad V_m(q) = \begin{bmatrix} -L_1 L_2 M_2 \dot{q}_2 \sin q_2 & -L_1 L_2 M_2 (\dot{q}_1 + \dot{q}_2) \sin q_2 \\ L_1 L_2 M_2 \dot{q}_1 \sin q_2 & 0 \end{bmatrix} \tag{12.1.17}$$

Skew symmetry follows from (12.1.16) and (12.1.17) as

$$\begin{aligned} & y^T(\dot{M}(q) - 2V_m(q, \dot{q}))y \\ &= L_1 L_2 M_2 \sin q_2 [y_1 \quad y_2] \left(\begin{bmatrix} -2\dot{q}_2 & -\dot{q}_2 \\ -\dot{q}_2 & 0 \end{bmatrix} - 2 \begin{bmatrix} -\dot{q}_2 & -\dot{q}_1 - \dot{q}_2 \\ \dot{q}_1 & 0 \end{bmatrix} \right) \begin{bmatrix} y_1 \\ y_2 \end{bmatrix} \\ &= L_1 L_2 M_2 \sin q_2 [y_1 \quad y_2] \left(\begin{bmatrix} 0 & 2\dot{q}_1 + \dot{q}_2 \\ -2\dot{q}_1 - \dot{q}_2 & 0 \end{bmatrix} \right) \begin{bmatrix} y_1 \\ y_2 \end{bmatrix} = 0 \end{aligned} \tag{12.1.18}$$

Assuming exact knowledge of the parameters in Table 12.1, but ignoring all dependencies on joint positions, we can find upper and lower bounds for the elements of the matrices $M(q)$ and $N(q, \dot{q})$ in (12.1.15); they are listed in Table 12.2. Using the 2-norm according to (12.1.7) results in upper and lower bounds for mass matrix $M(q)$ as described in equation (12.1.5): $M^- = 0.957\,\text{kg m}^2$ and $M^+ = 204.511\,\text{kg m}^2$. Matrix $N(q, \dot{q})$ can be upper bounded as $N^+ = 2(|\dot{q}_1 \dot{q}_2| + \dot{q}_2^2) \times 1\,\text{kg m}^2$.

Table 12.2 Lower and upper bounds of matrix elements

$m_{11}^- = 12$ kg m^2	$m_{11}^+ = 14$ kg m^2
$m_{12}^- = m_{21}^- = 1$ kg m^2	$m_{12}^+ = m_{21}^+ = 2$ kg m^2
$m_{22}^- = 1$ kg m^2	$m_{22}^+ = 1$ kg m^2
$n_1^+ = (2|\dot{q}_1 \dot{q}_2| + \dot{q}_2^2) \times 1$ kg m^2	
$n_2^+ = \dot{q}_2^2 \times 1$ kg m^2	

12.1.3 Nonholonomic robots: model of wheel set

Robots whose motion is subjected to a set of p nonintegrable constraints involving time derivatives of the configuration vector q are classified as nonholonomic systems (Neimark and Fufaev, 1972). The constraints usually take the form

$$G(q)\dot{q} = 0 \tag{12.1.19}$$

with the $(n - p)$ independent columns of the $p \times n$ matrix $G(q)$ forming the base for the nonholonomic constraint condition

$$\dot{q} = K(q)u \tag{12.1.20}$$

Note that the number of control inputs is less than the dimension of the system, i.e. underactuation with $u \in \Re^{n-p}$ follows from (12.1.19). Consider a set of wheels with a common axle, but independent wheel actuators as an example of a mobile robot with nonholonomic kinematics, as shown in Figure 12.2. Assuming no slip at the tires, the motion of each wheel is restricted to its longitudinal direction with velocities v_R and v_L, respectively, by a single nonholonomic constraint ($p = 1$). In other words, no motion can occur along the lateral robot coordinate axis y_R. Also shown in Figure 12.2 is the robot configuration $q = (x, y, \phi) \in \Re^3$ in the world coordinate frame (x_w, y_w). Control inputs are the two wheel velocities v_R and v_L, which may be translated into the translational and rotational velocity variables $u = (v_C, \omega) \in \Re^2$ for convenience. The motion of the wheel set in the world coordinate frame is given by

$$\begin{aligned} \dot{x} &= v_C \cos\phi \\ \dot{y} &= v_C \sin\phi \\ \dot{\phi} &= \omega \end{aligned} \tag{12.1.21}$$

which form the forward kinematics for this case.

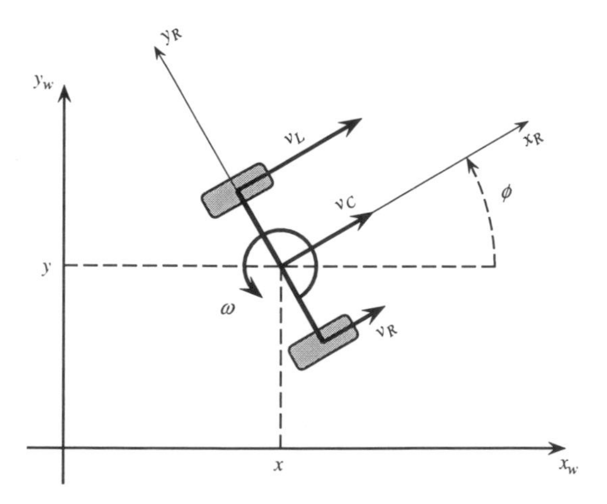

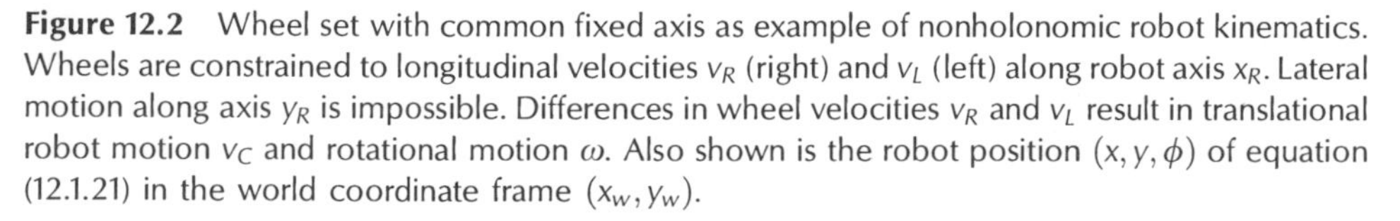

Figure 12.2 Wheel set with common fixed axis as example of nonholonomic robot kinematics. Wheels are constrained to longitudinal velocities v_R (right) and v_L (left) along robot axis x_R. Lateral motion along axis y_R is impossible. Differences in wheel velocities v_R and v_L result in translational robot motion v_C and rotational motion ω. Also shown is the robot position (x, y, ϕ) of equation (12.1.21) in the world coordinate frame (x_w, y_w).

For the wheel set shown in Figure 12.2, we assume inertial dynamics of the form

$$M\dot{v}_C + N_t(v_C, \omega) = \tau_t \tag{12.1.22}$$

$$J\dot{\omega} + N_r(v_C, \omega) = \tau_r \tag{12.1.23}$$

with positive scalars M and J denoting mass and rotational inertia about the vertical z_w-axis, $\tau = (\tau_t, \tau_r) \in \Re^2$ being the control inputs, and scalars $N_t(\dot{v}_C, \omega)$ and $N_r(\dot{v}_C, \omega)$ comprising all additional dynamics. A more complicated dynamic modelling was discussed by Bloch *et al.* (1992).

12.2 Trajectory tracking control

The control task commonly arising in robot control is to track a time-dependent trajectory described by

$$[q_d(t)\ \dot{q}_d(t)\ \ddot{q}_d(t)] \tag{12.2.1}$$

with bounded desired configurations

$$q_d(t) = [q_{d_1}(t), \ldots, q_{d_i}(t), \ldots, q_{d_n}(t)],$$

velocities

$$\dot{q}_d(t) = [\dot{q}_{d_1}(t), \ldots, \dot{q}_{d_i}(t), \ldots, \dot{q}_{d_n}(t)]$$

and accelerations

$$\ddot{q}_d(t) = [\ddot{q}_{d_1}(t), \ldots, \ddot{q}_{d_i}(t), \ldots, \ddot{q}_{d_n}(t)]$$

for each component of an n-dimensional system like (12.1.2). Control of a second-order mechanical system with a force/torque input as in (12.1.2) requires position and velocity feedback as a basis for stabilization, i.e. PD-type control. This requirement can either be met by measurement and feedback of both position and velocity variables or by a lead compensator for position measurements in the linear sense (Arimoto and Miyazaki, 1984). In sliding mode control designs, this requirement is reflected in the choice of the sliding manifold with a stable motion, designed as a (linear) combination of position and velocity variables.

Since the first set point sliding mode controller for robot manipulators suggested by Young (1978), numerous variations have been presented in the literature. In the sequel, we seek to outline the set of principal design choices to be made when designing a sliding mode tracking controller for a mechanical system like (12.1.2). First, the designer may choose between componentwise control and vector control. Second, a purely discontinuous controller has to be compared to a continuous feedback/feedforward controller with an additional discontinuity term to achieve robustness by generating sliding mode. For overviews see Table 12.6 and Table 12.7 in Section 12.2.4.

Example 12.2 Circular trajectory for planar two-link manipulator

For the control design examples in this section, we will take the planar two-link manipulator of Example 12.1 and require it to follow a circular trajectory in its workspace. The circle with centre (x_d, y_d) and radius r_d is given in world coordinates (x_w, y_w) by

$$x_d(t) = x_{d_0} + r_d \cos \psi_d$$

$$y_d(t) = y_{d_0} - r_d \sin \psi_d \tag{12.2.2}$$

$$\psi_d(t) = \frac{2\pi}{t_f} t - \sin\left(\frac{2\pi}{t_f} t\right) \qquad (0 \leq t \leq t_f)$$

where the operation is assumed to start at time $t = 0$ and to be completed at final time $t = t_f$. The parameters for the examples are listed in Table 12.3. Bounds over the time interval $0 \leq t \leq t_f$ can be obtained by using the inverse kinematics (12.1.13) as summarized in Table 12.4. The desired trajectory is depicted in Figure 12.3.

Table 12.3 Parameters of desired circular trajectory

$x_d = 1$ m	$y_d = 1$ m
$r_d = 0.5$ m	$t_f = 5$ s

Table 12.4 Bounds of desired circular trajectory

$\lvert\psi_d\rvert \leq \frac{2\pi}{5}$	$\lvert\dot{\psi}_d\rvert \leq \frac{4\pi}{5}$	$\lvert\ddot{\psi}_d\rvert \leq \frac{4\pi^2}{25}$
$\lvert\dot{q}_{d_1}\rvert \leq 1$ rad s^{-1}	$\lvert\dot{q}_{d_2}\rvert \leq 1$ rad s^{-1}	

12.2.1 Componentwise control

The first choice is mainly concerned with the structure of the sliding manifolds. Because in system (12.1.2), n force/torque inputs, are assumed to control n configuration outputs of the $2n$-dimensional dynamics, each component i of the n components of the output vector may be assigned its own sliding manifold and hence may be controlled independently. Alternatively, the n components are dealt with as a vector (Section 12.2.2). For componentwise control, the structure of the n sliding manifolds is

$$s_i = c_i q_{e_i} + \dot{q}_{e_i} \qquad (i = 1, \ldots, n) \tag{12.2.3}$$

where $c_i > 0$ are scalar gains to determine the rate of exponential convergence of the tracking error $q_{e_i} = q_{d_i} - q_i$ to zero after reaching the sliding manifold $s_i = 0$. Each component of the control input vector τ_i is responsible for ensuring sliding mode occurs along its respective manifold (12.2.3), e.g. by choosing

$$\tau_i = \tau_{0_i} \operatorname{sign}(s_i) \qquad (i = 1, \ldots, n) \tag{12.2.4}$$

Sliding mode is established independently in each manifold $s_i = 0$ and finally in the intersection $s = [s_1, \ldots, s_n]^T = 0_{n\times 1}$. The controller gains τ_{0_i} are to be determined from the stability analysis sketched in Theorem 12.1.

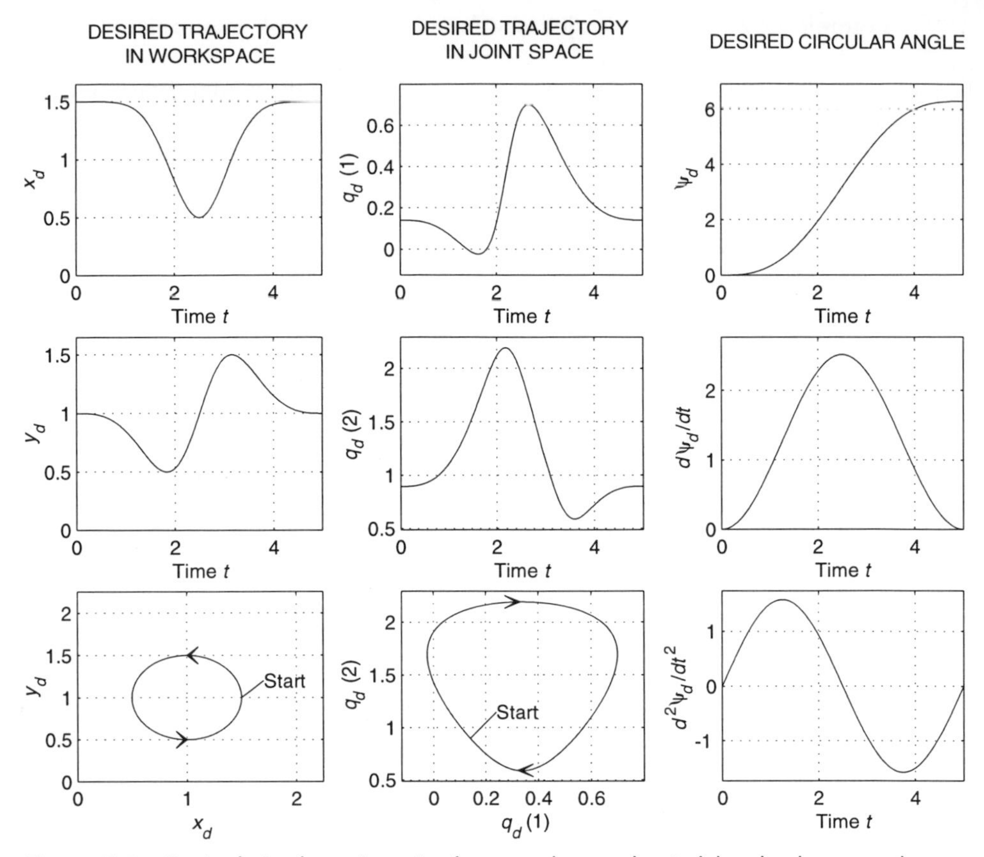

Figure 12.3 Desired circular trajectories for control examples. Left-hand column: workspace trajectory in world coordinates (x_w, y_w). Center column: associated joint space trajectory in (q_1, q_2) coordinates. Right-hand column: time trajectory of angle ψ_d.

Theorem 12.1

The states of system (12.2.5) with components (12.2.6) under control (12.2.8) will reach the sliding manifolds (12.2.7) in finite time:

Total system
$$\ddot{q} = M^{-1}(\tau - N(q, \dot{q})) \tag{12.2.5}$$

System component
$$\ddot{q}_i = (M^{-1}(\tau - N(q, \dot{q})))_i = \frac{1}{m_{ii}}\left(\tau_i - n_i - \sum_{j=1, j\neq i}^{n} m_{ij}\ddot{q}_j\right) \tag{12.2.6}$$

Manifolds
$$s_i = c_i q_{e_i} + \dot{q}_{e_i} = c_i(q_{d_i} - q_i) + (\dot{q}_{d_i} - \dot{q}_i) \tag{12.2.7}$$

Controls
$$\tau_i = \tau_{0_i}\ \text{sign}(s_i) \tag{12.2.8}$$

□

Proof

Consider a Lyapunov function candidate for each component,

$$V_i = \frac{1}{2} s_i^2 \tag{12.2.9}$$

with its time derivative along the system trajectories

$$\begin{aligned} \dot{V}_i &= s_i \left(c_i \dot{q}_{e_i} + \ddot{q}_{d_i} - \frac{1}{m_{ii}} \left(\tau_{0_i} \operatorname{sign}(s_i) - n_i - \sum_{j=1, j\neq 1}^{n} m_{ij} \ddot{q}_j \right) \right) \\ &\leq - \frac{\tau_{0_i}}{m_{ii}^{+}} |s_i| + |s_i| \left(c_i |\dot{q}_{e_i}| + |\ddot{q}_{d_i}| + \frac{1}{m_{ii}^{-}} \left(n_i^{+} + \sum_{j=1, j\neq 1}^{n} m_{ij}^{+} |\ddot{q}_j| \right) \right) \end{aligned} \tag{12.2.10}$$

Boundedness of the desired trajectory in (12.2.1), of the elements m_{ij} of the mass matrix (12.1.4) and of the elements n_i of vector $N(q)$ by (12.1.9) to (12.1.11), ensures the existence of a

$$\tau_{0_i} > m_{ii}^{+} \left(c_i |\dot{q}_{e_i}| + |\ddot{q}_{d_i}| + \frac{1}{m_{ii}^{-}} \left(n_i^{+} + \sum_{j-1, j\neq 1}^{n} m_{ij}^{+} |\ddot{q}_j| \right) \right) \tag{12.2.11}$$

to yield

$$\dot{V}_i \leq -\xi_i |s_i| \tag{12.2.12}$$

for some scalar $\xi_j > 0$. Consequently, finite convergence of each system component (12.2.6) to the manifold $s_i = 0$ in (12.2.7) is established. ■

The assumption of boundedness of the terms $(m_{ji}/m_{ii})\ddot{q}_j, j = 1, \ldots, n, j \neq i$, implies bounded coupling forces/torques due to bounded accelerations of the other masses. In practical mechanical systems with one mass m_i being associated with each control input τ_i, a hierarchical boundedness of coupling terms is inherent. In particular, robot manipulators tend to be constructed with stronger and hence heavier links and joints near the base, and with increasingly lighter links and joints towards the tip and the end effector. For an explicit stability analysis, the hierarchy of masses has to be solved in reverse, a process which may become tedious for higher-dimensional systems. The details of the hierarchical design method may be found in Utkin (1992).

As an alternative to the individual design of components τ_i of control vector τ pursued in Theorem 12.1, the design may be based on a Lyapunov function constructed for the whole system instead of for each subsystem. The main advantage of such a closed representation in vector form is the avoidance of hierarchical mass requirements due to the positive definite property of the mass matrix M. It was shown in Chapter 2 that using a Lyapunov function $V = s^T \operatorname{sign}(s)$ to derive the control vector τ likewise enforces sliding mode in the manifold $s = 0$.

Example 12.3 Componentwise control of two-link manipulator

Solving (12.2.6) for the planar two-link manipulator with dynamics (12.1.14) yields

$$\begin{aligned} \ddot{q}_1 &= \frac{1}{m_{11}}(\tau_1 - n_1 - m_{21}\ddot{q}_2) \\ \ddot{q}_2 &= \frac{1}{m_{22}}(\tau_2 - n_2 - m_{12}\ddot{q}_1) \end{aligned} \tag{12.2.13}$$

According to (12.2.7), the sliding manifolds are defined as

$$\begin{aligned} s_1 &= c_1 q_{e_1} + \dot{q}_{e_1} = 0 \qquad q_{e_1} = q_{d_1} - q_1 \\ s_2 &= c_2 q_{e_2} + \dot{q}_{e_2} = 0 \qquad q_{e_2} = q_{d_2} - q_2 \end{aligned} \tag{12.2.14}$$

leading to a control vector

$$\tau = \begin{bmatrix} \tau_1 \\ \tau_2 \end{bmatrix} = \begin{bmatrix} \tau_{0_1} \text{ sign } s_1 \\ \tau_{0_2} \text{ sign } s_2 \end{bmatrix} \tag{12.2.15}$$

To prove stability, a Lyapunov function candidate is used for each component. As an example, consider the first joint controller

$$V_1 = \frac{1}{2} s_1^2 \tag{12.2.16}$$

The time derivative along the system dynamics (12.2.13) to (12.2.14) under control (12.2.15) is given by

$$\begin{aligned} \dot{V}_1 &= s_1 \dot{s}_1 \\ &= s_1(c_1 \dot{q}_{e_1} + \ddot{q}_{d_1} - \ddot{q}_1) \\ &= s_1\left(-\frac{\tau_{0_1} \text{ sign } s_1}{m_{11}} + c_1 \dot{q}_{e_1} + \ddot{q}_{d_1} + \frac{n_1}{m_{11}} + \frac{m_{21}}{m_{11}}\ddot{q}_2\right) \end{aligned} \tag{12.2.17}$$

The requirement $\dot{V}_1 < -\xi_1 |s_1|$ leads to a condition for the required control resources τ_{0_1} as

$$\tau_{0_1} \geq m_{11}^+ \left(c_1 |\dot{q}_{e_1}| + |\ddot{q}_{d_1}| + \frac{n_1^+}{m_{11}^-} + \frac{m_{21}^+}{m_{11}^-} |\ddot{q}_2| \right) + \xi_1 \qquad (\xi_1 > 0) \tag{12.2.18}$$

The parameter bounds can be found from Tables 12.2 and 12.4. The second joint controller can be treated in a similar manner. Often the link accelerations $\ddot{q}_i$ are not available. In this case the desired link accelerations $\ddot{q}_{d_i}$ can be substituted under the assumption of close tracking.

For robots with on-off control inputs, i.e. control torques which may only take two values $\tau_i = -\tau_{0_i}$ or $\tau_i = +\tau_{0_i}$, the maximum value of (12.2.18) over the whole desired trajectory in the interval $0 \leq t \leq t_f$ should be taken. If the required control resources calculated from (12.2.18) exceed the actual control resources of a given robot, the desired trajectory should be modified to decrease the required joint accelerations. Also, since the starting point of the robot manipulator often coincides with the starting

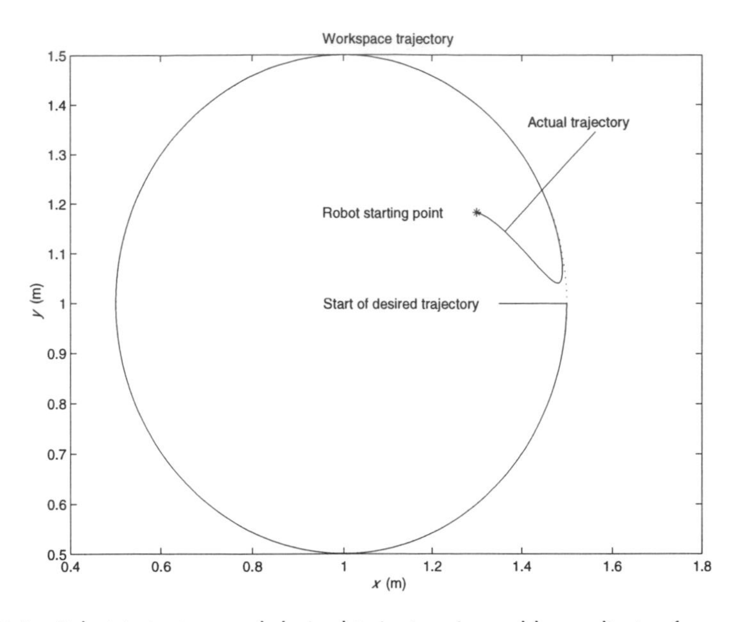

Figure 12.4 Robot trajectory and desired trajectory in world coordinates for componentwise control design.

point of the desired trajectory, the initial errors are zero and sliding mode may occur immediately at the start. Thus, all errors and their derivatives are zero throughout the entire operation.

Figures 12.4 and 12.5 show the time trajectory in the robot workspace, the distances to the sliding manifolds and the required control resources for the circular trajectory (12.2.2). The initial conditions were chosen to be nonzero for illustration purposes. Convergence to the sliding manifolds in finite time is illustrated by Figure 12.5(a). Note the difference in the required control resources for the two joints in Figure 12.5(b).

12.2.2 Vector control

Componentwise control is most suitable for systems with dominant terms in the diagonal of the mass matrix $M(q)$ and with truly discontinuous control inputs, i.e. inputs τ_i which may only take two values $\tau_i = -\tau_{0_i}$ and $\tau_i = +\tau_{0_i}$. For other systems, in particular for those with control inputs $-\tau_{0_i} \leq \tau_i \leq +\tau_{0_i}$, vector control is an elegant alternative.

In contrast to the set of n one-dimensional sliding manifolds $s_i = 0$ for componentwise control design, vector control is based on a single n-dimensional vector sliding manifold

$$s = Cq_e + \dot{q}_e \qquad (s \in \Re^{n\times 1}) \tag{12.2.19}$$

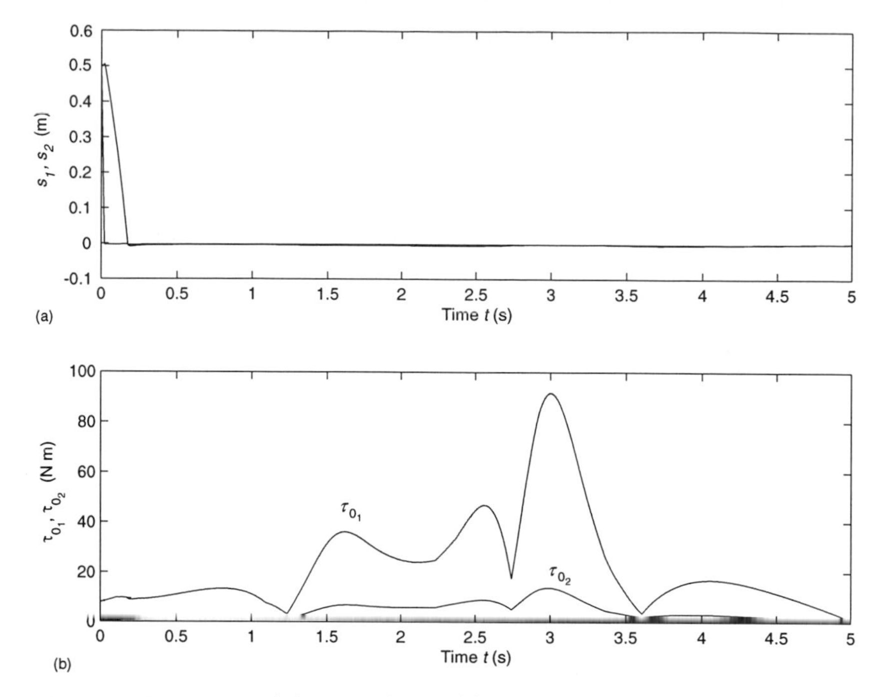

Figure 12.5 (a) Distance to sliding manifolds $s_1(t)$ and $s_2(t)$. (b) Required control resources τ_{0_1} and τ_{0_2} for each joint.

where $C \in \Re^{n \times n}$ is a Hurwitz and preferably diagonal gain matrix, $q_e(t) = q_d(t) - q(t)$ is the tracking error vector, and input vector τ is defined as

$$\tau = \tau_0 \frac{s}{\|s\|_2} \qquad (\tau \in \Re^{n \times 1}) \tag{12.2.20}$$

Sliding mode only occurs when all components of s in (12.2.19) are equal to zero instead of occurring in each component separately as for componentwise control. Likewise, $s = 0_{n \times 1}$ is the only discontinuity in control (12.2.20), whereas (12.2.4) features n control discontinuities in vector $\tau = [\tau_1, \tau_2, \ldots, \tau_n] \in \Re^{n \times 1}$ in (12.2.4). Note that the control vector always has length τ_0, which led to the name *unit control* for this approach (Ryan and Corless, 1984; Dorling and Zinober, 1986). The stability analysis in the following theorem is also vector-based.

Theorem 12.2

The system (12.2.21) with bounds (12.1.5), (12.1.9), (12.1.10) and (12.1.11) under control (12.2.23) will reach the sliding manifold (12.2.22) in finite time:

System $$\ddot{q} = M^{-1}(\tau - N(q, \dot{q})) \tag{12.2.21}$$

Manifold $$s = Cq_e + \dot{q}_e = C(q_d - q) + (\dot{q}_d - \dot{q}) = 0 \tag{12.2.22}$$

Control $$\tau = \tau_0 \frac{s}{\|s\|_2} \tag{12.2.23}$$

□

Proof

Consider the Lyapunov function candidate

$$V = \frac{1}{2} s^T s \tag{12.2.24}$$

with its derivative along the system trajectories (12.2.21) under control (12.2.23):

$$\begin{aligned} \dot{V} &= s^T\left(C\dot{q}_e + \ddot{q}_d - M^{-1}(q)\left(\tau_0 \frac{s}{\|s\|_2} - N(q,\dot{q})\right)\right) \\ &\leq -\frac{\tau_0}{M^+}\|s\| + \|s\|\left(C\|\dot{q}_e\| + \frac{N^+}{M^-} + \|\ddot{q}_d\|\right). \end{aligned} \tag{12.2.25}$$

Using the boundedness assumptions on (12.2.1) and using (12.1.5), (12.1.9), (12.1.10) and (12.1.11), we can find a sufficiently large τ_0 to guarantee

$$\dot{V} \leq -\xi\|s\|_2 \tag{12.2.26}$$

and henceforth finite converge to the manifold $s = 0_{n\times 1}$ in (12.2.22). ∎

In comparison to componentwise control design as in Theorem 12.1, vector control exploits matrix–vector norms for (12.1.5), (12.1.9), (12.1.10) and (12.1.11) in the stability analysis rather than single parameter bounds. Although matrix–vector norms yield a more concise and therefore more elegant mathematical formulation, they are known to be conservative and may result in overestimation of the required control resources. In particular, since $\tau_0 \geq \tau_{0i}$, $\forall i = 1, \ldots, n$ in (12.2.8) and (12.2.23), all components τ_i of the control input vector τ in (12.2.21) are required at least control resources with torques τ_0. For mechanical systems with inhomogeneous masses and actuators, such a requirement may considerably exceed the system capabilities, so componentwise control design is advisable. An example is a multilink robot manipulator having heavy base links with strong joint actuators, but smaller links near the end effector with preferably less powerful actuators.

Example 12.4 Vector control of two-link manipulator

Adaptation of the control design in Theorem 12.2 to the planar two-link manipulator example of (12.1.14) yields a sliding manifold

$$s = \begin{bmatrix} c_1 & 0 \\ 0 & c_2 \end{bmatrix}\begin{bmatrix} q_{e_1} \\ q_{e_2} \end{bmatrix} + \begin{bmatrix} \dot{q}_{e_1} \\ \dot{q}_{e_2} \end{bmatrix} \tag{12.2.27}$$

which in fact is similar to the componentwise control design (12.2.14) for this choice of gain matrix

$$C = \begin{bmatrix} c_1 & 0 \\ 0 & c_2 \end{bmatrix}$$

The difference in control design becomes apparent when defining the vector controller according to (12.2.23) as

$$\tau = \tau_0 \frac{s}{\|s\|_2} = \frac{\tau_0}{[(c_1 q_{e_1} + \dot{q}_{e_1})^2 + (c_2 q_{e_2} + \dot{q}_{e_2})^2]^{1/2}} \left(\begin{bmatrix} c_1 & 0 \\ 0 & c_2 \end{bmatrix} \begin{bmatrix} q_{e_1} \\ q_{e_2} \end{bmatrix} + \begin{bmatrix} \dot{q}_{e_1} \\ \dot{q}_{e_2} \end{bmatrix} \right) \tag{12.2.28}$$

compared to (12.2.15).

Stability can be established using Lyapunov function candidate

$$V = \frac{1}{2} s^T s = \frac{1}{2}(s_1^2 + s_2^2) \tag{12.2.29}$$

which also differs from the componentwise Lyapunov function (12.2.16). Differentiation along the system trajectories (12.1.14) with manifold (12.2.27) and under control (12.2.28) yields

$$\begin{aligned} \dot{V} &= s_1 \dot{s}_1 + s_2 \dot{s}_2 \\ &= -\frac{\tau_0}{M^+} \|s\| + \|s\| \left(\sqrt{c_1^2 \dot{q}_{e_1}^2 + c_1^2 \dot{q}_{e_1}^2} + \frac{N^+}{M^-} + \sqrt{\ddot{q}_1^2 + \ddot{q}_2^2} \right) \end{aligned} \tag{12.2.30}$$

Substituting the matrix bounds of the example in Section 12.1.2 leads to an upper bound on the required control resources as

$$\tau_0 \geq M^+ \left(\sqrt{c_1^2 \dot{q}_{e_1}^2 + c_1^2 \dot{q}_{e_1}^2} + \frac{N^+}{M^-} + \sqrt{\ddot{q}_1^2 + \ddot{q}_2^2} \right) + \xi \qquad (\xi > 0) \tag{12.2.31}$$

Again, the desired link acceleration $\|q_d\| = \sqrt{\ddot{q}_{d_1}^2 + \ddot{q}_{d_2}^2}$ may be substituted for the actual link acceleration $\|q\| = \sqrt{\ddot{q}_1^2 + \ddot{q}_2^2}$ under the assumption of exact tracking. Note that vector control design requires *both* joints to provide resources according to (12.2.31).

Simulation results are shown in Figures 12.6 and 12.7. Since the matrix bounds are more conservative than the bounds on the single matrix elements in Table 12.2, the requirement for τ_0 is considerably larger than for τ_{0_1} and τ_{0_2} in the case of componentwise control design; compare Figures 12.5 and 12.7. Due to the increase of control resources, convergence to the sliding manifold from the initial conditions is significantly faster. Also note that sliding mode occurs simultaneously in both components of vector $s = [s_1 \quad s_2]^T$.

12.2.3 Continuous feedback/feedforward plus a discontinuity term

The two designs in Theorems 12.1 and 12.2 yield purely discontinuous controllers. All system dynamics are treated as unknown disturbances and are suppressed by the control inputs. Only bounds on the system parameters are assumed to be known and are necessary for guaranteeing stability of the design. This approach is well suited for systems with mostly unknown or highly uncertain parameters and with direct implementability of a sliding mode controller. A more complex feedback/feedforward structure is advisable when the parameters are at least partially known and in the

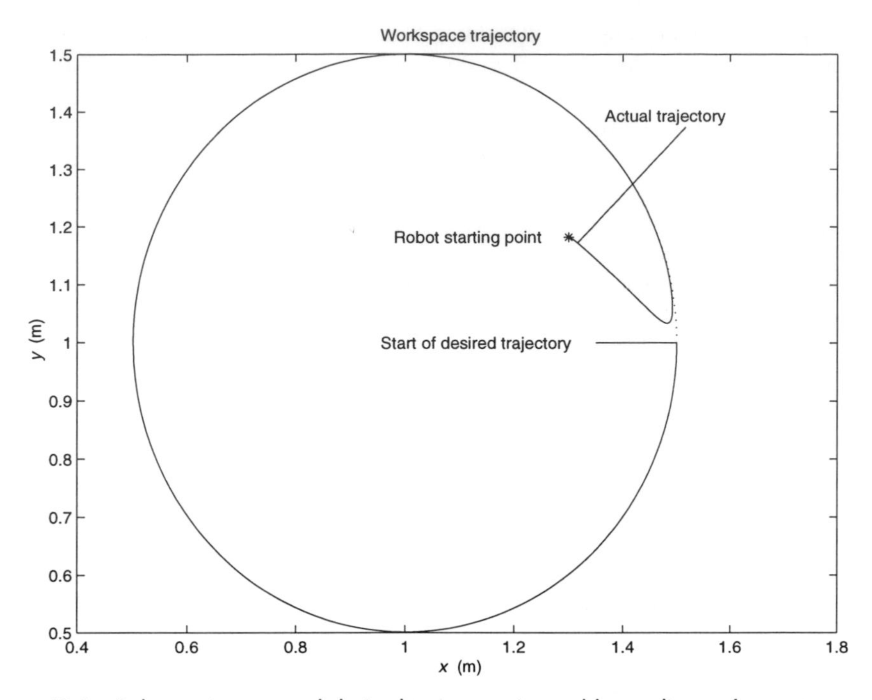

Figure 12.6 Robot trajectory and desired trajectory in world coordinates for vector control design.

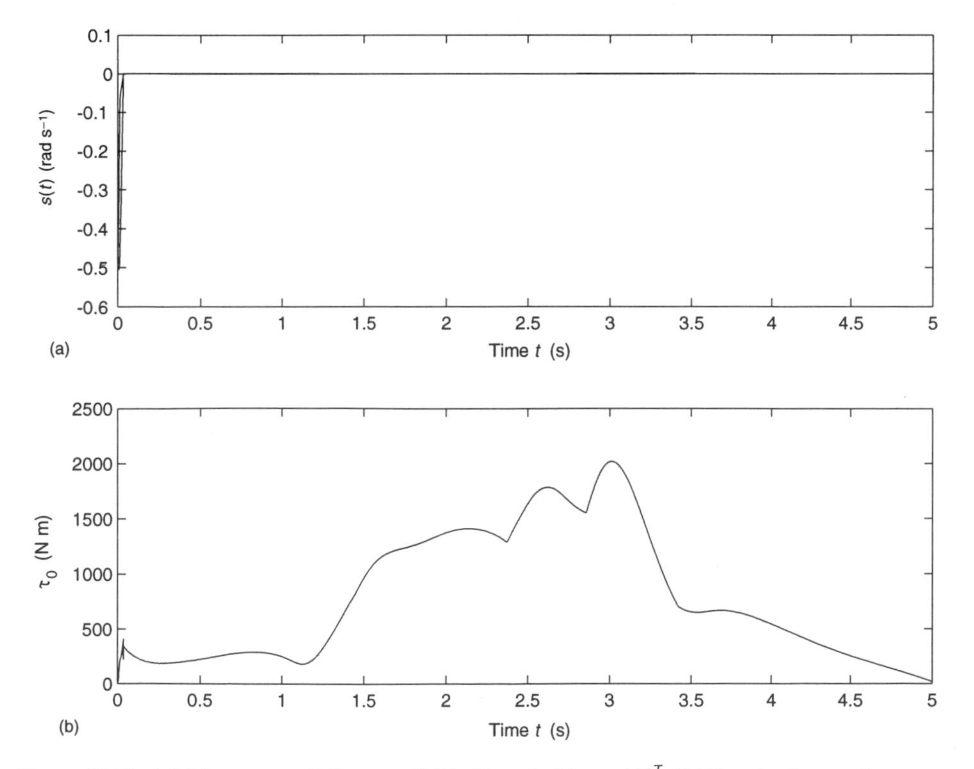

Figure 12.7 (a) Distance to sliding manifold $s(t) = [s_1(t) \quad s_2(t)]^T$. (b) Required control resources τ_0 for both joints.

presence of additional unmodelled dynamics, requiring one of the methods of Chapter 8 to prevent chattering. Another advantage of this alternative is the possibility of exploiting the physical properties of the system to benefit the control performance. In particular, for robot manipulators with dynamics described by (12.1.3), the skew symmetry property (12.1.8) can be used. Although this approach is suitable both for componentwise control and for vector control, we focus on vector control, for ease and clarity of presentation. Componentwise control is discussed by Slotine (1985) and by Chen *et al.* (1990).

Assume there exist estimates for the matrices and vectors in (12.1.3) with similar structures and estimated parameters, denoted by $\hat{M}(q)$, $\hat{V}_m(q, \dot{q})$, $\hat{F}(\dot{q})$ and $\hat{G}(q)$. In addition to (12.2.19) we define

$$\dot{\lambda} = Cq_e + \dot{q}_d = s + \dot{q} \tag{12.2.32}$$

and present the following control design.

Theorem 12.3

The system (12.2.33) with bounds (12.1.5), (12.1.9), (12.1.10) and (12.1.11) under control (12.2.35) will reach the sliding manifold (12.2.34) in finite time:

System $\quad \ddot{q} = M^{-1}(q)(\tau - V_m(q, \dot{q})\dot{q} - F(\dot{q}) - G(q)) \quad$ (12.2.33)

Manifold $\quad s = \dot{\lambda} - \dot{q} = 0 \qquad \lambda = C(q_d - q) + \dot{q}_d \quad$ (12.2.34)

Control $\quad \tau = \hat{M}(q)\ddot{\lambda} + \hat{V}_m(q, \dot{q})\dot{\lambda} + \hat{F}(\dot{q}) + \hat{G}(q) + \tau_0 \frac{s}{\|s\|_2} \quad$ (12.2.35)

□

Proof

Consider the Lyapunov function candidate

$$V = \frac{1}{2} s^T M(q) s \tag{12.2.36}$$

which is positive semidefinite with respect to q since the mass matrix $M(q)$ is positive definite. Differentiation along the system trajectories under control (12.2.35) and using the skew symmetry property (12.1.8) yields

$$\begin{aligned} \dot{V} &= s^T \left(-\tau_0 \frac{s}{\|s\|_2} + \overline{M}(q)\ddot{\lambda} + \overline{V}_m(q, \dot{q})\dot{\lambda} + \overline{F}(\dot{q}) + \overline{G}(q) \right) \\ &\leq -\tau_0 \|s\| + \|s\| (\overline{M}^+ \|\ddot{\lambda}\| + \overline{V}_m^+ \|\dot{\lambda}\| + \overline{F}_0 + \overline{F}^+ \|\dot{q}\| + \overline{G}^+) \end{aligned} \tag{12.2.37}$$

Since estimates $\hat{M}(q)$, $\hat{V}_m(q, \dot{q})$, $\hat{F}(\dot{q})$, and $\hat{G}(q)$ have similar structures as their equivalents in (12.1.3), they also fulfill the boundedness properties described in Section 12.1.2. Hence the estimation errors $\overline{M}(q) = M(q) - \hat{M}(q)$, $\overline{V}_m(q, \dot{q}) = V_m(q, \dot{q}) - \hat{V}_m(q, \dot{q})$, $\overline{F}(\dot{q}) = F(\dot{q}) - \hat{F}(\dot{q})$ and $\overline{G}(q) = G(q) - \hat{G}(q)$ may be bounded similar to (12.1.5)–(12.1.11) by appropriate scalars $\overline{M}^-$, $\overline{M}^+$, $\overline{V}_m^+$, $\overline{F}_0$, $\overline{F}^+$ and $\overline{G}^+$. For a bounded trajectory with $|q_d(t)| \leq q_d^+$, $|\dot{q}_d(t)| \leq \dot{q}_d^+$ and $|\ddot{q}_d(t)| \leq \ddot{q}_d^+$, and bounded link positions $q(t)$, velocities $\dot{q}(t)$, $\dot{\lambda}(t)$ and $\ddot{\lambda}(t)$ are also bounded and there exists a finite

$$\tau_0 > \overline{M}^+\|\ddot{\lambda}\| + \overline{V}_m^+\|\dot{\lambda}\| + \overline{F}_0 + \overline{F}^+\|\dot{q}\| + \overline{G}^+ \tag{12.2.38}$$

such that the sliding condition is fulfilled, i.e.

$$\dot{V} \leq -\xi\|s\|_2 \tag{12.2.39}$$

and it is guaranteed that the sliding manifold (12.2.34) is reached, in finite time. ■

The first four continuous terms in (12.2.35) fulfill both feedback and feedforward tasks by compensating for dynamic terms in (12.2.33) according to the current error and the desired trajectory as comprised in the variable λ of (12.2.32). Note that $\ddot{\lambda}(t)$ in controller (12.2.35) does not require measurement of link acceleration $\ddot{q}(t)$; see also definition (12.2.32). In the case of exact knowledge of all model parameters, i.e. $\overline{M}(q) = 0_{n\times n}$, $\overline{V}_m(q, \dot{q}) = 0_{n\times n}$, $\overline{F}(\dot{q}) = 0_{n\times 1}$, and $\overline{G}(q) = 0_{n\times 1}$, a small discontinuous term $\tau_0(s/\|s\|_2)$ would suffice to guarantee stability. In practice the sliding mode term also has to cope with uncertainties in the estimates $\hat{M}(q)$, $\hat{V}_m(q, \dot{q})$, $\hat{F}(\dot{q})$ and $\hat{G}(q)$, and possibly also with additive external disturbance forces/torques not accounted for in model (12.1.3).

The feedback/feedforward terms follow the ideas of feedback linearization: to obtain a basically linearized outer control loop by appropriate cancellation of all nonlinear terms in an inner control loop. In robotic applications this technique is known as 'computed torque control' (Hunt *et al.*, 1983; Gilbert and Ha, 1984). Related approaches directly utilize energy considerations derived from the Euler–Lagrange formulation (Takegaki and Arimoto, 1981) or the 'natural motion' of the mechanical system (Koditschek, 1991). An advantage over classical feedback linearization is the possibility to exploit the dynamics of the system such as the skew symmetry property of robot manipulators, rather than cancelling all nonlinear terms via feedback regardless of their possibly beneficial influence in closed loop. Various other approaches in this class of controllers have been proposed in the literature to improve the accuracy of the feedback/feedforward terms, e.g. by reducing the estimation error using adaptive or robust adaptive control. The underlying idea is to minimize the control action in the outer loop. The interested reader is encouraged to study textbooks on robot control (Craig, 1988; Spong and Vidyasagar, 1989; Lewis *et al.*, 1993) for a more detailed treatment of feedback/feedforward strategies.

Example 12.5 Feedback/feedforward control of two-link manipulator

In order to design a feedback/feedforward controller, estimates of the model parameters are required. For clarity, the parameter estimates in this example are chosen as the average of the lower and upper bounds of the model parameters as given in Table 12.2. The estimated parameters are denoted with hats; their numeric values are given in Table 12.5. Note that the estimates for Coriolis/centripetal effects are set to zero due to the contained trigonometric functions.

Table 12.5 Estimates for the matrix elements

$\hat{m}_{11} = 13\ \mathrm{kg\ m^2} quad$	$\hat{m}_{22} = 1.5\ \mathrm{kg\ m^2}$
$\hat{m}_{12} = \hat{m}_{21} = 1.5\ \mathrm{kg\ m^2}$	$\hat{V}_m = 0_{2\times 2}$

Furthermore, auxiliary variable $\dot{\lambda}(t)$ needs to be calculated in addition to sliding manifold (12.2.34):

$$\dot{\lambda} = \begin{bmatrix} c_1 & 0 \\ 0 & c_2 \end{bmatrix} \begin{bmatrix} q_{d_1} & -q_1 \\ q_{d_2} & -q_2 \end{bmatrix} + \begin{bmatrix} \dot{q}_{d_1} \\ \dot{q}_{d_2} \end{bmatrix} \tag{12.2.40}$$

Given the parameter estimates in Table 12.5 and definition (12.2.40), a control vector according to (12.2.35) can be defined as

$$\tau = \begin{bmatrix} \hat{m}_{11} & \hat{m}_{12} \\ \hat{m}_{21} & \hat{m}_{22} \end{bmatrix} \begin{bmatrix} \ddot{\lambda}_1 \\ \ddot{\lambda}_2 \end{bmatrix} + \hat{V}_m(q, \dot{q}) \begin{bmatrix} \dot{\lambda}_1 \\ \dot{\lambda}_2 \end{bmatrix} + \tau_0 \frac{s}{\|s\|_2} \tag{12.2.41}$$

Note that since the example manipulator operates in the plane without the influence of gravity and since no friction was included in model (12.1.14), $N(q, \dot{q}) = V_m(q, \dot{q})\dot{q}$ holds.

The stability proof exploits Lyapunov function candidate (12.2.36) with time derivative (12.2.37). Matrix bounds for the estimation errors are obtained in a similar manner as in Example 12.1. For example, the estimation error of the mass matrix, $\overline{M} = M - \hat{M}$, can be bounded by $\overline{M}^- = 3.177\ \mathrm{kg\ m^2}$ and $\overline{M}^+ = 5.323\ \mathrm{kg\ m^2}$. In particular, the value for the upper bound is considerably smaller than the corresponding value of $M^+ = 205.511\ \mathrm{kg\ m^2}$. Consequently, the amplitude of the control discontinuity term, τ_0 in (12.2.41), can be chosen significantly smaller than in (12.2.28).

Figures 12.8 and 12.9 illustrate the performance of the continuous feedback/feedforward controller with a small discontinuity term in a similar manner as in Examples 12.3 and 12.4. However, since the continuous feedback/feedforward part also requires control resources, the bottom graph of Figure 12.9 depicts a lowpass-filtered average of the total control torques τ_1 and τ_2. The simulation example suggests that the required control resources have the same order of magnitude as for componentwise control in Section 12.2.1. Note that convergence to the sliding manifold is slower than in the previous examples, but both components reach the manifold simultaneously due to the unit vector characteristic of the discontinuity term in (12.2.41).

12.2.4 Comparing the design choices

The previous three sections presented three control design alternatives for mechanical systems based on two choices: componentwise control or vector control, and purely discontinuous control versus continuous feedback/feedforward control with an additional discontinuity term. The fourth possible control alternative, continuous feedback/feedforward control with an additional, componentwise discontinuity term, is omitted here for brevity. The interested reader is encouraged to study a simple design example like the planar two-link manipulator given in Section 12.1.2 for this combination of choices.

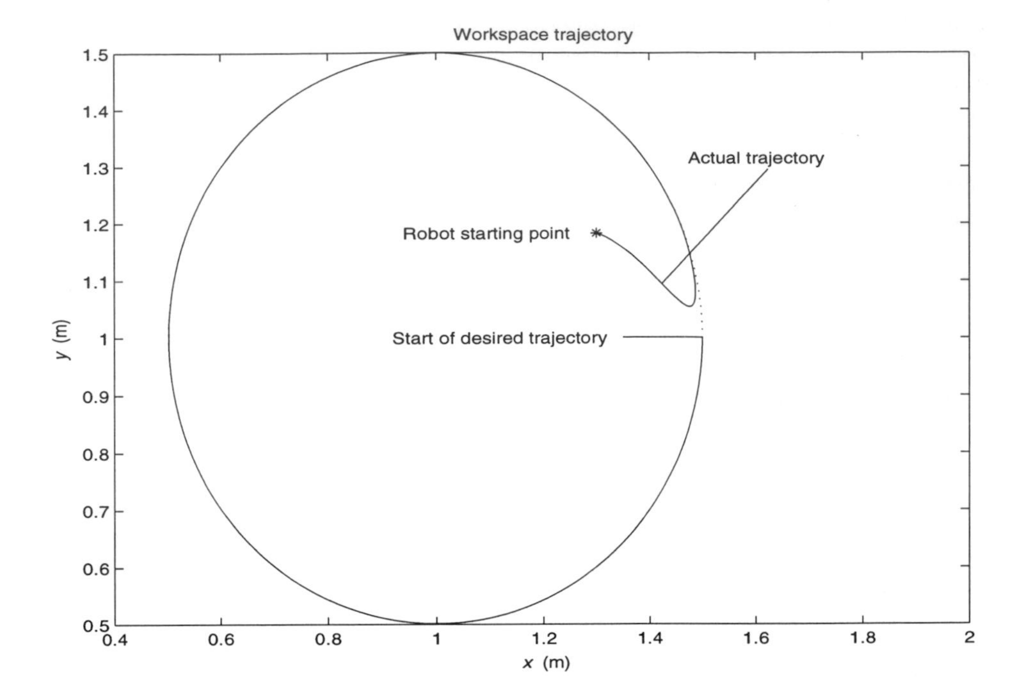

Figure 12.8 Robot trajectory and desired trajectory in world coordinates for continuous feedback/feedforward control design with additional discontinuity term.

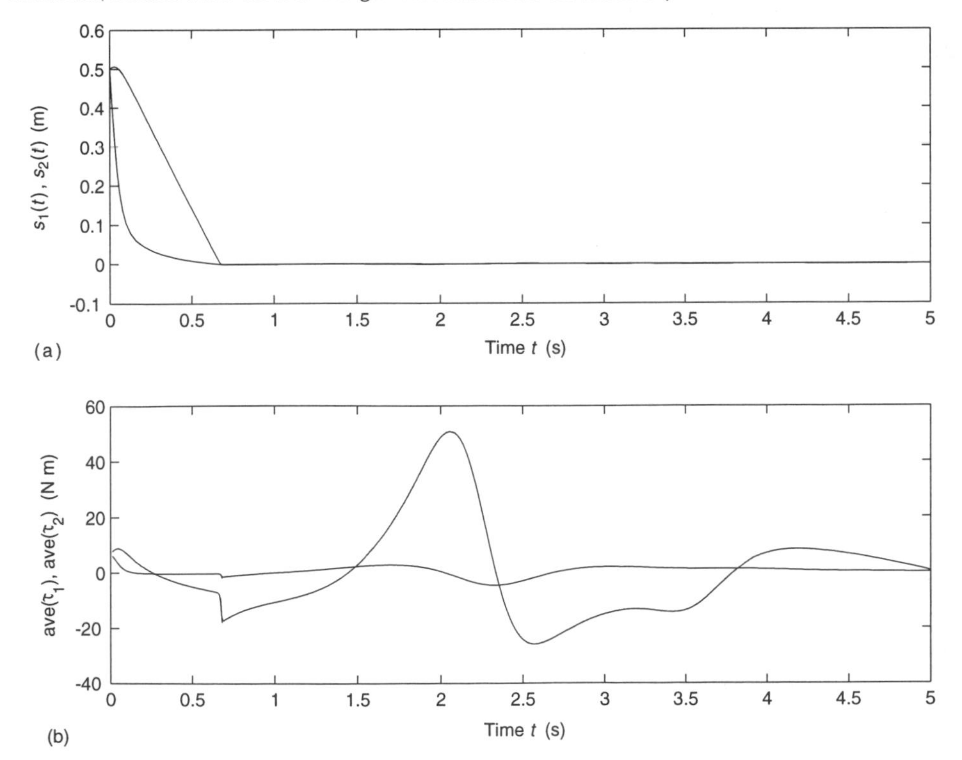

Figure 12.9 (a) Distances to sliding manifolds $s_1(t)$ and $s_2(t)$. (b) Required average control resources τ_1 and τ_2 for each joint.

All four control design alternatives have one feature in common: they enforce sliding mode in some manifold and thus achieve exact tracking of the desired trajectory. Besides the differences in control design methodologies, the four alternatives differ in the required control resources for each joint. Componentwise control seeks to determine the necessary resources for each joint but requires tedious derivations; vector control is more concise at the expense of higher control resource requirements.

In all four alternatives the control discontinuities are employed to suppress model uncertainties and, in practice, external disturbances. Continuous feedback/feedforward control reduces the amount of model uncertainty to be suppressed by the control discontinuity. However, when choosing between componentwise control and vector control, the uncertainty to be suppressed remains the same, and the difference in amplitude of the discontinuity terms arises from the stability proof rather than from system requirements.

For ideal sliding mode control systems with direct implementation of the discontinuity term, purely discontinuous control as in Theorem 12.2 is equivalent to the continuous feedback/feedforward control with additional discontinuous control in Theorem 12.3. After reaching the sliding manifold, the motions of the two systems are identical. This can be established using the equivalent control method discussed in Section 2.3. In fact, the equivalent control torques during sliding mode are similar for all four alternatives; on average, all the control algorithms apply the necessary torques to follow the desired trajectory. Higher amplitudes of the discontinuity only shorten the transient phase for approaching the sliding manifold.

For sliding mode designs in systems with unmodelled dynamics, the feedback/feedforward structure enables 'smaller' sliding gains τ_0 to be employed. Recall that the sliding gain was thought to be the cause of chattering. But even with a reliable solution to the chattering problem, a similar qualitative argument holds for most chattering prevention schemes, in particular for the saturation function method in Section 8.2 and the observer-based approach in Section 8.3. These methods avoid excitation of unmodelled dynamics by replacing the infinite gain of an ideal sliding mode controller with a finite gain in the linear zone of the saturation function or in the observer loop, respectively.

For a given system, the nature of the unmodelled dynamics, especially their frequency range, determines a practical upper bound for the finite gain. This bound on the saturation feedback gain implies bounded ability to suppress uncertainties and disturbances. In general, the smaller the possible gains, the larger the errors, i.e. the larger the boundary zone introduced by the saturation function or the larger the observer errors. Consequently, feedback of estimated dynamics and feedforward of the desired trajectory improves the performance of the control system by reducing the amount of uncertainty to be suppressed by the sliding mode term. Tables 12.6 and 12.7 contrast the general choices for sliding mode control design in mechanical systems.

12.3 Gradient tracking control

Trajectory tracking control as discussed in the previous section implies an a priori trajectory design q_d to be given for the entire robot operation. Expansion of the range of robot tasks and an increase in robot autonomy created a need to generate

Table 12.6 Comparison between componentwise sliding mode control design and vector control/unit control[a]

	Componentwise control	Vector control/unit control
Preferred systems	Truly discontinuous inputs in each component which can take only two values, $-\tau_{0_i}$ or $+\tau_{0_i}$	Ability to vary inputs in a continuous range, i.e. $-\tau_{0_i} \leq \tau_i \leq \tau_{0_i}$
Sliding manifold(s)	n $(2n-1)$-dimensional manifolds $s_i = 0, i = 1, \ldots, n$ in $2n$-dimensional state space $\Re^{2n\times 1}$	One n-dimensional manifold $s = 0, s \in \Re^{n\times 1}$ for the entire system vector $q \in \Re^{n\times 1}$
Control	Discontinuous in each component, $\tau_i = \tau_{0_i}\ \mathrm{sign}(s_i), i = 1, \ldots, n$	Unit vector with one discontinuity at the origin, $\tau = \tau_0\ s/\|s\|_2$, $\tau \in \Re^{n\times 1}$
Stability analysis	Separate Lyapunov functions $V_i = \frac{1}{2}s_i^2$ for each component $i = 1, \ldots, n$; often tedious analysis of coupling terms for higher-dimensional or complex systems based on bounds for each system parameter	One Lyapunov function $V = \frac{1}{2}s^T s$ for the entire system; often conservative analysis based on matrix–vector bounds for the whole system model

[a] Equations are given for purely discontinuous control.

Table 12.7 Comparison between purely discontinuous control and continuous feedback/feedforward control with additional discontinuity term[b]

	Purely discontinuous control	Feedback/feedforward with additional discontinuity term
Preferred systems	Ideal systems with truly discontinuous inputs and no unmodelled dynamics	Mechanical systems with continuous inputs and additional unmodelled dynamics
Sliding manifold	$s = Cq_e + \dot{q}_e = 0, q_e = q_d - q$	$s = Cq_e + \dot{q}_e = 0, q_e = q_d - q$, $\dot{\lambda} = s + \dot{q}_e$
Control	$\tau = \tau_0(s/\|s\|_2)$	$\tau = \hat{M}(q)\ddot{\lambda} + \hat{V}_m(q, \dot{q})\dot{\lambda} + \hat{F}(\dot{q})$ $+\hat{G}(q) + \tau_0(s/\|s\|_2)$
Stability analysis	Straightforward Lyapunov analysis with $V = \frac{1}{2}s^T s$; discontinuous term suppresses all system dynamics	Includes system dynamics in a Lyapunov function like $V = \frac{1}{2}s^T M(q)s$; discontinuous term suppresses uncertainty in estimated feedback/feedforward terms

[b] Equations are given for the unit control approach using generic robot manipulator dynamics as an example.

trajectories on-line; for example, robots need to adjust trajectories on-line to avoid collisions with obstacles in the workspace while approaching a given goal point. The artificial potential field method is a popular tool for on-line trajectory generation with inherent collision avoidance; it was pioneered by Khatib (1986) and refined by many others. The key idea is to design an artificial potential field in the control computer with a global minimum at the goal point of the robot operation and local maxima at workspace obstacles. Robot motion is then guided by the gradient of the computer-generated artificial potential field instead of by a predetermined trajectory.

This method is an early form of virtual reality, since the potential field is generated 'artificially' in the control computer without any physical relevance. However, many approaches exploit physical phenomena to design the artificial potential field, e.g. electrostatics, fluid dynamics and stress mechanics. To avoid additional local minima away from the goal point, which may trap the robot and cause premature termination of the operation, harmonic Laplace fields are often utilized, using either the Neumann boundary conditions (n) or the Dirichlet boundary conditions.

A number of methods to guide the robot motion using artificial potential fields have been presented in the literature:

- *Coupling of the gradient of the artificial potential field into the robot dynamics via the control input.* The control force/torques are applied collinear to the gradient, implementing a dynamic relationship between the robot and its environment in the general framework of impedance control of mechanical systems (Hogan, 1985). Effectively, the robot acceleration vector $\ddot{q}(t)$ is oriented along the gradient, which obviously does not lead to tracking of the gradient due to the robot's inertial dynamics. However, safe collision avoidance can be guaranteed for potentials tending to infinity at obstacle boundaries, even for bounded actuator resources (Rimon and Koditschek, 1992).
- *Accounting for the robot dynamics and actuator limitations in the design of the potential field.* Krogh's 'generalized potential field' (Krogh, 1984) considers the *time* needed to reach the goal point or an obstacle rather than the respective distance. The control design is based on a double integrator to model a mobile robot and takes into account limited actuator resources, but no additional dynamics.
- *Using feedback for tracking the gradient of the potential field.* Khatib (1986) proposed a feedback linearization strategy, similar to the continuous feedback/feedforward control discussed in Section 12.2.3, assuming exact parametric knowledge of all robot dynamics. This rather restrictive requirement was relieved by Koditschek's (1991) 'natural motion' approach for dissipative mechanical systems and a related, energy-based approach by Takegaki and Arimoto (1981). However, in Koditschek's approach, the gradient of the artificial potential field implicitly takes into account the robot mass/inertia matrix for generating 'natural motion' and hence cannot be designed independently of the robot dynamics.
- *Using sliding mode control for exact tracking of a gradient.* Utkin *et al.* (1991) proposed to orient the robot's velocity vector along the gradient of the artificial potential field to achieve exact tracking. The method was later generalized by Guldner and Utkin (1995) and will be the basis for control design in this section. The primary advantage is the exact gradient tracking property, which allows an artificial gradient field to be designed independently.

Let the gradient of an artificial potential field $U(q)$ be denoted by

$$\boldsymbol{E}(q) = -\nabla U(q) \qquad (\boldsymbol{E}(q) \in \Re^n) \tag{12.3.1}$$

For each point $q \in \Re^n$ in the robot workspace, the gradient (12.3.1) defines a vector of a desired direction of motion. Integration along the gradient vector via

$$\frac{\partial q(t)}{\partial t} = \boldsymbol{E}(q(t)) \tag{12.3.2}$$

with integration variable t (not necessarily denoting time) yields a continuous trajectory from the starting point $q_0 \in \Re^n$, called a gradient line. A set of gradient lines is depicted in Figure 12.10 for a planar example of the harmonic dipole potential method (Guldner *et al.*, 1997) with a single obstacle being protected by a circular security zone. All gradient lines starting in the robot freespace continuously approach the goal point without entering the security zone. However, a number of gradient lines come close to the security circle, which can only result in safe operation if exact tracking is guaranteed via robust sliding mode control.

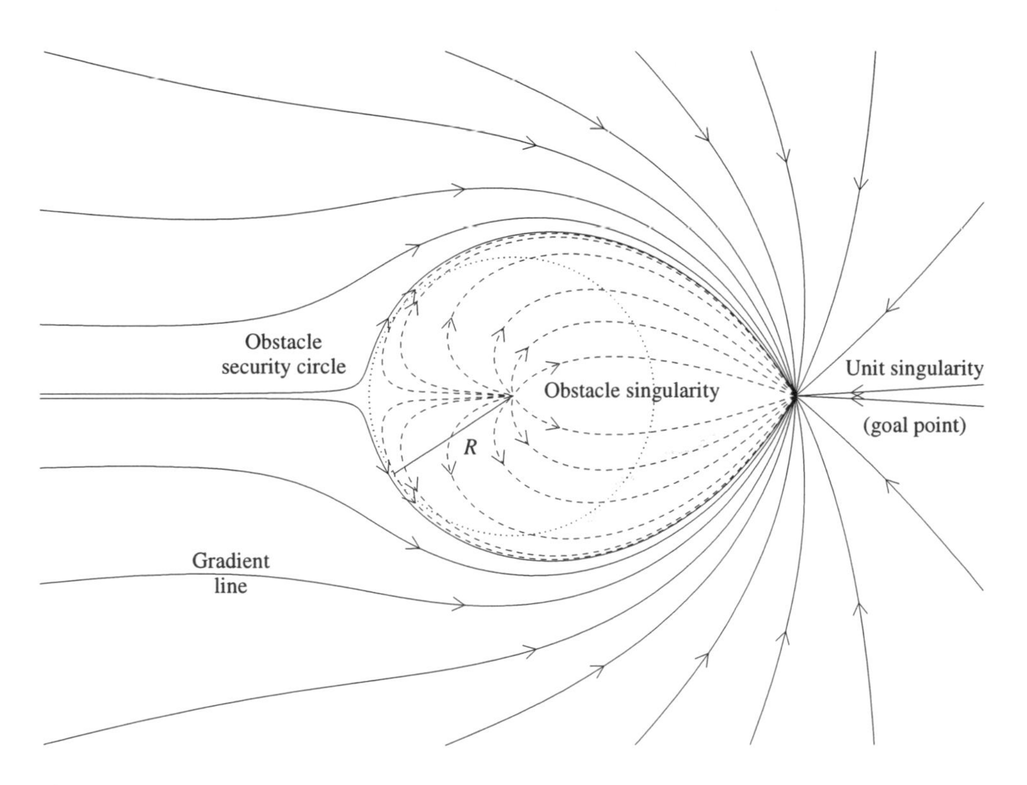

Figure 12.10 Gradient lines of harmonic dipole potential field (solid and dashed lines) with a negative unit singularity at the goal point and a positive obstacle singularity at the centre of the obstacle security circle (dotted line). For details see Sections 12.4.2 and 12.4.3.

12.3.1 Control objectives

The goal of control is to track the gradient of an arbitrary artificial potential field (12.3.1) with a robot manipulator described by (12.1.3) or a mobile robot such as (12.1.22) to (12.1.24). Exact tracking can be achieved by orienting the velocity vector $\dot{q}(t)$ of (12.1.3) collinear to the gradient vector $\boldsymbol{E}$ in (12.3.1). In order to avoid singularities at equilibrium points $\boldsymbol{E} = 0_{n\times 1}$, we define the desired motion vector $\dot{q}_d(q) \in \Re^n$ as

$$\dot{q}_d(q) = v_d(q, t)\frac{\boldsymbol{E}(q)}{\max(\|\boldsymbol{E}(q)\|, \varepsilon)} \tag{12.3.3}$$

for some small scalar $\varepsilon > 0$. The task of the controller to be designed in Section 12.3.2 is to guarantee $\dot{q}(t) = \dot{q}_d(q(t))$ at all times.

In the previous section, a sliding mode controller was designed for tracking a given trajectory. According to the characteristics of this desired trajectory, the required control resources were calculated from the stability analysis. Here we choose a different path: the control resource limitations are assumed to be given. Then the motion profile (velocity and acceleration) is derived on-line together with the desired trajectory, i.e. the gradient of the artificial potential field. The desired scalar tracking velocity $v_d(q, t)$ should therefore be adjusted to enable exact tracking with limited control resources. Given the maximum possible acceleration a_0, to be determined later from the available control resources, suitable starting and goal-approach phases are given by

$$v_d(q, t) = \min(a_0 t, v_0(q, t), \sqrt{2a_0\zeta}) \tag{12.3.4}$$

where $v_0(q, t)$ is the desired 'travelling' velocity, to be determined below, and ζ is the remaining distance to the goal point q_G, i.e.

$$\zeta = \|q(t) - q_G\| \tag{12.3.5}$$

Starting from initial time $t = 0$, the desired velocity $v_d(q, t)$ is increased using maximum acceleration a_0 until the travelling velocity $v_0(q, t)$ is reached. The required acceleration a_0 during 'normal travelling' with $v_d(q, t) = v_0(q, t)$ is found from (12.3.3) as

$$\begin{aligned}\ddot{q}_d(q, t) &= \dot{v}_0(q, t)\frac{\boldsymbol{E}(q)}{\|\boldsymbol{E}(q)\|} + v_0(q, t)\frac{\partial}{\partial q}\left(\frac{\boldsymbol{E}(q)}{\|\boldsymbol{E}(q)\|}\right)\dot{q}(t) \\ &= \left(\dot{v}_0(q, t) + v_0^2(q, t)\frac{\partial}{\partial q}\left(\frac{\boldsymbol{E}(q)}{\|\boldsymbol{E}(q)\|}\right)\right)\frac{\boldsymbol{E}(q)}{\|\boldsymbol{E}(q)\|}\end{aligned} \tag{12.3.6}$$

assuming $\|\boldsymbol{E}(q)\| > \varepsilon$ and exact tracking with $\dot{q} = v_0(q, t)(\boldsymbol{E}(q)/\|\boldsymbol{E}(q)\|)$. An upper bound for desired acceleration $\ddot{q}_d(q, t)$ in (12.3.6) is given by

$$\|\ddot{q}_d(q, t)\| = |\dot{v}_0(q, t)| + v_0^2(q, t)\left\|\frac{\partial}{\partial q}\left(\frac{\boldsymbol{E}(q)}{\|\boldsymbol{E}(q)\|}\right)\right\| \leq a_0 \tag{12.3.7}$$

For given maximum acceleration a_0, determined by the available control resources in the subsequent section, a suitable travelling velocity $v_0(q, t)$ is dynamically determined by the solution of

$$\dot{v}_0(q, t) = a_0 - v_0^2(q, t)\left\| \frac{\partial}{\partial q}\left(\frac{\boldsymbol{E}(q)}{\|\boldsymbol{E}(q)\|}\right)\right\| \tag{12.3.8}$$

Since $\boldsymbol{E}(q)$ is time independent, its derivatives with respect to vector q are known in closed form and dynamic equation (12.3.8) can be solved on-line. A similar procedure can be followed for $\|\boldsymbol{E}(q)\| \leq \varepsilon$. Note that $v_0(t)$ is automatically decreased in areas of high curvature of the gradient $\boldsymbol{E}(q)$, i.e. for large values of

$$\left\| \frac{\partial}{\partial q}\left(\frac{\boldsymbol{E}(q)}{\|\boldsymbol{E}(q)\|}\right)\right\|$$

In practice the travelling velocity $v_0(t)$ will be bounded by $0 \leq v_0(t) \leq v_{0_{\max}}$, where $v_{0_{\max}}$ is selected according to physical limitations of the robot and the maximum change of curvature of the gradient encountered for the specific robot task.

When approaching the goal point q_G, the trajectory can be approximated by the remaining scalar distance $\zeta(t)$ under the assumption of an approach with $q(t) = q_d(q(t))$ as proposed by Utkin *et al.* (1991):

$$\dot{\zeta}(t) = -\sqrt{2a_0\zeta(t)} \tag{12.3.9}$$

Using (12.3.4) and (12.3.9), a bound on scalar $\dot{v}_d(q, t)$ can be established for the goal-approach phase as

$$|\dot{v}_d(q, t)| = \left|\frac{d}{dt}\sqrt{2a_0\zeta(t)}\right| = \left|\frac{1}{2}\frac{2a_0\zeta(t)}{\sqrt{2a_0\zeta(t)}}\right| \leq a_0 \tag{12.3.10}$$

The choice of $v_d(q, t) = \sqrt{2a_0\zeta(t)}$ during the goal-approach phase also prevents overshoot at the goal point and results in a finite reaching time. By virtue of definition (12.3.5), $\zeta \geq 0$ holds at all times. Integrating (12.3.9) from $\zeta_s = \zeta(q(t_s)) = \zeta(t_s)$ at some starting position q_s at time t_s,

$$\int_{\zeta_s}^{\zeta} \frac{d\zeta}{\sqrt{2a_0\zeta}} = -\int_{t_s}^{t} dt \tag{12.3.11}$$

directly yields

$$\sqrt{2a_0\zeta} - \sqrt{2a_0\zeta_s} = -a_0(t - t_s) \tag{12.3.12}$$

Since the right-hand side of (12.3.12) is negative for $t > t_f$, $\zeta < \zeta_s$ is decreasing monotonously to reach $\zeta_f = 0$ after finite time

$$t_f = t_s + \sqrt{\frac{2\zeta_s}{a_0}} \tag{12.3.13}$$

12.3.2 Holonomic robots

Among the four choices of control design outlined for trajectory tracking control in Tables 12.6 and 12.7, purely discontinuous vector control features the most concise mathematical formulation and hence will be used here to reveal the basic structure of gradient tracking control design. The design presented below can be easily reformulated in terms of one of the three other methods described in Section 12.2.

The control objective defined in the preceding section is to guarantee $\dot{q}(t) = \dot{q}_d(t)$ given in (12.3.3) and (12.3.4) at all times. The main difference between trajectory tracking control and gradient tracking control is that gradient tracking control uses velocity vector $\dot{q} \in \Re^n$ as the control variable, rather than position vector $q \in \Re^n$. Desired velocity $\dot{q}_d(t)$ implicitly depends on position vector $q \in \Re^n$ via the gradient vector $\boldsymbol{E}(q) \in \Re^n$ and hence can be regarded as an outer control loop encompassing the robot and its environment. For gradient tracking control on a velocity scale, an n-dimensional sliding variable $s \in \Re^n$ is chosen as

$$s(\dot{q}, \dot{q}_d) = \dot{q}_d(t) - \dot{q}(t) \tag{12.3.14}$$

The following theorem exemplifies the gradient tracking control design for purely discontinuous vector control.

Theorem 12.4

The system (12.3.15) with bounds (12.1.5), (12.1.9), (12.1.10) and (12.1.11) under control (12.3.17) will reach the sliding manifolds (12.3.16) in finite time:

System $\quad \ddot{q} = M^{-1}(\tau - N(q, \dot{q}))$ (12.3.15)

Manifold $\quad s = \dot{q}_d(t) - \dot{q}(t) = 0$ (12.3.16)

$$\dot{q}_d(q) = v_d(q, t)\frac{\boldsymbol{E}(q)}{\max(\|\boldsymbol{E}(q)\|, \varepsilon)}$$

Control $\quad \tau = \tau_0 \dfrac{s}{\|s\|_2}$ (12.3.17)

□

Proof

Consider the Lyapunov function candidate

$$V = \frac{1}{2}s^T s \tag{12.3.18}$$

with the following time derivative along (12.3.15), (12.3.16) and control (12.3.17):

$$\begin{aligned} \dot{V} &= s^T(\ddot{q}_d(t) - \ddot{q}(t)) \\ &\leq -\frac{\tau_0}{M^+}\|s\| + \|s\|\left(a_0 + \frac{N^+}{M^-}\right) \end{aligned} \tag{12.3.19}$$

where bounds (12.3.7) and (12.3.10) for $\ddot{q}_d(t)$ helped to reduce (12.3.19). For given control resources τ_0, the maximum acceleration a_0 for determining travelling velocity $v_0(t)$ in (12.3.8) and the goal approach in (12.3.9) is calculated as

$$a_0 < \frac{\tau_0 - N^+}{M^-} \tag{12.3.20}$$

such that $\dot{V} \leq -\xi \|s\|_2$ is ensured. Stability and finite approach of the sliding manifold $s(\dot{q}, \dot{q}_d) = 0$ in (12.3.14) then follows along standard arguments for (12.3.19), guaranteeing exact tracking of the gradient lines. ■

12.3.3 Nonholonomic robots

Gradient tracking control for nonholonomic robots also follows the control objectives given in Section 12.3.1. As an example, we will discuss control design for a mobile robot modelled as a nonholonomic wheel set (Figure 12.2). The kinematic model is given in (12.1.21) and the dynamic model in (12.1.22), (12.1.23). The task of orienting the motion vector of the wheel set along the gradient vector according to (12.3.3) to (12.3.5) can be split into two subtasks due to the definition of the input vector u in (12.1.21); the two subtasks are orientation control using ω and velocity control using v_C. Hence componentwise control design is most suitable. The sliding manifold for velocity control along the gradient lines, the translational component, is defined as

$$s_t = v_d(q, t) - v_C = 0 \tag{12.3.21}$$

for the case $\|\boldsymbol{E}(q)\| > \varepsilon$.

The orientation error for control of the rotational component is given by

$$\phi_e = \phi_d - \phi, \; \phi_d = \text{Atan}\frac{E_y(q)}{E_x(q)} \tag{12.3.22}$$

where the arctangent function Atan(.) returns angles in all four quadrants, i.e. $\phi_d \in [-\pi, \pi[$, and $\boldsymbol{E} = [E_x(q) \quad E_y(q)]^T$ is the gradient vector in the plane. Once again, control of the robot's orientation is a position control, so the associated sliding manifold is defined as for second-order systems:

$$s_r = C\phi_e + \dot{\phi}_e = 0 \tag{12.3.23}$$

The control designs are summarized in the following theorem.

Theorem 12.5

Systems (12.3.24) and (12.3.25) with bounds $0 < M^- \le M \le M^+$, $0 < J^- \le J \le J^+$, $0 < |N_t(v_C, \omega)| \le N_t^+$ and $0 < |N_r(v_C, \omega)| \le N_r^+$ under controls (12.3.28) and (12.3.29) will reach the sliding manifolds (12.3.26) and (12.3.27) in finite time:

System
$$M\dot{v}_C + N_t(v_C, \omega) = \tau_t \tag{12.3.24}$$

$$J\dot{\omega} + N_r(v_C, \omega) = \tau_r \tag{12.3.25}$$

Manifolds
$$s_t = v_d(q, t) - v_C = 0 \tag{12.3.26}$$

$$s_r = C\phi_e + \dot{\phi}_e = 0 \qquad \phi_e = \phi_d - \phi \qquad \phi = \text{Atan}\frac{E_y}{E_x} \tag{12.3.27}$$

Control
$$\tau_t = \tau_{t0} \text{ sign } s_t \tag{12.3.28}$$

$$\tau_r = \tau_0 \text{ sign } s_r \tag{12.3.29}$$

□

Proof

Consider the Lyapunov function candidates

$$V_t = \frac{1}{2} s_t^T s_t \tag{12.3.30}$$

and

$$V_r = \frac{1}{2} s_r^T s_r \tag{12.3.31}$$

Differentiation of (12.3.30) along (12.3.24) and (12.3.26) under control (12.3.28) with the above bounds yields

$$\begin{aligned} \dot{V}_t &= s\left(\dot{v}_d(t) + \frac{N_t(v_C, \omega)}{M} - \frac{\tau_t}{M}\right) \\ &\le |s|\left(a_0 + \frac{N_t^+}{M^-}\right) - |s|\frac{\tau_{t0}}{M^+} \end{aligned} \tag{12.3.32}$$

where (12.3.7) was used to reduce the expression. Stability follows from (12.3.32) for sufficiently large τ_{t0} and a suitable choice of a_0.

Similarly, stability of the orientation controller with manifold (12.3.27) is shown by differentiation of (12.3.31) with system (12.3.25) under control (12.3.29):

$$\begin{aligned} \dot{V}_t &= s\left(C\dot{\phi}_e + \ddot{\phi}_d + \frac{N_r(v_C, \omega)}{J} - \frac{\tau_r}{J}\right) \\ &\le |s|\left(C|\dot{\phi}_e| + |\ddot{\phi}_d| + \frac{N_r^+}{J^-}\right) - |s|\frac{\tau_{r0}}{J^+} \end{aligned} \tag{12.3.33}$$

Boundedness of $\dot{\phi}_e = \dot{\phi}_d - \omega$ and $\ddot{\phi}_d$ is established for $\|\boldsymbol{E}\| > \varepsilon$ due to

$$\dot{\phi}_d = \frac{d}{dt}\left(\text{Atan}\frac{E_y(q)}{E_x(q)}\right) = H_0 v_C \tag{12.3.34}$$

$$\ddot{\phi}_d = \frac{d^2}{dt^2}\left(\text{Atan}\frac{E_y(q)}{E_x(q)}\right) = H_1 + \frac{H_0}{M}\tau_t \tag{12.3.35}$$

where

$$\begin{aligned}
H_0 &= \frac{E_x^2[D_x + D_y](E_y) - E_xE_y[D_x + D_y](E_x)}{\|\boldsymbol{E}\|^2} \\
H_1 &= v_C^2 H_2 - \frac{N_t(v_C, \omega)}{M} H_0 \\
H_2 &= E_x\left\{\frac{1}{\|\boldsymbol{E}\|^4}[2E_x[D_x](E_x)[D_x + D_y](E_y) + E_x^2[D_{xx} + D_{xy}](E_y)\right. \\
&\quad - E_y[D_x^2 + D_xD_y](E_x) - E_x[D_x](E_y)[D_x + D_y](E_x) - E_xE_y[D_{xx} + D_{xy}](E_x)] \\
&\quad \left. - \frac{3}{\|\boldsymbol{E}\|^6}[E_x^2[D_x + D_y](E_y) - E_xE_y[D_x + D_y](E_x)][E_x[D_x](E_x) + E_y[D_x](E_y)]\right\} \\
&\quad + E_y\left\{\frac{1}{\|\boldsymbol{E}\|^4}[2E_x[D_y](E_x)[D_x + D_y](E_y) + E_x^2[D_{xy} + D_{yy}](E_y)\right. \\
&\quad - E_y[D_xD_y + D_y^2](E_x) - E_x[D_y](E_y)[D_x + D_y](E_x) - E_xE_y[D_{xy} + D_{yy}](E_x)] \\
&\quad \left. - \frac{3}{\|\boldsymbol{E}\|^6}[E_x^2[D_x + D_y](E_y) - E_xE_y[D_x + D_y](E_x)][E_x[D_y](E_x) + E_y[D_y](E_y)]\right\}.
\end{aligned} \tag{12.3.36}$$

$[D_x]$ denotes the first-order differential operator $\partial/\partial x$ and $[D_{xy}]$ denotes the second-order differential operator $\partial^2/\partial x\, \partial y$. Sliding mode was assumed to exist in the translatory manifold (12.3.26), enabling the equivalent control methodology for τ_t to be used.

Solving $\dot{s}_t = 0$ for the translational control input yields

$$(\tau_t)_{eq} = N_t(v_C, \omega) + M\dot{v}_d(t) \tag{12.3.37}$$

which was substituted for τ_t into (12.3.34) to reduce the expression. Bounds for the above terms can be determined to calculate the necessary control input τ_{r0} for (12.3.29). ■

Two application examples for gradient tracking control can be found in Sections 12.4.2 and 12.4.3.

12.4 Application examples

In this section we present a variety of examples of the sliding mode philosophy in advanced robotic systems. First, torque control of a flexible robot joint using integral sliding mode is presented. Next we discuss the application of the gradient tracking control derived in Section 12.3 for collision avoidance of mobile robots and robot manipulators in known workspaces. Finally, we introduce fully automatic steering control for passenger cars as an extension of robotic systems.

12.4.1 Torque control for flexible robot joints

In many practical applications, robot joints and robot links exhibit flexibilities which need to be considered when designing advanced control algorithms. As an example (Shi and Lu, 1996), consider a single revolute link with a flexible joint, i.e. a robot consisting of just one link actuated by one drive with flexibilities in the joint (Figure 12.11). Extension of this model to multilink robot arms follows along the lines of (12.1.3), but is omitted here for ease of notation. Neglecting actuator dynamics, a flexible joint model can be derived as

$$\begin{aligned} J_l \ddot{q}_l + d_l &= \tau_l \\ J_m \ddot{q}_m + d_m &= \tau_m - \frac{\tau_l}{n} \end{aligned} \tag{12.4.1}$$

where $J_l^- \leq J_l \leq J_l^+$ is the link inertia, q_l is the joint angular position, n is the gear ratio, $J_m^- \leq J_m \leq J_m^+$ is the motor/gear inertia, q_m is the angular motor position, $k^- \leq k \leq k^+$ denotes the joint stiffness, $\tau_l = k(q_m/n - q_l)$ is the link torque at the joint, and the motor torque τ_m represents the control input. Also, unknown but bounded torque disturbances $|d_l| \leq d_l^+$ and $|d_m| \leq d_m^+$ were added to (12.4.1) at the link and motor sides, respectively.

Since we have already introduced trajectory tracking control (position scale) in Section 12.2 and gradient tracking control (velocity scale) in Section 12.3, we will study torque tracking control (acceleration scale) in this example, i.e. the problem of controlling link torque τ_l to track a known desired profile $[\tau_d, \dot{\tau}_d, \ddot{\tau}_d]$.

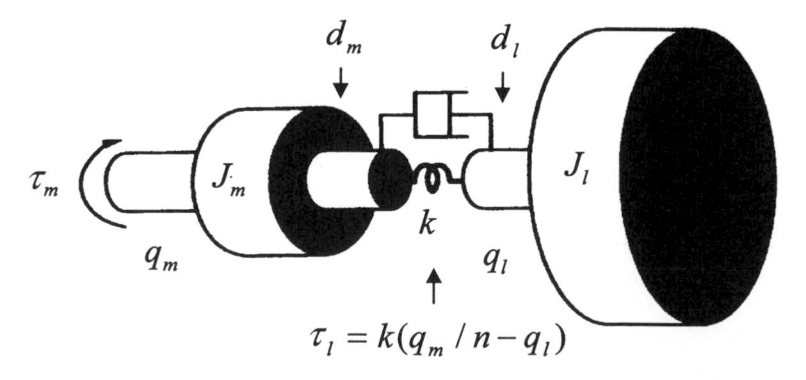

Figure 12.11 Definition of variables for flexible joint control problem.

A linear input/output representation of (12.4.1) is given by

$$\ddot{\tau}_l + a\tau_l + d = b\tau_m$$
$$\text{with} \quad a = \frac{k}{n^2 J_m} + \frac{k}{nJ_l} \qquad b = \frac{k}{nJ_m} \qquad d = \frac{k}{nJ_m} d_m + \frac{k}{J_l} d_l \tag{12.4.2}$$

This example will exploit both the continuous feedback/feedforward technique discussed in Section 12.2.3 and integral sliding mode control presented in Chapter 7 to reject the unknown disturbance $|d| < D^+$ and uncertainties in a and b. The control law for τ_m is therefore composed of a continuous controller of feedback linearizing type with pole placement τ_f, and a disturbance rejecting controller τ_r, i.e.

$$\tau_m = \tau_f + \tau_r \tag{12.4.3}$$

Define a continuous feedback/feedforward controller as

$$\tau_f = \hat{b}^{-1}(\hat{a}\tau_l + \ddot{\tau}_d + c_1\dot{\tau}_e + c_0\tau_e) \tag{12.4.4}$$

where $\hat{a}$ and $\hat{b}$ are estimates of a and b, respectively. The first term $(\hat{a}/\hat{b})\tau_l$ in (12.4.4) seeks to compensate the term $a\tau_l$ on the left-hand side of (12.4.2) as a type of feedback linearization. The second term in (12.4.4), $\ddot{\tau}_d/\hat{b}$, is the acceleration of the desired torque; it is a feedforward term. Finally, the last two terms in (12.4.4), $(c_1\dot{\tau}_e + c_0\tau_e)/\hat{b}$, are a linear pole placement controller. Substitution of (12.4.4) into (12.4.2) yields error dynamics for $\tau_e = \tau_d - \tau_l$, $\dot{\tau}_e = \dot{\tau}_d - \dot{\tau}_l$ and $\ddot{\tau}_e = \ddot{\tau}_d - \ddot{\tau}_l$ as

$$\ddot{\tau}_e + c_1\dot{\tau}_e + c_0\tau_e = \tau_p - \hat{b}\tau_r \tag{12.4.5}$$

The left-hand side of equation (12.4.5) is linear with poles determined by $c_1, c_0 > 0$, but is subject to the perturbation torque

$$\tau_p = \left(\frac{\hat{b}}{b} - 1\right)\ddot{\tau}_l + \left(\frac{\hat{b}}{b}a - \hat{a}\right)\tau_l + \frac{\hat{b}}{b}d + (b - \hat{b})\tau_r \tag{12.4.6}$$

which may be simplified by substitution of $\ddot{\tau}_l = b\tau_m - a\tau_l - d$ from (12.4.2) to

$$\tau_p = (a - \hat{a})\tau_l + (\hat{b} - b)\tau_m + d \tag{12.4.7}$$

In order to improve the control performance, the disturbance (12.4.7) should be compensated by the additional disturbance rejection term τ_r in (12.4.3). Since τ_p is not measurable, an estimate is obtained using a sliding mode observer of the form

$$\dot{\hat{z}} = \ddot{\tau}_d + c_1\dot{\tau}_e + c_0\tau_e + \hat{b}\tau_r - u \tag{12.4.8}$$

with $\hat{z}$ being an estimate for $\dot{\tau}_l$. Basically, observer (12.4.8) is a copy of (12.4.5) with $\ddot{\tau}_l = \dot{\hat{z}}$ and observer feedback u as a replacement for τ_p being defined as

$$u = (\bar{a}^+|\tau_l| + \bar{b}^+|\tau_m| + \bar{d}^+) \operatorname{sign} \bar{z} \tag{12.4.9}$$

Here $\bar{a}^+$, $\bar{b}^+$ and $\bar{d}^+$ denote upper bounds for $\bar{a} = a - \hat{a}$, $\bar{b} = b - \hat{b}$ and $\bar{d} = d - \hat{d}$, respectively, obtained from the bounds on the system parameters given above, e.g.

$$\bar{a}^+ = \max\left\{\left(\frac{k^+}{n^2 J_m^-} + \frac{k^+}{nJ_l^-}\right) - \hat{a}, \hat{a} - \left(\frac{k^-}{n^2 J_m^+} + \frac{k^-}{nJ_l^+}\right)\right\} \tag{12.4.10}$$

The control discontinuity is introduced along the observation error $\bar{z} = \hat{z} - \dot{\tau}_l$. Stability of the observer system is ensured via Lyapunov function candidate $V = \frac{1}{2}\bar{z}^2$; differentiation along the system trajectories (12.4.8) with control (12.4.9) yields

$$\begin{aligned} \dot{V} &= (\ddot{\tau}_d + c_1\dot{\tau}_e + c_0\tau_e + \hat{b}\tau_r - (\bar{a}^+|\tau_l| + \bar{b}^+|\tau_m| + \bar{d}^+)\ \text{sign}\ \bar{z} - \ddot{\tau}_l)\bar{z} \\ &= \tau_p\bar{z} - \tau_p^+|\bar{z}| \end{aligned} \tag{12.4.11}$$

where $\tau_p^+ = \bar{a}^+|\tau_l| + \bar{b}^+|\tau_m| + \bar{d}^+ + \xi > \max(\tau_p)$ ensures the existence of sliding mode via

$$\dot{V} < -\xi|\bar{z}| \tag{12.4.12}$$

for some small scalar $\xi > 0$.

Since (12.4.8) is generated in the control computer, the initial conditions can be set as

$$\hat{z}|_{t=0} = \dot{\tau}_l|_{t=0} \tag{12.4.13}$$

such that $\bar{z} = 0$ for all $t \geq 0$, i.e. sliding mode is initiated immediately at $t = 0$. In order to estimate the disturbance torque τ_p to be compensated via disturbance rejection term τ_r in (12.4.3), the equivalent control method (Section 2.3) is exploited. In sliding mode, $\dot{\bar{z}} = \dot{\hat{z}} - \ddot{\tau}_l = 0$, i.e.

$$\dot{\bar{z}} = \ddot{\tau}_e + c_1\dot{\tau}_e + c_0\tau_e + \hat{b}\tau_r - u = 0 \tag{12.4.14}$$

The control signal u in (12.4.9) contains two components: a high-frequency switching component resulting from the discontinuous sign term and a low-frequency component, i.e. the equivalent control u_{eq}. As discussed in detail in Utkin (1992), the equivalent control is equal to the average of u, perhaps obtained by a lowpass filter. With this in mind, $u_{eq} = u_{ave} = \tau_p$ follows from comparing (12.4.7) and (12.4.14). Consequently, defining

$$\tau_r = \hat{b}u_{eq} = \hat{b}u_{ave} \tag{12.4.15}$$

leads to closed-loop error dynamics

$$\ddot{\tau}_e + c_1\dot{\tau}_e + c_0\tau_e = 0 \tag{12.4.16}$$

Hence the sliding mode estimator τ_r successfully rejects the uncertainties in parameters a and b, as well as additive disturbance d in (12.4.2), and it allows controller (12.4.4) to perform exact pole placement. Note that the lowpass filter time constant or bandwidth has to be carefully chosen to be faster than the perturbation dynamics (12.4.7) for successful rejection of τ_p, but slow enough not to excite unmodelled dynamics in the system like the neglected actuator dynamics producing torque τ_m.

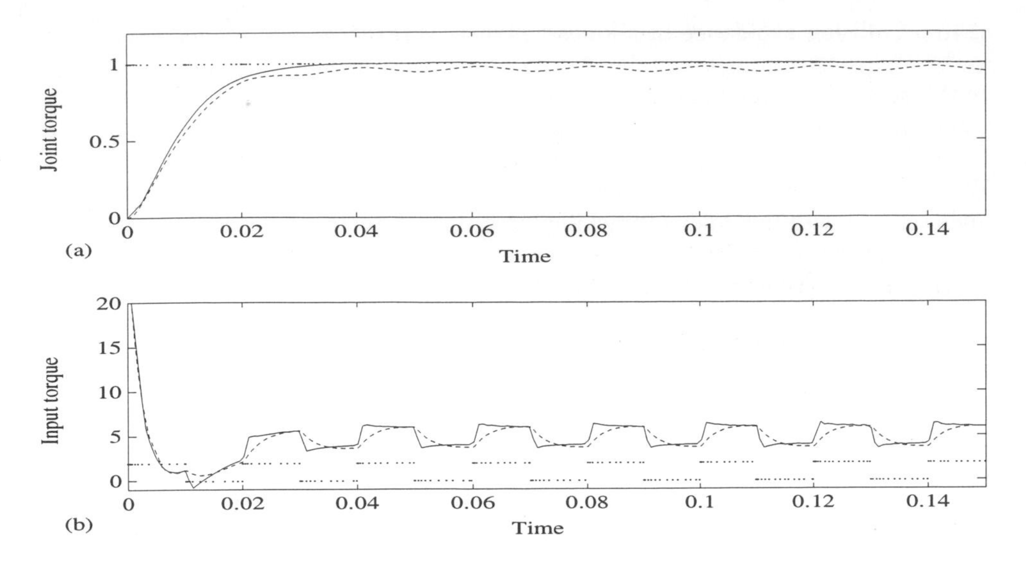

Figure 12.12 Dynamic responses. (a) Output torque: (——) with perturbation compensation, (- - -) without perturbation compensation, (····) the torque command. (b) Input torque: (——) with perturbation compensation, (- - -) without perturbation compensation, (····) the disturbance torque on the motor side.

In particular, the extreme case of direct implementation without averaging as $\tau_r = u$ with u defined in (12.4.9) led to chattering in experiments reported by Lu and Chen (1993). The interested reader is referred to Shi and Lu (1996) for a more detailed discussion on the selection of filter time constant and bandwidth.

A simulation study of the controller discussed above is shown in Figure 12.12. The system parameters were chosen as

$$J_m = 1.0 \times 10^{-5} \qquad n = 600 \qquad J_l = nJ_m/4 \qquad k = 10 \tag{12.4.17}$$

and subject to a square wave disturbance of the form

$$d_m = 1 + \mathrm{sqw}(50t) \tag{12.4.18}$$

on the motor side. Assuming perfect parameter knowledge, i.e. known J_m, J_l, n and k, the poles were both assigned to -200 by defining $c_1 = 400$ and $c_0 = 40\,000$. The disturbance was upper bounded by $d^+ = (k/nJ_m) \max d_m = 5000$. The averaging of u in (12.4.9) is performed by a simple first-order lowpass:

$$\frac{\dot{u}_{ave}}{2000} + u_{ave} = u \tag{12.4.19}$$

i.e. 10 times faster than the closed loop. Simulation results in Figure 12.12 are shown for step responses of $\tau_d = 1$ and demonstrate the significant performance improvement achieved by the perturbation compensation.

12.4.2 Collision avoidance in a known planar workspace

In this section we present an example of planar collision avoidance for mobile robots using the artificial potential field approach of Section 12.3. A mobile robot is to be guided to a specified goal point through a planar workspace with known obstacles. Workspaces of higher dimension are discussed in the next section. The obstacle avoidance scheme with the gradient tracking algorithm of Section 12.3 form the core of a hierarchical path control scheme (Guldner *et al.*, 1995).

The known obstacles in the robot workspace are protected by security zones, which form the basis for the design of an artificial potential field. A set of diffeomorph transformations given by Rimon and Koditschek (1992) provides a mapping between arbitrary star-shaped security zones and security circles. Hence the design of an artificial potential field may concentrate on obstacle security circles in transformed space. After calculating the artificial potential field and its gradient in the transformed space, an inverse mapping provides the back transformation into the real robot workspace. The obstacle security zones are assumed to be nonoverlapping with the goal point lying outside of all security zones.

The main advantage of the transformation procedure is the possibility to concentrate the design of the artificial potential field on security circles for the obstacles rather than on security zones of complex shape. Thanks to the exact tracking capability of the gradient tracking controller discussed in Section 12.3, the artificial potential field can be designed independently of the robot dynamics, enabling a wide variety of applications.

An example of a possible choice of an artificial potential field is the harmonic dipole potential in $\Re^2$ suggested by Guldner and Utkin (1995). For a singularity at the origin, the harmonic potential is given by

$$U = \rho \ln \frac{1}{r} \tag{12.4.20}$$

where ρ denotes the 'strength' of the singularity and r is the distance from the origin. For convenience, a polar coordinate system (r, φ) is adopted here. The planar gradient is given by

$$\boldsymbol{E} = -\nabla U = \begin{cases} -\dfrac{\partial U(r,\varphi)}{\partial r} = \dfrac{\rho}{r} \\[2ex] -\dfrac{1}{r}\dfrac{\partial U(r,\varphi)}{\partial \varphi} = 0 \end{cases} \tag{12.4.21}$$

Consider a dipole for a single obstacle security circle formed by a unit singularity in the goal point and a singularity of opposite polarity in the centre of the obstacle security circle with strength

$$0 < \rho = \frac{R}{R+D} < 1 \tag{12.4.22}$$

where R is the radius of the security circle and D is the distance between the two singularities. The total artificial potential field in a polar coordinate system with the origin in the goal point with the negative unit singularity is given by

$$U(r, \varphi) = \frac{\rho}{2} \ln\left(\frac{1}{r^2 - 2Dr\cos\varphi + D^2}\right) - \ln\frac{1}{r} \tag{12.4.23}$$

with the associated gradient derived from (12.4.21) as

$$\boldsymbol{E} = \begin{cases} -\dfrac{\partial U(r, \varphi)}{\partial r} = \rho\left(\dfrac{r - D\cos\varphi}{r^2 - 2Dr\cos\varphi + D^2}\right) - \dfrac{1}{r} \\[2ex] -\dfrac{1}{r}\dfrac{\partial U(r, \varphi)}{\partial \varphi} = \rho\dfrac{2Dr\sin\varphi}{r^2 - 2Dr\cos\varphi + D^2} \end{cases} \tag{12.4.24}$$

The resulting gradient lines are shown in Figure 12.10. Note two properties of harmonic dipoles:

- All gradient lines beginning outside the security circle remain outside.
- Along any gradient line outside the security circle, the distance to the goal point decreases monotonically.

These properties are preserved under the above diffeomorph transformations. For details and proofs, see Guldner and Utkin (1995).

In order to reduce the computational complexity, many artificial potential field approaches only consider one obstacle at each time instant and switch between the potentials associated with different obstacles, e.g. when the robot gets closer to a different obstacle than the one currently considered. The switching between potentials of different obstacles led to oscillations in several approaches (Koren and Borenstein, 1991), which are caused by the robot inertia and time delays in the closed-loop system obstacle–potential field–robot. An example is shown in Figure 12.13 for a circular robot, oscillating around the equidistance line between two obstacle potentials. The obstacles themselves are omitted in this and the following figures.

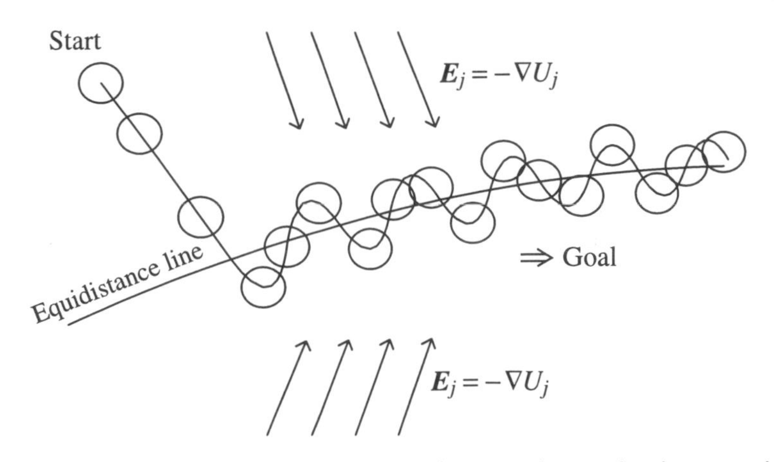

Figure 12.13 Robot (depicted by circles) oscillating along equidistance line between obstacles i and j. Obstacles and goal point (located to the right of the graph) are not shown. Planar gradients $\boldsymbol{E}_i = -\nabla U_i$ and $\boldsymbol{E}_j = -\nabla U_j$ are symbolized by arrows.

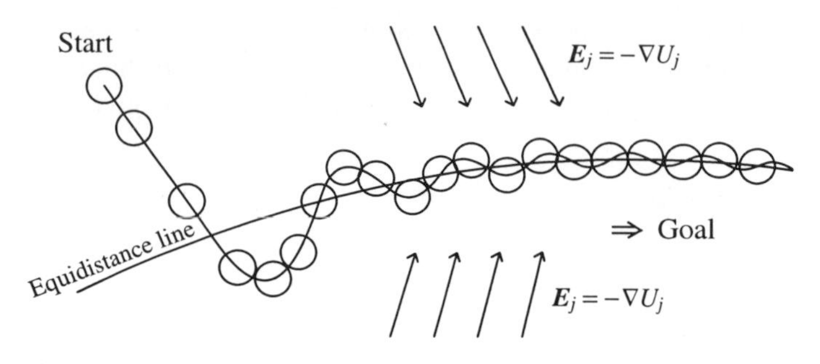

Figure 12.14 Robot (depicted by circles) oscillating along equidistance line between obstacles i and j. The oscillations are damped by a lowpass filter applied to the gradient. Obstacles and goal point (located to the right of the graph) are not shown. Planar gradients $\boldsymbol{E}_i = -\nabla U_i$ and $\boldsymbol{E}_j = -\nabla U_j$ are symbolized by arrows.

An intuitive solution to the oscillation problem is the introduction of a lowpass filter for the artificial potential field or its gradient, leading to a time-dependent field. However, a lowpass filter is reactive, i.e. it only starts to consider the new potential when the robot crosses the equidistance line between the obstacles. Thus the oscillations can only be damped, they cannot be eliminated completely, as illustrated in Figure 12.14.

A closer look at the oscillation phenomenon reveals it is similar to chattering in variable-structure systems (Chapter 8). In fact, the oscillations are caused by unmodelled dynamics in the closed-loop system obstacle–potential field–robot: the switching between the potential fields of different obstacles neglects the robot dynamics. In the vicinity of the equidistance line, the gradient lines may be oriented towards it (Figure 12.15(a)) and the trajectory field resembles the state space of a dynamic system with sliding mode. Formally, sliding mode may appear along the equidistance line in our case and the robot trajectory should coincide with the equidistance line. In the course of this motion the robot avoids collisions with obstacles and approaches the goal. Unfortunately, this ideal motion cannot be implemented. Indeed, ideal sliding mode would require the speed vector to track the gradient lines of each of two adjoining potential fields, undergoing discontinuities on the equidistance line. This implies that the control force or torque should develop infinite acceleration, which is impossible due to physical constraints. As a result, the robot trajectory oscillates in the vicinity of the equidistance line (Figure 12.13) like in a system with chattering. This similarity means that a reliable solution is to apply one of the methods discussed in Chapter 8 to prevent chattering.

As an example, consider the boundary layer approach of Section 8.2. Instead of switching the potential field abruptly when the robot crosses the equidistance line between two obstacles, both potentials $U_i(q)$ and $U_j(q)$ are considered simultaneously in a boundary layer of width δ along the equidistance line. The resulting total potential field is

$$U_{res}(q) = \begin{cases} U_i(q) & \text{for } d < -\dfrac{\delta}{2} \\ \left(\dfrac{1}{2}+\dfrac{d}{\delta}\right)U_i(q) + \left(\dfrac{1}{2}-\dfrac{d}{\delta}\right)U_j(q) & \text{for } |d| < \dfrac{\delta}{2} \\ U_j(q) & \text{for } d > \dfrac{\delta}{2} \end{cases} \qquad (12.4.25)$$

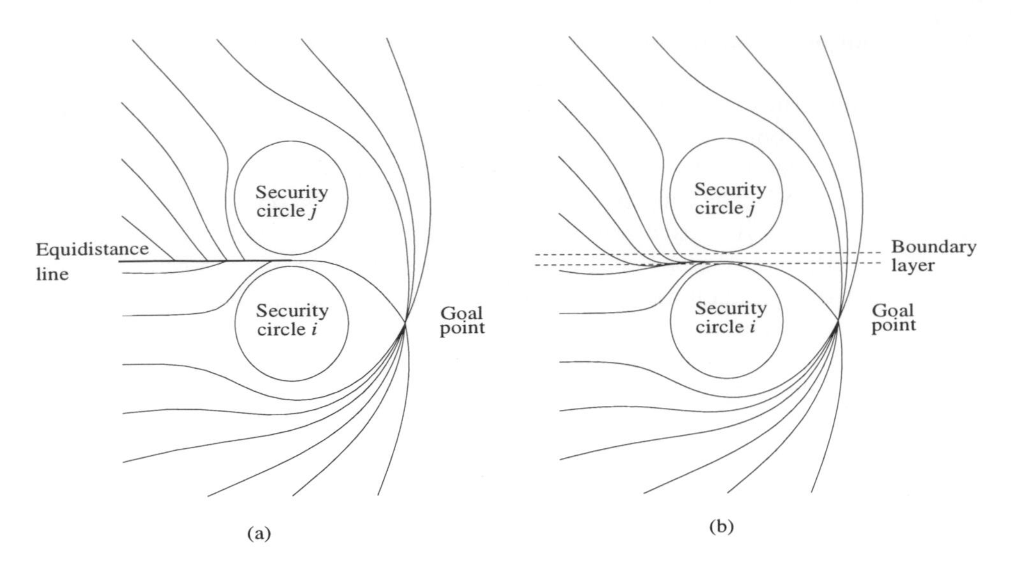

Figure 12.15 Effect of boundary layer for smoothing gradient lines along equidistance line between obstacle security zones: (a) discontinuous switching between gradients of obstacle security circles i and j; (b) smooth switching in boundary layer along equidistance line.

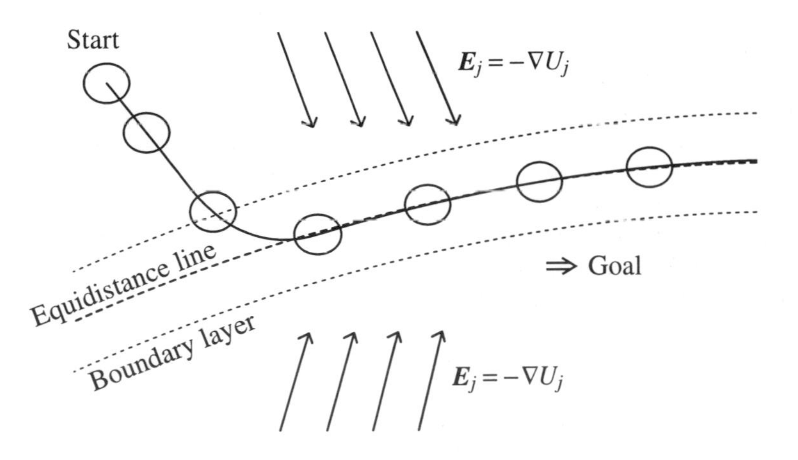

Figure 12.16 Robot (depicted by circles) follows the equidistance line without oscillations when gradients are smoothed inside boundary layer. Obstacles and goal point (located to the right of the graph) are not shown. Planar gradients $\boldsymbol{E}_i = -\nabla U_i$ and $\boldsymbol{E}_j = -\nabla U_j$ are symbolized by arrows.

where d denotes the distance of the robot from the equidistance line. Figure 12.15 compares the gradient lines for an example with two obstacle security circles. Note the difference of the gradient lines in the vicinity of the equidistance line $s_{equi} = 0$. The discontinuities along $s_{equi} = 0$ visible in Figure 12.15(a) are eliminated by the introduction of the boundary layer δ along $s_{equi} = 0$ in Figure 12.15(b). Figure 12.16 shows how this application of sliding mode theory to the collision avoidance problem successfully eliminates the oscillations; see also Guldner *et al.* (1994a).

12.4.3 Collision avoidance in known workspaces of higher dimension

This section extends the results of the previous section for mobile robots in planar workspaces to robots in higher-dimensional workspaces ($\Re^n, n > 2$), in particular to robot manipulators. Details of the development in this section may be found in Guldner *et al.* (1997). The concept of transforming star-shaped obstacle security zones into security circles is carried over to $\Re^n$ by using security spheres (for $n = 3$) and security hyperspheres (for $n > 3$). However, it is beyond this book to generalize the diffeomorph transformations given by Rimon and Koditschek (1992) to $n > 2$. As in the planar case, obstacles are assumed to be known, with nonoverlapping security zones and a reachable goal point outside any security zone.

Instead of redesigning the harmonic potential in (12.4.20) for $\Re^n, n > 2$, the properties of the planar gradient (12.4.21) are further exploited. Since in any space $\Re^n, n > 2$, three points uniquely define a subspace of dimension $n = 2$, a projection onto such a planar subspace allows us to continue using the planar harmonic dipole potential of Section 12.4.2. Let the goal point $q_G \in \Re^n$, the centre of the obstacle (hyper)sphere $q_S \in \Re^n$ and the robot position $q_R \in \Re^n$ define a plane

$$\wp(q) : q = q_R + \alpha_1 \frac{q_R - q_G}{\|q_R - q_G\|} + \alpha_2 \frac{q_R - q_S}{\|q_R - q_S\|} \qquad (\forall \alpha_1, \alpha_2 \subset \Re) \qquad (12.4.26)$$

The plane $\wp(q)$ serves as a design platform for the planar harmonic dipole potential (12.4.23) with gradient (12.4.24). In $\Re^n$ the gradient is directly found as

$$\boldsymbol{E} = \frac{q_R - q_G}{\|q_R - q_G\|^2} - \rho \frac{q_R - q_S}{\|q_R - q_S\|^2} \in \Re^n \qquad (12.4.27)$$

where ρ is found according to (12.4.22) with $D = \|q_R - q_S\|$ and $q_G, q_R, q_S \in \Re^n$ taken as vectors. A three-dimensional example is shown in Figure 12.17 with a sphere-like robot travelling in plane $\wp(q)$ towards the goal point at the origin of the coordinate system.

For multiple obstacles in the workspace, a switching procedure similar to the planar case is designed. Each obstacle is assigned its own security zone, each of which is separately transformed into a security (hyper)sphere. In $\Re^n$, $n > 2$, the switching between different artificial potentials for various obstacles takes place in subspaces of varying dimension. Consider a three-dimensional situation. Between two obstacles, one two-dimensional equidistance subspace can be found, i.e. an equidistance plane. Introducing a third obstacle in this three-dimensional workspace leads to a total of three equidistance planes for each combination of two out of the three obstacle security zones. Additionally, the points of equidistance to all three obstacles form a one-dimensional subspace, an equidistance line.

Generalizing this to arbitrary $\Re^n$, $n > 2$, yields subspaces $\Re^{n-1}$ equidistant to exactly two of the obstacle security zones, subspaces $\Re^{n-2}$ equidistant to exactly three of the obstacle security zones, and subspaces $\Re^{n-k+1}$ equidistant to exactly $1 < k < n$ of the obstacle security zones, with k being the number of obstacles among all workspace obstacles which may have to be considered simultaneously. The equidistance subspaces for various obstacle security zones are equivalent to *Voronoi planes* (Latombe, 1991), often employed for robot path planning. In contrast to Voronoi-based approaches, the equidistance subspaces do not have to be computed explicitly for the following switching strategy.

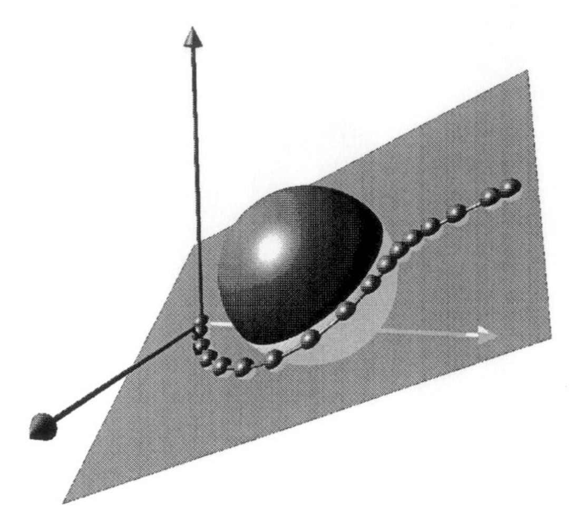

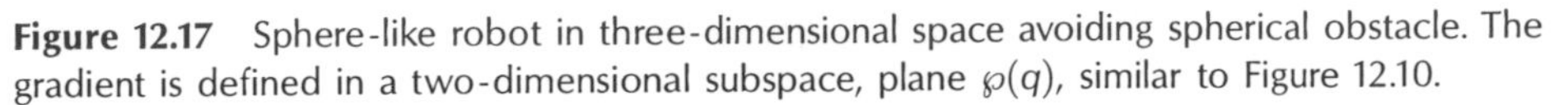

Figure 12.17 Sphere-like robot in three-dimensional space avoiding spherical obstacle. The gradient is defined in a two-dimensional subspace, plane $\wp(q)$, similar to Figure 12.10.

In order to avoid oscillations along the switching planes, i.e. the equidistance subspaces, boundary zones of appropriate dimension are introduced as an extension of (12.4.25). Consider an $\Re^n$ situation with k obstacles in which the gradients in all adjacent regions converge to equidistance subspace $\Theta_k^n \in \Re^{n-k+1}$. An appropriate boundary zone of width $\delta > 0$ is defined as

$$B_k^n = \{q \in \Re^n \,|\, \|q \perp \Theta_k^n\| \leq \delta\} \tag{12.4.28}$$

where $\|q \perp \Theta_k^n\|$ denotes the length of the normal projection of vector $q \in \Re^n$ onto subspace $\Theta_k^n \in \Re^{n-k+1}$. Within boundary zone B_k^n, all k obstacles contribute to the resulting potential field $U_{res} \in \Re^n$. Let U_G be the unit potential of the goal point, which is common to all single obstacle potentials. The individual weights η_i of each obstacle potential U_{S_i} with appropriate strength according to (12.4.22) are given by

$$\eta_i = \frac{1}{k}\left(1 - \frac{1}{\delta}\left((k-1)d_i - \sum_{\substack{l=1 \\ l \neq i}}^{n} d_l\right)\right) \qquad (i = 1, 2, \ldots, k) \tag{12.4.29}$$

with d_i being the distance of the robot position $q_R \in \Re^n$ from the equidistance subspace with respect to obstacle i. The resulting potential field is found as

$$U_{res} = U_G + \sum_{i=1}^{k} \eta_i U_{S_i} \tag{12.4.30}$$

The necessary condition $\sum_{i=1}^{k} \eta_i = 1$ ensures that the properties of the harmonic dipole potentials are preserved and smooth transition between different potentials enables tracking without oscillations. Note that the robot motion is only confined to the boundary zone B_k^n as long as all k gradients in adjacent regions are directed towards the equidistance subspace Θ_k^n. When the robot has passed an obstacle, the boundary zone may be left and new equidistance subspaces, with a different number of obstacles considered simultaneously, are found until the goal point is

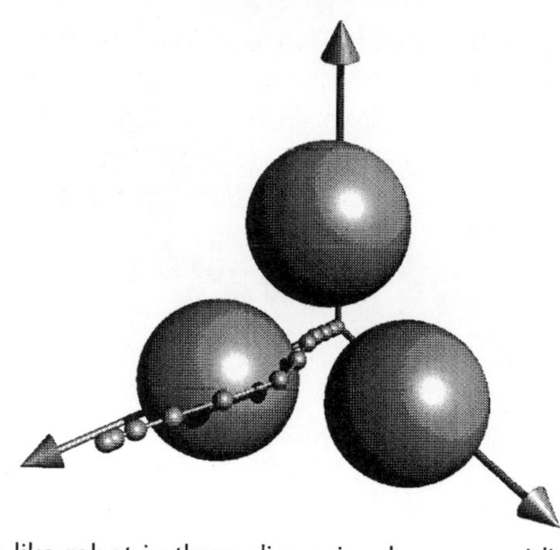

Figure 12.18 Sphere-like robot in three-dimensional space avoiding three spherical obstacles. The goal point is located at the origin of the coordinate system (see also Figure 12.19). Obstacle 1 is in the lower left corner; obstacle 2 is in the lower right corner; obstacle 3 is in the top centre.

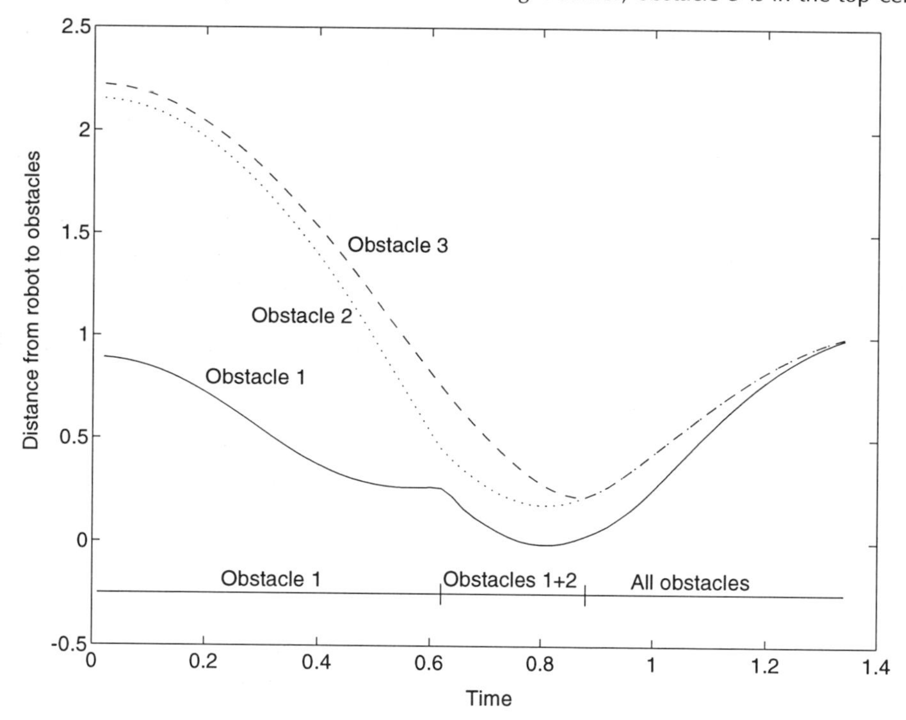

Figure 12.19 Distances between robot and obstacle security spheres for the simulation in Figure 12.18: (——) distance to obstacle 1 (lower left); (····) distance to obstacle 2 (lower right); (- - -) distance to obstacle 3 (top centre). The labels just above the time axis indicate the respective obstacles considered when defining potentials according to (12.4.29) and (12.4.30).

reached. There is no intention to create sliding motion along the equidistance subspaces, although the motion trajectories may temporarily resemble boundary layer sliding mode trajectories.

This section concludes with two examples in $\Re^3$. First, consider the spherical robot of Figure 12.17 in a workspace with three obstacles and the goal point at the origin of the coordinate system in Figure 12.18. The varying number of obstacles is illustrated in Figure 12.19 according to the distance of the robot to the respective obstacles. Initially, only obstacle 1 in the lower left corner is considered. After about 0.6 s, the robot enters the boundary plane between obstacles 1 and 2. Finally, when obstacle 3 is the same distance from the robot as obstacle 2, all three obstacles are considered in the calculation of the resulting potential and the robot travels in the boundary zone of the equidistance line towards the goal point.

Since the example of Figures 12.17 to 12.19 are rather abstract, based on 'free-flying' spherical robots, the next example applies the algorithm to a more realistic problem, a revolute robot manipulator that has three links and three joints and must avoid cylindrical obstacles in its three-dimensional workspace. The potential fields were calculated in the robot's $\Re^3$ configuration space rather than in the $\Re^3$ workspace. Figure 12.20 shows the robot motion in the actual workspace in nine snapshots. Details of this application are described in Guldner *et al.* (1997).

12.4.4 Automatic steering control for passenger cars

Automatic steering control is the last example for the application of advanced robotics. Automation of vehicles, e.g. for automated highways systems (AHS), has been discussed for several decades and is studied in various programmes worldwide in the framework of intelligent transport systems (Stevens, 1996; Tsugawa *et al.* 1996).

Two control subtasks arise for automated driving of 'robot cars': steering control to keep the vehicle in the lane (controlling the lateral motion) and throttle/brake control to maintain speed and proper spacing between vehicles (controlling the longitudinal motion). Both subtasks have been solved using sliding mode control. The focus in this section will be on automatic steering control. Longitudinal control has been studied by Hedrick *et al.* (1994) and Pham *et al.* (1994). The automatic steering system of an automated vehicle consists of a reference system to determine the lateral vehicle

Figure 12.20 Robot arm with three links avoiding two cylindrical obstacles. The obstacle security zones were transformed into configuration space and the potential fields were calculated in this three-dimensional configuration space.

position with respect to the lane centre, sensors to detect the vehicle motion (typically yaw rate and lateral acceleration), and a steering actuator to steer the front wheels. The reference systems that are employed range from look-ahead systems like machine vision or radar to look-down systems like electric wires or magnets embedded in the road surface. The lateral displacement of the vehicle with respect to the reference is measured at a point ahead of the vehicle (look-ahead) or directly down from the front bumper (look-down); Patwardhan *et al.* (1997) give a more detailed treatment. The control design below, however, is valid for any reference system.

Control design is usually based on the so-called single-track model, which concentrates on the main vehicle mass by lumping the two wheels at each axle into a single wheel. The road–tire interaction forces are responsible for generating planar lateral and yaw vehicle motions, with the front wheel steering angle δ_f being the input variable. A linearized second-order model for constant speed v is given by

$$\begin{bmatrix} \dot{\beta} \\ \ddot{\psi} \end{bmatrix} = \mu \begin{bmatrix} -\dfrac{c_f + c_r}{Mv} & -1 - \dfrac{c_f l_f - c_r l_r}{Mv^2} \\ -\dfrac{c_f l_f - c_r l_r}{J} & -\dfrac{c_f l_f^2 + c_r l_r^2}{Jv} \end{bmatrix} \begin{bmatrix} \beta \\ \dot{\psi} \end{bmatrix} + \mu \begin{bmatrix} \dfrac{c_f}{Mv} \\ \dfrac{c_f l_f}{J} \end{bmatrix} \delta_f \tag{12.4.31}$$

with states sideslip angle β and yaw rate $\dot{\psi}$. For a detailed derivation of (12.4.31) see Peng (1992) or Ackermann *et al.* (1993). The parameters are vehicle mass M and yaw inertia J, distances l_f and l_r of front and rear axles from the centre of gravity (CG), front and rear tire cornering stiffness c_f and c_r, and road adhesion factor μ. All parameters are uncertain within known bounds, e.g. $0 < \mu^- \leq \mu \leq \mu^+ \leq 1$.

When following a reference path with curvature ρ_{ref} as depicted in Figure 12.21, lateral vehicle displacement y_{es}, measured at some sensor position d_S ahead of CG, and angular error ψ_e can be described by linearized dynamic model

$$\begin{aligned} \dot{y}_{es} &= v(\beta + \psi_e) + d_S \dot{\psi} \\ \dot{\psi}_e &= \dot{\psi} - v\rho_{ref} \end{aligned} \tag{12.4.32}$$

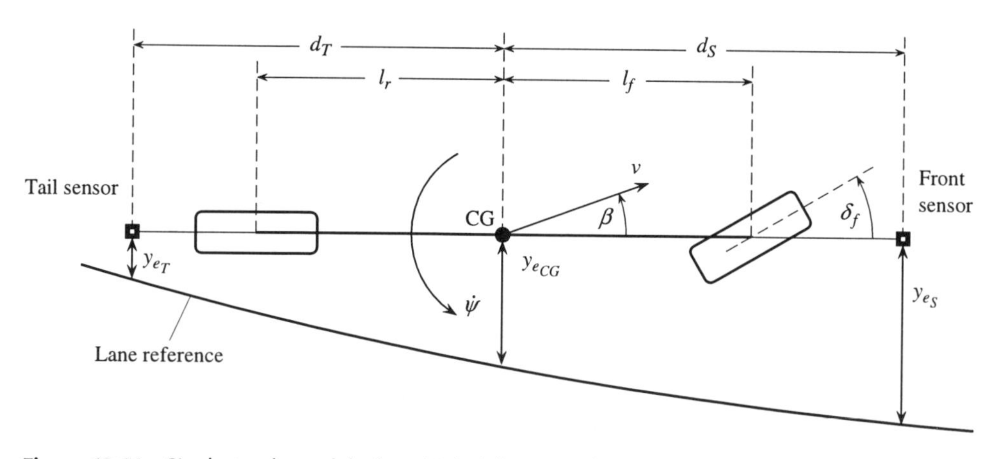

Figure 12.21 Single-track model of a vehicle following a lane reference. Sensors at the front and tail bumpers measure lateral displacements y_{e_S} and y_{e_T}, respectively. Also shown are vehicle states, sideslip angle β and yaw rate $\dot{\psi}$, input steering angle δ_f, and various distances from the centre of gravity (CG).

Given (12.4.31) and (12.4.32), various control design options are possible. As an example, we present a cascaded control design under the assumption that vehicle yaw rate $\dot{\psi}$ is measurable by a gyroscope. The control design follows the regular form methodology (Section 3.3) and considers subsystem (12.4.31) as the input to subsystem (12.4.32). Hence the first design step assumes yaw rate $\dot{\psi}$ to be a direct input to (12.4.32) and derives a desired yaw rate $\dot{\psi}_d$. The second step then ensures that the actual, measured vehicle yaw rate $\dot{\psi}$ follows $\dot{\psi}_d$ exactly via appropriate control design for steering angle δ_f in (12.4.31), the true system input. A suitable continuous feedback/feedforward 'yaw rate' controller to stabilize the first equation in (12.4.32) would be

$$\dot{\psi}_d = -\frac{1}{d_S}(v(\beta + \psi_e) + Cy_{es}) \tag{12.4.33}$$

with linear feedback gain $C > 0$. However, neither sideslip angle β nor yaw angle error ψ_e can be measured and hence have to be estimated by an observer (Guldner *et al.*, 1994b). Introducing auxiliary variable $z = \beta + \psi_e$, an observer is designed as

$$\begin{bmatrix} \dot{\hat{y}}_{es} \\ \dot{\hat{z}} \end{bmatrix} = \begin{bmatrix} 0 & v \\ 0 & 0 \end{bmatrix} \begin{bmatrix} \hat{y}_{es} \\ \hat{z} \end{bmatrix} + \begin{bmatrix} d_S \\ 0 \end{bmatrix} \dot{\psi} + \begin{bmatrix} c_1 \\ c_2 \end{bmatrix} \bar{y}_{es} \tag{12.4.34}$$

with feedback of the observation error $\bar{y}_{es} = y_{es} - \hat{y}_{es}$ via gains $c_1 > 0$ and $c_2 > 0$, chosen faster than the vehicle dynamics in (12.4.31). With the help of the observed auxiliary variable $\hat{z} = \hat{\beta} + \hat{\psi}_e$, a desired yaw rate is defined as

$$\dot{\psi}_d = -\frac{1}{d_S}(v(\hat{\beta} + \hat{\psi}_e) + Cy_{es}) \tag{12.4.35}$$

The second step of control design uses the steering angle δ_f as the input to (12.4.31) to drive yaw rate error $\dot{\psi}_e = \dot{\psi}_d - \dot{\psi}$ to zero, e.g. by purely discontinuous sliding mode control

$$\delta_f = \delta_0 \text{ sign } \dot{\psi}_e \tag{12.4.36}$$

The stability analysis follows the previously discussed Lyapunov approach and is omitted here for brevity. An alternative approach uses a combination of continuous feedback/feedforward and a discontinuity term:

$$\delta_f = \delta_1 \text{ sign } \dot{\psi}_e + \frac{1}{\hat{c}_f \hat{l}_f}\left(\left((\hat{c}_f \hat{l}_f - \hat{c}_r \hat{l}_r)\hat{\beta} + \frac{1}{v}(\hat{c}_f \hat{l}_f^2 + \hat{c}_r \hat{l}_r^2)\dot{\psi}\right) + \frac{\hat{J}}{\hat{\mu}}\ddot{\psi}_d\right) \tag{12.4.37}$$

where estimates of vehicle parameters are denoted by hats; the estimate $\hat{\beta}$ of sideslip angle β stems from an observer similar to (12.4.34). The derivative of the desired yaw rate, $\ddot{\psi}_d$, can be derived from (12.4.35) by virtue of known observer dynamics. Due to the continuous feedback/feedforward terms in (12.4.37), the gain of the discontinuity term can be reduced as compared to (12.4.36), i.e. $\delta_1 < \delta_0$.

This control design neglects the dynamics of the steering actuator, which will lead to chattering in practical implementations. In addition to the chattering prevention methods discussed in Chapter 8, the introduction of an integrator in the control loop proved to be a promising approach. Originally, the integrator was a physical model of the actuator dynamics (Ackermann *et al.*, 1993) with the steering rate as the system input rather than the steering angle, so

$$\dot{\delta}_f = u \tag{12.4.38}$$

instead of $\delta_f = u$. The additional integrator only requires to alter the outer control loop (12.4.36), (12.4.37). Define a second-order sliding variable

$$s = C_3 \dot{\psi}_e + \ddot{\psi}_e \tag{12.4.39}$$

leading to a control law

$$\delta_f = \delta_0 \text{ sign } s \tag{12.4.40}$$

instead of (12.4.36). The alternative feedback/feedforward controller term in (12.4.37) has to be adjusted accordingly. If the real steering actuator is not an integrator as in (12.4.38), but features more complex dynamics, e.g. for an electrohydraulic actuator, two design alternatives are left to the control engineer. Either a sliding mode controller is designed according to (12.4.36) or (12.4.37) with appropriate measures to prevent chattering as discussed in Chapter 8, or an integrator like (12.4.38) is introduced as part of the controller, i.e. realized in the controller software. The latter case follows the ideas of integral sliding mode by implementing sliding motion in an integral manifold rather than directly in the control input variable δ_f. Hence the switching action of the sliding mode discontinuity is first filtered by integrator (12.4.38) and thus

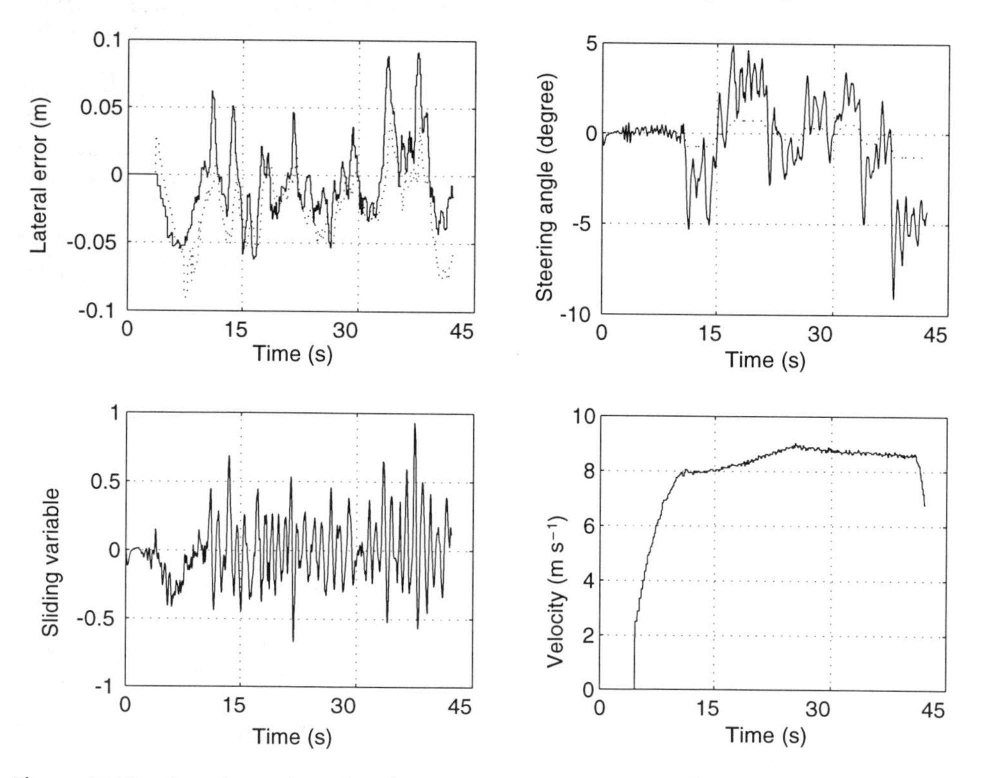

Figure 12.22 Experimental results of an automatic steering controller based on sliding mode design. (Courtesy Dr Hingwe and Professor Tomizuka, University of California of Berkeley, USA.)

does not directly reach the input δ_f, which inherently prevents chattering. A different integrator location in the control loop has been proposed by Pham *et al.* (1994), before rather than after the switching discontinuity. The interested reader is referred to Hingwe and Tomizuka (1995) for a comparison of different integrator locations in the controller loop. Experimental results from this work are displayed in Figure 12.22.

References

ACKERMANN, J., *et al.*, 1993, *Robust Control: Analysis and Design of Linear Control Systems with Uncertain Physical Parameters*, London: Springer-Verlag.

ARIMOTO, S. and MIYAZAKI, F., 1984, Stability and robustness of PID feedback control of robot manipulators of sensory capability, *Proceedings of the First International Symposium on Robotics Research*, pp. 783–99.

BLOCH, A.M., REYHANOGLU, M. and MCCLAMROCH, N.H., 1992, Control and stabilization of nonholonomic dynamic systems, *IEEE Transactions on Automatic Control*, **37**, 1746–57.

CHEN, Y.-F., MITA, T. and WAHUI, S., 1990, A new and simple algorithm for sliding mode control of robot arms, *IEEE Transactions on Automatic Control*, **35**, 828–29.

CRAIG, J., 1986, *Introduction to Robotics: Mechanics and Control*, Reading MA: Addison-Wesley.

CRAIG, J., 1988, *Adaptive Control of Mechanical Manipulators*, Reading MA: Addison-Wesley.

DORLING, C. and ZINOBER, A., 1986, Two approaches to hyperplane design in multivariable variable structure control systems, *International Journal of Control*, **44**, 65–82.

GILBERT, E. and HA, I., 1984, An approach to nonlinear feedback control with applications to robotics, *IEEE Transactions on Systems, Man, and Cybernetics,* **14**, 879–84.

GULDNER, J. and UTKIN, V.I., 1995, Sliding mode control for gradient tracking and robot navigation using artificial potential fields, *IEEE Transactions on Robotics and Automation*, **11**, 247–54.

GULDNER, J., UTKIN, V.I. and BAUER, R., 1994a, On the navigation of mobile robots in narrow passages: a general framework based on sliding mode theory, *Preprints of the IFAC Symposium on Robot Control*, Capri, Italy, pp. 79–84.

GULDNER, J., UTKIN, V.I. and ACKERMANN, J., 1994b, A sliding mode approach to automatic car steering, *Proceedings of the American Control Conference*, Baltimore MA, USA, pp. 1969–73.

GULDNER, J., UTKIN, V.I. and BAUER, R.,1995, A three-layered hierarchical path control system for mobile robots: algorithms and experiments, *Robotics and Autonomous Systems*, **14**, 133–47.

GULDNER, J., UTKIN, V.I. and HASHIMOTO, H., 1997, Robot obstacle avoidance in *n*-dimensional space using planar harmonic artificial potential fields, *ASME Journal of Dynamic Systems, Measurement, and Control*, **119**(2), 160–66.

HEDRICK, J. K., TOMIZUKA, M. and VARAIYA, P., 1994, Control issues in automated highway systems, *IEEE Control Systems*, **14**(6), 21–32.

HINGWE, P. AND TOMIZUKA, M., 1997, Experimental evaluation of chatter freee sliding mode control for lateral control in AHS, *Proceedings of the American Control Conference*, Albuquerque, NM, USA.

HOGAN, N., 1985, Impedance control: an approach to manipulation (parts I to III), *ASME Journal of Dynamic Systems, Measurement, and Control*, **107**, 1–24.

HUNT, L., SU, R. and MEYER, G., 1983, Global transformation of nonlinear systems, *IEEE Transactions on Automatic Control*, **28**, 24–31.

KHATIB, O., 1986, Real-time obstacle avoidance for robot manipulators and mobile robots, *International Journal of Robotics Research*, **5**, 90–98.

KODITSCHEK, D., 1991, The control of natural motion in mechanical systems, *ASME Journal of Dynamic Systems, Measurement, and Control*, **113**, 547–51.

KOREN, Y. and BORENSTEIN, J., 1991, 'Potential field methods and their inherent limitations for mobile robot navigation', *Proceedings of the International Conference on Robotics and Automation*, Sacramento CA, USA, pp. 1398–1404.

KROGH, B., 1984, A generalized potential field approach to obstacle avoidance control, *Proceedings of the SME Conference on Robotics Research*, Bethlehem PA, USA.

LATOMBE, J.-C., 1991, *Robot Motion Planning*, Dordrecht: Kluwer.

LEWIS, F., ABDALLAH, C. and DAWSON, D., 1993, *Control of Robot Manipulators*, New York, NY, USA: Macmillan.

LU, Y. and CHEN, J., 1993, Sliding mode controller design for a class of nonautonomous systems – an experimental study, *Proceedings of the IEEE Conference IECON*, NY, USA, pp. 2358–63.

NEIMARK, J. and FUFAEV, N., 1972, *Dynamics of Nonholonomic Systems*, Providence RI, USA: American Mathematical Society.

PATWARDHAN, S., TAN, H.-S. and GULDNER, J., 1997, A general framework for automatic steering control, *Proceedings of the American Control Conference*, Albuquerque NM, USA, pp. 1598–1602.

PENG, H., 1992, 'Vehicle lateral control for highway automation', PhD dissertation, University of California, Berkeley CA.

PHAM, H., HEDRICK, K. and TOMIZUKA, M., 1994, Combined lateral and longitudinal control of vehicles for IVHS, *Proceedings of the American Control Conference*, Baltimore MA, USA, pp. 1205–6.

RIMON, E. and KODITSCHEK, D.E., 1992, Exact robot navigation using artificial potential functions, *IEEE Transactions on Robotics and Automation*, **8**, 501–18.

RYAN, E.P. and CORLESS, M., 1984, Ultimate boundedness and asymptotic stability of a class of uncertain systems via continuous and discontinuous feedback control, *IMA Journal of Mathematics, Control and Information*, **1**, 223–42.

SHI, J. and LU, Y., 1996, Chatter-free variable structure perturbation estimator on the torque control of flexible robot joints with disturbance and parametric uncertainties, *Proceedings of the IEEE International Conference on Industrial Electronics, Control, and Instrumentation IECON*, pp. 238–43.

SLOTINE, J.-J.-E., 1985, The robust control of robot manipulators, *International Journal of Robotics Research*, **4**, 49–64.

SPONG, M. and VIDYASAGAR, M., 1989, *Robot Dynamics and Control*, New York, USA: J. Wiley.

STEVENS, W., 1996, The automated highway systems program: a progress report, *Preprints of the 13th IFAC World Congress* (plenary volume), San Francisco CA, pp. 25–34.

TAKEGAKI, M. and ARIMOTO, S., 1981, A new feedback method for dynamic control of manipulators, *ASME Journal of Dynamic Systems, Measurement, and Control*, **102**, 119–25.

TSUGAWA, S., AOKI, M., HOSAKA, A. and SEKI, K., 1996, A survey of present IVHS activities in Japan, *Preprints of the 13th IFAC World Congress* (Vol. Q), San Francisco CA, USA, pp. 147–52.

UTKIN, V.I., 1992, *Sliding Modes in Control and Optimization*, London: Springer-Verlag.

UTKIN, V.I., DRAKUNOV, S., HASHIMOTO, H. and HARASHIMA, F., 1991, Robot path obstacle avoidance via sliding mode approach, *Proceedings of the IEEE/RSJ International Workshop on Intelligent Robots and Systems (IROS)*, Osaka, Japan, pp. 1287–90.

YOUNG, K.-K.D., 1978, Controller design for a manipulator using theory of variable structure systems, *IEEE Transactions on Systems, Man, and Cybernetics*, **8**, 210–18.

Index

(α, β) coordinate frame 189–92, 200, 206, 213–17
(a, b, c) coordinate frame 189–92, 200, 206, 213–17
(d, q) coordinate frame 125, 189–92, 200, 206, 213–7, 253–4, 258

AC motor 125, 187–212
  see also PMSM motor
AC/DC converter 251–69
  see also boost AC/DC converter
Ackermann's formula 87, 90, 121
actuator
  see DC motor, induction motor or PMSM motor
AHS 313
angular acceleration 176, 203, 222
  observer 176–9
arm robot
  see robot manipulator
armature current 172, 189, 214–17
armature parameters (inductance, resistance) 125, 172, 189, 207, 213
artificial potential field 294–6, 306–12
asymptotic observer
  see observer
automated highway systems (AHS) 313
automatic steering control 313–17
auxiliary control loop 142, 146
  see also cascaded control
average of discontinuous control 11, 119, 151
  see also equivalent control

back-EMF
  see EMF
balance condition 193, 196, 259
bilinear system 232
block control 99, 145, 165
  see also regular form
block observable form 106–8
boost AC/DC converter 252–63
  current control 255–8
  model 252–4, 258
  observer-based control 262–8
  performance requirements 254, 259
  voltage regulation 258–61
boost DC/DC converter 236–40, 246–51
  current control 238–40, 246–51
  model 236–7
  observer-based control 246–51
  voltage regulation 238–40
bound
  desired trajectory 279
  mass matrix elements 290
  observation error 143
boundary layer 23, 139–41, 292, 308–9, 312
boundary value problem 167
boundary zone 311–2
brushless DC motor 187
  see also PMSM
buck DC/DC converter 232–6, 241–5
  model 234–5
  observer-based control 241–5
  voltage regulation 235–6
Butterworth filter 185, 209

canonical space 7–8
cascaded control 115–30, 144–51, 165, 173, 201, 219, 228, 234, 238–49, 255, 315
  AC/DC converter 255–62
  chattering prevention 145–51
  current control 125–9, 173–4, 193–201, 238–40, 246–51, 255–258
  DC motor 173–5
  DC/DC converter 233–9
  discrete-time 164–8
  example 145–7
  integral sliding mode 115–30, 147–51
  power converter 235–6, 238–40
  speed control 174–6, 179–1, 201–5, 219–23, 313–5
  vehicle lateral control 315
  see also regular form
centrifugal force 122
characteristic polynomial 42, 140, 143, 179, 242
characteristics of sliding mode 12, 158
chattering 6, 70, 118–20, **131–56**, 292, 308–9, 316–7
  causes 134–9, 155–6
  comparison of solutions 151–3
  discretization chatter 131, 155–8

disturbance rejection solution 118–20, 147–51, 302–5
example: first order system 132–9
observer-based solution 141–4, 181–2
prevention 118–20, 139, 141–54, 180, 292, 308–9, 311–2, 316–7
problem analysis 131–9
regular form solution 144–7
robot motion 307–9
solutions 141–54
chattering due to discretization 70, 156–8
chattering due to unmodelled dynamics 134–54
chopper control 193, 200
collision avoidance 294–7, 306–13
componentwise control 26, 45
mobile robot 299–301
robot manipulator 279–83
computed torque 122, 289
concept of sliding mode 31, 157–61
continuous control with discontinuity term
automatic steering control 314–7
chattering prevention 147–51
disturbance rejection 117–29, 303–5
integral sliding mode 117–29, 303–5
lateral vehicle control 314–7
perturbation estimation 117–29, 303–5
robot control 286–90
trajectory tracking 286–90
torque perturbation 303–5
vehicle control 314–7
control
AC/DC converter 3, 251–69
automatic steering 314–17
block control 99, 144, 165
see also regular form
chopper control 193, 200
componentwise control 279–83, 299–301
continuous with discontinuity term **117–29**, 148–51, **286–90**, 303–5, 316
current control
see current control
DC motor 124–5, 172–87
DC/DC converter 231–51
digital
see discrete-time
discrete-time and delay systems 155–69
equivalent
see equivalent control
feedback/feedforward 286–90, 303–5, 316
flexible joint 302–5
gradient tracking 292–301
high gain 116, 119, 240
hysteresis control 193, 200, 234
induction motor 218–28
integral sliding mode **115–30**, 240, 302–5, 316
lateral vehicle control 314–317
linear systems 79–102
output feedback 94–102
PMSM/AC motor or brushless DC motor 125–9, 193–212
relay 1–5
robots 278–301
sliding mode
see sliding mode control
speed control
see speed control
trajectory tracking 278–292
unit control 43–6, 80, 294
vector control 283–6, 298–9
vehicle control 314–17
converter 231–69
see also power converter
coordinate frame 189–93, 213–5, 234, 252–3
Coriolis force 122
cornering stiffness 314
Coulomb friction 2, 9
Cuk converter 232
current control
AC motor or brushless DC motor 125–9, 193–201
AC/DC converter 255–258
DC motor 173–4
DC/DC converter 238–40, 246–51
PMSM motor 125–9, 193–201
current observer
AC/DC converter 263–4
EMF components 207–8
PMSM/AC motor or brushless DC motor 205–8
current phase angle of AC/DC converter 260
curvature
gradient of potential field 297
road 314

DC generator 4
DC motor 124, 172–187
current control 173–4
example 124–5
field oriented control 124–5
integral sliding mode 124–5
model 124, 172–3, 179
observer design 176–9, 182–5
PWM via sliding mode 124–5
sensorless control 182–5
speed control 174–6, 179–1
DC/DC converter 231–51
see also boost converter and buck converter
definition
discrete-time sliding mode 161
equivalent control 25
ideal sliding mode 118
matching condition 43
sliding manifold 31
unit control/vector control 44
delay systems 21, **164–8**
diffeomorph transformation 306–7, 310
difference equation 164
differential-difference equation 164, 167
digital control
see discrete-time control
dipole potential 295, 306–12
direct speed control
DC motor 175–6
induction motor 219–23
PMSM/AC motor or brushless DC motor 201–5
Dirichlet boundary condition 294
discrete-time
design procedure 159
equivalent control 162

example 156–61
sliding mode control 155–69
discretization chatter 70, **155–8**, 168
dissipative mechanical system 294
distributed systems 164–8
disturbance 42
estimator 112, 119, 147
model 83
rejection 117, 119, 133–4, 147–51, 303
torque 122, 302–5
drive system
see DC motor, induction motor or PMSM motor
dynamic compensator 83–5
dynamics
holonomic 273
nonholonomic 277–8
pendulum 51, 55, 62, 90
unmodelled 132, **136**, 292, 308–9, 316
see also model

eigenvalue placement 79–80, 87, 96–7
see also pole placement
eigenvector 88–9
electric drive
see DC motor, induction motor or PMSM motor
electric wire for automatic steering 314
EMF (electromotive force) 172, 190, 207–9, 266
energy conservation 272, 289, 294
equidistance line 307–12
equidistance subspace 310–2
equilibrium point of potential field 296
equivalent control 11, **24**-8, 117, 119, 124, 134, 143, 150, 162, 183–5, 196, 207–10, 224–5, 238, 241–2, 247, 264, 301–4
average of discontinuous control 11, 28, 119, 151
definition 25
discrete-time systems 162
gradient tracking 301
integral sliding mode **119**, 123, 150–1, 183
observer design **105**, **109**, 143, 183, 209, 242,
physical meaning 27–8
PWM via sliding mode 196–8
estimate
desired trajectory bounds 279
mass matrix elements 290
rotor angle and speed 209, 224
rotor flux 224
see also observer
estimation
boundary layer 139–41
disturbance 112, 118–20, 128, 151
load torque 184
perturbation 118–24, 303–5
rotor flux 218–28
rotor speed 183, 209, 223–228
shaft speed 183
torque perturbation 122–4, 303–5
uncertainty 118–20, 128
see also observer
Euler-Lagrange formulation 51, **272–3**, 275, 289
Example
AC/DC converter 3, 252–63
Ackermann's formula 90, 121
automatic steering 314–17
cart pendulum 51–5, 90–4
cascaded control 145–7
see also cascaded control
chattering in 1st order system 132–9
collision avoidance for robots 306–9, 310–3
componentwise control 282–3
continuous control with discontinuity term 148–51, **286–90**, 303–5, 316
DC generator 4
DC motor 124–5
delay system 165–6
discrete-time sliding mode 156–61
distributed system 166–8
disturbance rejection 147–51, 302–5
dynamic compensator 85–7
first-order system 2–4
flexible robot joint torque control 302–5
flexible shaft 166–8
friction system 2, 9–10
ideal sliding mode 132–4
integral sliding mode 120–30
inverted double pendulum 60–2
lateral vehicle control 314–17
multilink manipulator 35–6
observer-based chattering prevention 141–4
observer for time-varying system 110–3
output feedback sliding mode control 97–9
pendulum 51–78, 90–4
planar potential field 309
PMSM/AC motor or brushless DC motor 125–9
relay system 4–5
rotational inverted pendulum 68–78
saturation function 139–41
second order system 4–6, 9, 15, 30, 85–7, 89–90
ship control 4–5
three-link manipulator 312–13
torque control 302–5
two-dimensional sliding mode 16–18, 30–1
two-link manipulator 274–6, 279, 282–3, 285–6, 298–9
variable structure system 7
vector control 285–6
vehicle lateral control 314–7

feedback linearization 289, 294, 303
feedback/feedforward control **286–90**, 303–5, 316
field oriented control
DC motor 124–5
induction motor 217
PMSM/AC or brushless DC motor 125–9, 187, 193–4
field oriented frame **189**-92, 200, 206, 213–17, 253–4
field weakening 200–12
Filippov's method 22–6
finite time convergence 29, 33, 45
see also reaching phase
flexible joint 302–5
flexible shaft 166–8
flux
electrical motor 124, 126, 189, 216
frame 189–92, 200, 206, 213–17
observer 218–28
forward kinematics 274

frames of reference
  see coordinate frames
free-flying robot 313
frequency shaped sliding mode 240
friction system 2, 9–10
full-order observer 143, 178
full-order sliding mode 121, 240

Gauss method 121
  see also LQG
general sliding mode concept 157–61, 168
generalized potential field 294
global minimum 294
global sliding mode 240
globally uniform ultimate boundedness (GUUB) 140
goal approach phase 297
goal point 294–7, 306, 310–12
gradient
  line 295, 299, 307
  of potential field 294–9, 306–12
  tracking 292–301, 306–12
gravity force 122
gyroscope 315

harmonic dipole potential 295, 306–12
harmonic potential field (Laplace field) 294, 306
hierarchical path control scheme 306
high gain control 11, 116, 119, 240
holonomic dynamics 273
holonomic robot
  see robot manipulator
Hurwitz stability 81, 140
hypersphere 310–12
hysteresis control 19–21, 193, 200, 234

ideal sliding mode 6, 19–21, 24, 28, 132–4, 142, 156
IGBT 251
impedance control 294
imperfection 18, 19, 21
  in mechanical systems 47–50
  decoupling 36
induced 2-norm 274
induced back EMF
  see EMF
induced matrix norm 274
  see also norm
inductance in electrical motor armature 172, 189, 198, 207, 215
induction motor 212–29
  field oriented control 217
  model 213–18
  observer design 218, 223–8
  sensorless control 227–8
  speed control 219–23
inductor 233–4, 252
inertia matrix 36, 47, 122
integral sliding mode **115–30**, 147–51, 240, 302–5, 316
integrator backstepping 145, 235
intelligent transport systems 313
interaction force between road and tire 314
interconnected system 82
invariance property 5–7, 41–3, 82, 158
inverse kinematics 275
inversion 251
  see also power converter
inverted pendulum
  cart (trolley)-inverted pendulum 51–5, 90–4
  double inverted pendulum 55–62
  rotational inverted pendulum 62–78

joint angle 122
joint stiffness 302
joint torque 122

Kalman filter 105, 208, 228
kinetic energy 272

Lagrange formulation 51, 272–3, 275, **289**
Laplace field 294, 306
lateral vehicle control 313–17
linear
  delay systems 165–8
  discrete time sliding mode control 161–8
  integral sliding mode 120–1
  pole placement 79–80, 87, 303–5
  quadratic Gauss method 121
    see also LQG
  time-invariant systems (LTI) 19, 79–82, 105, 120–1, 163–8
  time varying system 99–102, 106–13
link of manipulator 273–6, 302
  see also robot
link voltage 173, 189, 214–7
Lipschitz condition 18
load inertia 172, 215
load torque 172, 176, 193, 215
  estimation 184
  observer 176
  variation 222
local maxima 294
longitudinal vehicle control 313–4
loop inductor 233–4, 252
LQG 121
LQR 60–2, 69
LTI (linear time-invariant system) 120–1, 163–8
Luenberger observer 141–4, 178
  see also observer

magnets for automatic steering 314
manipulator (arm) 122–3
  see robot manipulator
mass matrix
  see inertia matrix
mass of a vehicle 314
matching condition 42–3, 117
MCT 251
mobile robot
  collision avoidance 306–12
  componentwise control 299–301
  gradient tracking control 299–301
  model 277–8
model following sliding mode 240
model
  actuator (generic) 132, 146, 316
  bicycle 314
  bilinear systems 232
  boost AC/DC converter 252–4, 258
  boost DC/DC converter 236

buck DC/DC converter 234–6
car: single-track 314
chattering example 132
delay system 165
discrete time system 159
distributed system 164
DC motor 124, 172–3, 179
DC/DC converter 234–7
Euler-Lagrange 51, 272
flexible joint 302
flexible shaft 166
holonomic robot 273
ideal sliding mode 132
induction motor 213–18
Lagrangian 51, 272
manipulator 122
mechanical system 47, 272–3
mobile robot 277–8
nonholonomic robot 277–8
pendulum 51, 55, 62, 90
see also inverted pendulum
PMSM/AC motor or brushless DC motor 125, 188–93
reduced-order DC motor 179
rigid body robot manipulator 122
single-link robot with flexible joint 302
single-track (car) 314
techniques 272–3
torsion bar 166
two-link manipulator 274–6
wheel set 277
vehicle: single-track 314
Moore-Penrose inverse 191
see also pseudo inverse
MOSFET 251
motion separation 180, 184, 238
see also singular perturbation
motor
see DC motor, induction motor or PMSM motor
multi-dimensional sliding mode 9–10
multiple-phase converter 251
see also boost AC/DC converter

natural motion 289, 294
nature of sliding mode 158
Neumann boundary condition 294
neutral point 188–9, 193, 213, 252
nonholonomic constraint 277
nonholonomic robot
collision avoidance 306–12
componentwise control 299–301
gradient tracking 299–301
model 277–8
nonminimum phase 233
norm 274, 285

observer **103–13**, 141–4, 176–9, 182–5, 205–9, 240–261, 292
AC/DC converter 262–8
angular acceleration 176–9
chattering prevention 141–4, 292
current of PMSM motor 205–8
current of source phase voltage 263
DC motor 176–9
DC/DC converter 240–51
disturbance rejection 303–5
EMF components of AC/DC converter 266–8
EMF components of PMSM motor 208–9
flux 218–28
induction motor 223–8
load torque 176–9, 181, 184–5
rotor flux of induction motor 218–28
rotor speed
DC motor 181, 183–4
induction motor 223–8
PMSM motor 206
sensorless control
AC/DC converter 162–8
DC motor 182–5
PMSM motor 206–11
sliding mode observer 103–13
source voltage 264
torque disturbance 303–5
observer-based chattering prevention 141–4, 292
obstacle avoidance 294–5, 306–12
ordinary differential equation 145
orientation control 299–301
oscillations 6, 19–20, **135–9**
robot trajectory 307–9
see also chattering
output feedback 94–102

pendulum
see inverted pendulum
permanent magnet synchronous motor 125, 187–212
see also PMSM motor
perturbation
compensator 117
estimation 303–5
see also disturbance
phase current 190, 207, 215
phase frame 189–92, 200, 206, 213–17, 253
phase voltage 190, 214–17, 252
planar manipulator
see two-link manipulator
planar workspace 306–9
PMSM (AC motor or brushless DC motor) 125–9, 187–212
current control 125–9, 193–201
current observer 205–8
example 125–9
field oriented control 125–9, 187, 193–5
integral sliding mode 125–9
model 125, 188–93
observer for speed sensorless control 206–11
PWM via sliding mode 127–9
sensorless control 206–9
speed control 201–5
pole-assignable system 95–6
pole placement 121, 179, 303–5
see also eigenvalue placement
potential energy 272
potential field method 294–6, 306–12
power balance 193, 196, 259
power converters 231–69
boost type AC/DC 251–69
DC/DC 231–51
multiple phase 251

prevention of chattering 139, **141–54**, 180, 292, 308–9, 311–12, 316–17
properties of harmonic dipole potential 307, 311
properties of sliding mode 12, 158
pseudo inverse 101, 109, **190–2**, 202, 215–17, 220
PWM (pulse width modulation)119, **124–9**, 141, 173, 187, 194–8, 234, 251

radar for automatic steering 314
reaching phase 2, 17–8, **29–30**, 45, 115–16, 133, 150, 158–9, 184, 296–7
rectification 251
  see also power converter
reduced-order
  model of DC motor 180
  motion equation 8
  observer 104, 178
  speed control of DC motor 181–3
reference current 174, 194–200
regenerative capability 254
regular form **39–41**, 47, 63, 79–80, 82, 95, 144–7, 164, 235, 315
regularization 18–24
relay control 1–5, 15–16, 19–20
  DC motors 179
    see also sliding mode
reluctance torque 189, 193
resistance in electrical motor armature 172, 189, 207, 215
ripple-free output voltage 254
road-tire interaction force 314
robot
  arm 35–6, 122–4, 273–92, 298–9, 310–3
  collision avoidance 306–13
  gradient tracking control 292–301, 306–12
  holonomic model 273
  manipulator 35–6, 122–4, 273–92, 298–9, 310–3
  mobile robot 277–8, 299–301, 306–12
  nonholonomic model 277
  sphere-like robot 310–2
  three-link manipulator 312–3
  trajectory tracking control 278–293
  two-link manipulator **274–6**, 282–3, 285–6, 289–90
  wheel-set 277–8
rotor
  flux angle 212, 222
  flux observer with known speed 218–23
  flux observer with unknown speed 223–8
  frame 189–92, 200, 206, 213–7
  inertia 172, 215

saturation function 70, **139**, 292
security sphere 310–2
security zone 295, 306
sensorless control
  AC/DC converter 262–8
  DC motor 182–5
  induction motor 227–8
  PMSM/AC motor or brushless DC motor 206–9
shaft speed 172, 189–90, 213–7
ship steering control 4–5
side slip angle 314
single-link robot with flexible joint 302–5
single-track model 314
singular perturbation **134–8**, 180, 184, 234, 238, 258
singularity in potential field 295–6, 306
skew symmetry 274, 276, 288–90
sliding manifold 31
sliding mode 1–5
  AC/DC converter 251–69
  bilinear system 232–3
  characteristics 12, 158
  chattering
    due to discretization 156–8
    due to unmodelled dynamics 134
  componentwise control 29, **279–83**, 299–301
  concept 157–61, 168
  continuous control with discontinuous term 117–29, 148–51, **286–90**, 303–5, 316
  chopper control 193, 200
  current control
    see current control
  DC motor 124–5, 172–87
  DC/DC converter 233–51
  delay system 164–8
  design procedure 36–45
  discrete-time design procedure 159
  discrete-time sliding mode 155–64, 159–69
  distributed systems 164–8
  disturbance rejection 115–30, 147–51
  domain 31
  equivalent
    see equivalent control
  existence condition 28–34
  feedback/feedforward 286–90, 303–5, 316
  flexible joint 302–5
  frequency shaped 240
  full-order 121, 240
  global 240
  gradient tracking 296–301
  hysteresis control 193, 200, 234
  ideal 6, 19–21, 24, 28, 132–4, 142, 156
  induction motor 218–28
  integral sliding mode control **115–30**, 147–51, 240, 302–5, 316
  linear delay systems 165–8
  linear discrete-time 161–4
  linear time invariant 19, **82**, 120–1, 163–8
  linear time varying 99–102
  model following 240
  multi-dimensional 9–10
  nature of sliding mode 158
  observer 103–13
    see observer
  output feedback 94–102
  PMSM/AC motor or brushless DC motor 125–9, 193–212
  properties 12, 158
  PWM 127–9, 195–8
  regularization 18–24
  speed control
    see speed control
  time invariant 19, **82**, 120–1, 163–8
  time varying 99–102
  torque control 302–5
  two-dimensional 9–10, 16–8
  two-link manipulator 282–3, 285–7, 289–90

vector control 283–6, 298–9
see also vector control
slip of induction motor 212
source voltage observer 264
spacing control for vehicles 313–4
speed control
AC motor or brushless DC motor 201–5
cars 313–4
DC motor 174–6, 179–81
induction motor 219–23
mobile robot 299–301
PMSM motor 201–5
reduced-order DC motor 179–81
sensorless 206–11
vehicles 313–14
speed observer
DC motor 181–4
induction motor 223–8
PMSM/AC or brushless DC motor 206–11
star-shapes 306, 310
state plane 5, 7, 86
state-space averaging method 231
stator current and voltage 125, 190, 213–17
stator frame 189–92, 200, 206, 213–7
steering angle 314–17
steering control 313–17
stored error energy 244–50
switching **1–8**, 118, 128, 131, 157–8, 173, 188, 194–7, 200–2, 219, 231–2, 234–6, 251, 307
synchronous motor
see PMSM

terminal voltage 172, 213–15
three-link manipulator 313
time-invariant system (linear) LTI 19, 79, 120–1, 163–8
time-varying system 99–102
torque constant of electrical motor 172, 193, 214–6
torque control 302–5
torque disturbance 122–4
torsion bar 166
trajectory tracking 1–2, 278–293
two-dimensional sliding mode 9–10
two-link manipulator 282–3, 285–7, 289–90
uncertainty
estimation 118–24, 147–51
parametric 147
underactuation **47**, 271
unit control 43–6, 80, 294
see also vector control
unit singularity 306
unity power factor 254
unmodelled dynamics 132, **136**, 292, 308–9, 316

variable structure system (VSS) 4–7, 115
see also sliding mode
vector control 283–6, 298–9
example: robot manipulator 285–6
gradient tracking 298–9
permanent magnet synchronous motor (PMSM) 187
robot 283–6, 298–9
trajectory tracking 283–6
see also field oriented control and unit control
velocity control of mobile robot 299–301
vibration control 4
virtual potential field 294, 306
see also potential field
vision system for automatic steering 314
voltage modulation 251
voltage regulation
AC/DC converter 258–60
boost DC/DC converter 238–40, 246–51
buck DC/DC converter 235–6, 241–5
von-Neumann boundary condition 294
Voronoi plane 310
VSS (variable structure system) 4–7, 115
see also sliding mode

wheel-set 277–8
see also mobile robot

yaw motion 314

zero dynamics 48, 64
zero speed problem 227